OBJECTIVE
BIOCHEMISTRY

OBJECTIVE BIOCHEMISTRY

N ARUN
Assistant Professor
Department of Biochemistry
Thanthai Hans Roever College of Arts and Science
Perambalur, TN

S KARTHICK
Researcher in Biochemistry
Dravidian University
Kuppam, AP

MJP PUBLISHERS

Chennai Trichy Tirunelveli NewDelhi

MJP PUBLISHERS

ISBN 978-81-8094-086-6

MJP PUBLISHERS
No.44, Nallathambi Street
Triplicane, Chennai 600 005

MJP 082

Publisher : C. Janarthanan

Mathru devo bhava, Pithru devo bhava, Acharya devo bhava (Upanishad)

To

Our beloved Parents, Teachers and Friends.

PREFACE

Multiple choice questions constitute an important method for objective assessment of an examinee's knowledge. Looking at the pitfalls of traditional question papers comprising a small number of long questions, educationists are laying stress on objective testing to broaden the coverage of syllabi, to test the analytical thinking of examinees and to reduce the element of subjectivity in the system of assessment. Many common eligibility tests and competitive exams are likely to follow suit. Moreover postgraduate competitive exams are generally based on objective type alone. Considering the importance of objective type questions in all the examinations, the present book has been written to provide students with a comprehensive question bank covering nearly the entire syllabus in biochemistry. This book would help students in preparing themselves for many examinations and in quick revision of the subject.

We have received help and encouragement from many quarters while writing this book. Firstly we thank our family members for extending their moral and financial support during the preparation of this book. Dr.P. Subramanian, Reader, Department of Biochemistry and Biotechnology, Annamalai University, and Dr.V.N. Janakarajan, Asstistant Professor, Department of Medical Biochemistry, IRT-Perundurai Medical College, have been a source of inspiration and encouragement to us. We are grateful to them for their help. We have to thank the Management and the Principal, THRC, for their kind support. We have had the benefit, wise counsel, scholarly criticism and constructive suggestions from many friends and professional colleagues. We are thankful to them for reviewing various chapters of the book.

We also thank our friends at Sri Kumaran Computers, for their support in preparing the manuscript. Finally, we thank MJP Publishers and their editorial team for bringing out our work in a book form. However for the errors still present, the responsibility is ours. Please send your suggestions and criticism to aamaravian@gmail.com, to enable us to improve the contents in the next editions.

N Arun

S Karthick

CONTENTS

1

WATER

I. CHOOSE THE CORRECT ANSWER

1. The pH of gastric juice of infants is
   (a) 2.0 (b) 4.0 (c) 4.5 (d) 5.0
2. The pH of water is 7.0 at the following temperature in centigrade
   (a) 22 (b) 30 (c) 37 (d) 40
3. If the pH of blood is 7.4, the ratio between $NaHCO_3$ and H_2CO_3 is
   (a) 10 : 1 (b) 20 : 1 (c) 25 : 1 (d) 30 : 1
4. The difference in pH between arterial and venous blood is rarely more than
   (a) 0.02 (b) 0.03 (c) 0.04 (d) 0.06
5. The pH of buffer is determined by $pH = pka + \log\frac{[salt]}{[acid]}$ which is also known as the equation of
   (a) Henderson–Joules (b) Henderson–Smith
   (c) Henderson–Harris (d) Henderson–Hasselbalch
6. The chief buffering system in the blood is
   (a) K_2HPO_4 and KH_2PO_4 (b) B-protein and H-protein
   (c) $NaHCO_3$ and H_2CO_3 (d) haemoglobin and H-haemoglobin
7. During severe muscular exercise, when the blood lactic acid content rises over 100 mg per 100 ml, the pH of blood
   (a) slightly increases (b) highly increases
   (c) slightly decreases (d) markedly decreases

8. In case of non-homogeneous solution, diffusion is expressed as $\frac{ds}{dt} = DA\frac{dc}{dx}$ as per
   (a) Fick's law (b) Peterson's law (c) Faraday's law (d) Dick's law
9. The average pH of urine is
   (a) 5.6 (b) 6.0 (c) 6.4 (d) 6.8
10. The osmotic pressure of a solution increases with rise in
    (a) temperature (b) cold (c) humidity (d) rancidity
11. The osmotic pressure of a solution relating to solute molecules depends on the
    (a) size (b) shape (c) number (d) volume
12. If a cell is immersed in a concentrated solution, it follows the phenomenon of
    (a) turgor (b) plasmolysis (c) haemolysis (d) paralysis
13. Osmosis is opposite to
    (a) effusion (b) affusion (c) confusion (d) diffusion
14. The intracellular fluid of red cells and the red cell membrane in 0.92% NaCl solution maintains ________ relation.
    (a) hypertonic (b) hypotonic (c) isotonic (d) none of these
15. The purgative action of Epsom ($MgSO_4$) salt follows
    (a) diffusion (b) osmosis (c) adsorption (d) absorption
16. Haemolysis is caused by the dilution of RBC by
    (a) diffusion (b) osmosis (c) effusion (d) imbibition
17. The surface tension of a solution is decreased by
    (a) calcium sulphate (b) barium sulphate
    (c) magnesium phosphate (d) potassium permanganate
18. The surface tension of a solution is lowered by
    (a) ammonia (b) sodium hydroxide
    (c) potassium hydroxide (d) aluminium hydroxide
19. The surface tension of a solution is increased by
    (a) bile salts (b) bile acids
    (c) conc.sulphuric acid (d) none of these
20. Lipids and proteins which are both effective in lowering surface tension are found concentrated in the cell wall following the principle of
    (a) Johnson–John (b) Gibbs–Thomson (c) Peterson–Pollen (d) none of these

21. Adsorption decreases with rise in
   (a) cold (b) humidity (c) dryness (d) temperature
22. The process of adsorption is applied in the purification of
   (a) enzymes (b) hormones (c) vitamins (d) coenzymes
23. Which of the following is a hydrotropic substance?
   (a) hydrochloric acid (b) nitric acid (c) hippuric acid (d) salicylic acid
24. Bile salts make emulsification with fat for the action of
   (a) amylase (b) lipase (c) pepsin (d) trypsin
25. The viscosity of a liquid increases due to the presence of
   (a) suspended particles (b) soluble particles
   (c) small sized particles (d) none of these
26. The size of each colloidal particle in nm is
   (a) 4 to 40 (b) 6 to 60 (c) 8 to 80 (d) 10 to 100
27. The serum colloidal protein particle may be precipitated by the addition of large amount of
   (a) calcium sulphate (b) ammonium sulphate
   (c) barium sulphate (d) magnesium sulphate
28. Some of the calcium phosphate of blood is held in colloidal suspension by the protective action of
   (a) lipids (b) carbohydrate (c) proteins (d) minerals
29. Water is not expelled by squeezing in
   (a) imbibition (b) precipitation (c) combination (d) dilution
30. In simple diffusion $\frac{ds}{dt}(C_0 - C_1)$ can be expressed by the modification of
   (a) Rongent's law (b) Fabry's law (c) Pollinger's law (d) Fick's law
31. Fatty acids can be transported into and out of mitochondria through
   (a) active transport (b) facilitated transfer
   (c) non-facilitated transfer (d) none of the above
32. The absorption of intact protein from the gut in the foetuses and newborn animals takes place by
   (a) pinocytosis (b) passive diffusion
   (c) simple diffusion (d) active transport
33. Thyroid function is determined by the use of which of the following isotopes?
   (a) Na^{24} (b) K^{42} (c) Ca^{45} (d) I^{131}

34. Pernicious anaemia is diagnosed using
    (a) Cl^{15} (b) P^{32} (c) Co^{60} (d) Fe^{59}
35. The protein biosynthesis has been studied by using
    (a) N^{15} (b) Ca^{45} (c) Na^{24} (d) Au^{198}
36. Of the total body weight, the average percentage of body water is
    (a) 55 to 65 (b) 50 to 60 (c) 60 to 70 (d) 70 to 80
37. Intracellular fluid contains how many per cent of the total body weight?
    (a) 45 (b) 50 (c) 55 (d) 60
38. The per cent of plasma in the extracellular fluid is
    (a) 3.5 (b) 3.8 (c) 4.2 (d) 4.5
39. The per cent of transcellular fluid in extracellular fluid is
    (a) 1.4 (b) 1.5 (c) 1.6 (d) 1.8
40. The organic compound of small molecular size is
    (a) urea (b) uric acid (c) creatinine (d) phosphates
41. The organic substance of large molecular size is
    (a) starch (b) inulin (c) lipids (d) proteins
42. 100 g of fat on combustion produces ______ of water.
    (a) 107 ml (b) 105 ml (c) 102 ml (d) 98 ml
43. 100 g of protein on combustion yields ______ of water.
    (a) 35 ml (b) 38 ml (c) 41 ml (d) 44 ml
44. The daily intake and output of water through different sources is
    (a) 2500 ml (b) 2700 ml (c) 2800 ml (d) 3000 ml
45. The amount of internal secretion of water per day in litres is
    (a) 107 ml (b) 105 ml (c) 102 ml (d) 98 ml
46. The amount of water excreted per day in faeces is
    (a) 90 ml (b) 95 ml (c) 100 ml (d) 105 ml
47. The regulatory mechanism of body water is influenced by the hormone
    (a) oxytocin (b) ACTH (c) FSH (d) epinephrine
48. All chemical reactions in the body proceed in the presence of
    (a) organic salts (b) inorganic salts (c) water (d) alkali
49. The symptom of water intoxication is
    (a) muscular weakness (b) anaemia (c) paralysis (d) fever

50. Dehydration may be ordinarily corrected by parenteral ingestion of a solution of
(a) $ZnCl_2$ (b) $MgCl_2$ (c) $CaCl_2$ (d) NaCl

51. A mixture of two-thirds isotonic saline solution and one-third lactate solution should be administered intravenously in case of
(a) constipation (b) prolonged diarrhoea
(c) malaria (d) jaundice

52. Dehydration is a symptom of
(a) vomiting (b) typhoid (c) uremia (d) hot climate

53. The principal mineral elements are also said to be
(a) macronutrients (b) micronutrients
(c) semi-micronutrients (d) none of these

54. The trace elements are subdivided into
(a) two groups (b) three groups (c) four groups (d) none of these

55. The essential trace elements include
(a) lead (b) nickel (c) boron (d) cobalt

56. The number of principal mineral elements is
(a) five (b) six (c) seven (d) eight

57. Zinc is a constituent of
(a) carbonic anhydrase (b) malate dehydrogenase
(c) aldolase (d) amylase

58. Haemoglobin formation needs both
(a) iron and zinc (b) iron and calcium
(c) iron and copper (d) iron and magnesium

59. Cobalt is a constituent of
(a) folic acid (b) vitamin B_{12} (c) niacin (d) biotin

60. Calcium is required for the activation of the enzyme
(a) isocitrate dehydrogenase (b) fumarase
(c) succinate thiokinase (d) ATPase

61. The absorption of calcium is increased by the higher dietary levels of
(a) fats (b) proteins (c) cereals (d) vitamin A

62. Calcium absorption is interfered by
(a) fatty acids (b) amino acids (c) vitamin D (d) vitamin B_{12}

63. Calcium in the complex form is about
    (a) 8 mg% (b) 6 mg% (c) 4 mg% (d) 2 mg%
64. The percentage of calcium in milligrams in nonionized form is about
    (a) 3 (b) 4 (c) 5 (d) 6
65. In man, the number of grams of calcium filtered in 24 hours by the renal glomeruli is about
    (a) 8 (b) 10 (c) 12 (d) 14
66. The percentage of calcium excreted in the faeces is
    (a) 40 to 60 (b) 50 to 70 (c) 60 to 80 (d) 70 to 90
67. The daily loss of calcium in mg in sweat is about
    (a) 12 (b) 13 (c) 14 (d) 15
68. The concentration of serum calcium may drop below 7 mg/100 ml in
    (a) hyperparathyroidism (b) hypoparathyroidism
    (c) tetany (d) rickets
69. Serum phosphate concentration is 1 to 2 mg/100 ml in
    (a) rickets (b) tetany (c) osteoporosis (d) hyperthyroidism
70. Under optimal conditions, the percentage of calcium absorbed is
    (a) 75 (b) 80 (c) 90 (d) 95
71. Succinate dehydrogenase requires ______ for its activation.
    (a) calcium (b) magnesium (c) sodium (d) none of these
72. In rickets, serum alkaline phosphatase activity is
    (a) increased (b) decreased (c) same (d) uncertain
73. Phosphorus is essential for the formation and development of bones and teeth along with
    (a) calcium (b) magnesium (c) sodium (d) none of these
74. Co-carboxylase is provided with
    (a) magnesium (b) calcium (c) sodium (d) none of these
75. Parathyroid hormone ______ the absorption of magnesium.
    (a) increases (b) decreases (c) unaltered (d) none of these
76. In certain stages of pregnancy, steroid hormones cause the retention of which of the following which results in gain in weight?
    (a) sodium (b) water (c) both a and b (d) none of these
77. Protein biosynthesis by ribosomes requires ______.
    (a) potassium (b) magnesium (c) calcium (d) sodium

78. Total iron content of the normal adult is about
    (a) 4 to 5 g (b) 3 to 4 g (c) 1 to 2 g (d) 5 to 6 g
79. What per cent of the total iron is present in haemoglobin?
    (a) 60 to 70 g (b) 50 to 60 g (c) 40 to 50 g (d) 30 to 40 g
80. How much per cent of the red blood cell copper is present as superoxide dismutase (erythrocuprein)?
    (a) 50 (b) 60 (c) 70 (d) 80
81. The normal inorganic phosphate of plasma per 100 ml in adults is
    (a) 3 to 4.5 mg (b) 2 to 3 mg (c) 1 to 2 mg (d) none of these
82. The osmotic pressure of serum protein is about
    (a) 30 mm Hg (b) 20 mm Hg (c) 10 mm Hg (d) 15 mm Hg
83. Calcium exists in the plasma in ______ form.
    (a) diffusible (b) nondiffusible (c) complexed (d) all of these
84. The cell membrane consists of small water-filled pores of radius about
    (a) 0.4 nm (b) 0.3 nm (c) 0.2 nm (d) 0.5 nm
85. The nonhaem iron is completely protein-bound and exists in the
    (a) unbound form (b) T state (c) both (a) and (b) (d) none of these
86. In regulation of acid–base equilibrium, sodium is largely associated with
    (a) chloride (b) bicarbonate (c) both (a) and (b) (d) none of these
87. In man, erythrocytes contain little or no
    (a) sodium (b) potassium (c) magnesium (d) calcium
88. Hypernatraemia occurs in prolonged treatment of
    (a) cortisone (b) ACTH (c) sex hormones (d) all of these
89. The symptoms of hyperkalaemia are corrected by the administration of deoxycorticosterone which help in the excretion of
    (a) potassium (b) sodium (c) magnesium (d) none of these
90. Prolonged hypokalaemia causes injury to
    (a) myocardium (b) kidneys (c) both (a) and (b) (d) none of these
91. Hypochloraemic alkalosis may develop in Cushing's disease or after the administration of
    (a) ACTH (b) cortisone (c) both (a) and (b) (d) none of these
92. Sulphur is present primarily in the cell protein in the form of ______ and ______.
    (a) cysteine (b) methionine (c) both (a) and (b) (d) none of these

II. Fill in the Blanks

1. Solution of mixed indicators having a number of colour changes over a wide range of pH are called _______ indicators.
2. The logarithm of the reciprocal of hydrogen ion concentration is called _______.
3. Temperature cannot alter the pH of a solution of _______.
4. The mixture of weak acids and their salts of strong bases are called _______.
5. The spreading of solute molecules throughout water is called _______.
6. Diffusion takes place against _______.
7. The most selective artificial membrane is _______.
8. Soaps have _______ osmotic pressure.
9. If a cell is kept in a hypotonic solution, the cell is called _______.
10. The force with which the surface molecules are held together is called _______.
11. Surface tension is involved in the process of _______.
12. The process by which water-insoluble substances are made water-soluble by hydrotropic substances is called _______.
13. The resistance experienced by one layer of a liquid in moving over another is called _______.
14. The sparingly soluble cholesterol and the calcium salt of bilirubin is kept in colloidal solution in the protective colloids by _______.
15. The process of separation of crystalloids from colloids by diffusion through a membrane by osmotic force is called _______.
16. Dialysis is applied in medicine in the _______.
17. An emulsoid may be changed into a suspensoid by _______.
18. Atoms having the same atomic number but different atomic weights are said to be _______.
19. Absorption of fat is studied by using _______.
20. The lifespan of RBC has been determined by tagging _______ to RBC.
21. The absorption, mobilization and transport of iron have been studied by the use of _______.
22. Plasma volume is measured by using _______ labelled serum albumin.
23. The formation of antibody in the reticuloendothelial system has been studied by using _______.
24. _______ is the end product of Au^{198} after emitting β- and γ-rays.

25. Some of the carrier proteins are called ______ and others are called ______.
26. Na^+–K^+ ATPase catalyses the hydrolysis of ______ to ______.
27. There is virtually no calcium in ______.
28. The active transport process of calcium is regulated by ______ a metabolite of vitamin D which is produced in the kidney in response to low plasma ______ concentration.
29. Ionized calcium is required in blood for the ______ process.
30. Phytic acid and oxalates ______ with iron absorption.
31. Decreased ionized fraction of serum calcium causes ______.
32. ______ which occurs in cereals forms the insoluble salt ______ with calcium and magnesium resulting in the impaired absorption of calcium.
33. An increase in the ______ calcium levels in the plasma is the stimulus for the production of ______ which then causes deposition of calcium in bone.
34. Iron absorption is enhanced by proteins of ______ digestive products forming ______.
35. Most of the iron in food occurs in the ______ state.
36. Renal rickets is more accurately designated as ______ which is an inherited disease.
37. Vitamin D ordinarily does not relieve renal rickets. Hence, it is sometimes referred to as ______.
38. Phosphorus is required in the absorption of glucose by ______.
39. High calcium diet and phytic acid ______ phosphorus absorption.
40. The reabsorption of phosphorus is inhibited by ______.
41. Magnesium is the principal ______ of the soft tissue.
42. Marked sulphate retention in advanced glomerulonephritis causes the development of______.
43. In case of Kwashiorkor, the serum magnesium level is ______ causing ______.
44. In uraemia, the serum magnesium level is ______.
45. In the absorption of glucose and galactose ______ ion plays an important role.
46. ______ increases plasma sodium level.
47. Loss of sodium by excessive sweating causes ______ with the intense and painful contraction of ______ of men working hard in hot humid climates.
48. The metabolism of sodium is regulated by ______.
49. An increase in the blood urea concentration may occur in ______.
50. Pyruvate kinase requires ______ for maximal activity.

51. A small amount of sulphide if absorbed into the bloodstream is rapidly oxidized to _______.
52. _______ decreases serum potassium level.
53. In man, _______ contain large amounts of potassium.
54. The serum potassium level _______ during increased carbohydrate utilization following administration of _______.
55. Hypokalaemia is exhibited in heart failure treatment with _______.
56. Chloride ion is important as an activator of _______.
57. Excessive consumption of sodium salt causes _______.
58. Sodium concentration in sweat is decreased by _______.

III. State if True or False

1. Chromium potentiates the action of insulin in accelerating utilization of glucose in humans.
2. Chromium deficiency is characterized by impaired growth, disturbances in glucose, lipid and protein metabolism.
3. About 1.0 per cent of the total iron is carried in the plasma in combination with β globulin transport protein transferrin.
4. Iron is not involved in cellular respiration.
5. The richest sources of iron are egg yolk, fish, figs and apples.
6. A defect in haemoglobin synthesis in anaemic condition is commonly due to copper deficiency.
7. In humans, haem is broken down in the mucosa and iron appears in the plasma transferrin.
8. Vitamin D and alkali increase iron absorption.
9. Apoferritin, in the mucosal cells, is the controlling factor for iron absorption.
10. Transferrin or siderophilin is a specific iron-binding protein.
11. Only the reticulocytes can utilize the ferric ion bound to transferrin, although reticulocytes and mature erythrocytes can take up unbound ferric ion.
12. In hepatic diseases both the bound iron and the total iron-binding capacity of the plasma is high.
13. Ferritin is not present in the intestine and also in liver.
14. Iron deficiency anaemia is not prevalent among children and adolescent girls.
15. Copper is required for melanin formation, phospholipid synthesis and collagen synthesis.

16. The daily biliary excretion of copper is 0.5 to 1.3 mg.
17. Wilson's disease is a rare hereditary disorder of copper metabolism.
18. The absorption of copper from the intestine is very low.
19. Excessive deposition of copper in the liver causes hepatic cirrhosis.
20. The excess of tissue copper can be removed by administering the copper chelating agent penicillamine.
21. Ceruloplasmin is a copper-binding plasma protein.
22. Triiodothyronine is less active than thyroxine in many respects.
23. The body normally contains about 10 to 20 g of iodine.
24. All the iodine in saliva is organic.
25. Muscles contain less amount of iodine.
26. Erythrocytes contain no inorganic iodine.
27. Most tissues contain less amount of iodine in the inorganic form and most of the iodine is present in the organic form.
28. 90 per cent of the iodine of the thyroid gland is in organic combination.
29. The urinary elimination of iodine is largest when the intake is high.
30. Urinary iodine is increased by exercise and other metabolic factors.
31. In children, severe iodine deficiency results in the extreme retardation of growth which is known as cretinism.
32. Fluorine in combination with vitamin C is required for the treatment of osteoporosis.
33. Sodium fluoride is a powerful inhibitor of phosphoglycerate kinase.
34. Fluoride ions inhibit the metabolism of oral bacterial enzymes and diminish the local production of acids which cause the production of dental caries.
35. The retina contains a zinc metalloenzyme, retinene reductase, which is required for the formation of retinene.
36. Zinc is not concerned with the healing of wounds.
37. High concentration of zinc is present in spermatozoa, prostate, and epididymis.
38. The lowest concentration of copper occurs in the choroids of the eye.
39. Zinc present in cereals is highly absorbed owing to the presence of phytic acid.
40. The deficiency of zinc causes hepatosplenomegaly, delayed closure of the epiphysis of the long bones and anaemia.

41. Methyltetrahydrofolate oxidoreductase and homocysteine methyltransferase require vitamin B_{12} for activity.
42. Cobalt in large doses develops a condition known as polyerythemia.
43. Manganese ions inhibit lipid peroxidation reactions.
44. The heart and the brain are the main storage organs for manganese.
45. Manganese toxicity develops hepatolenticular degeneration resembling Parkinson's disease.
46. Liver and kidney contain less amount of molybdenum than other tissues.
47. Selenium is essential for normal growth and fertility of human beings.
48. Glutathione peroxidase, a selenoprotein, catalyses the peroxidation of glutathione.
49. Selenium is involved in immune mechanism and ubiquinone synthesis.

I. ANSWERS

1. (d)	2. (a)	3. (b)	4. (c)	5. (d)	6. (c)	7. (d)
8. (a)	9. (b)	10. (a)	11. (c)	12. (b)	13. (d)	14. (c)
15. (a)	16. (b)	17. (d)	18. (a)	19. (c)	20. (b)	21. (d)
22. (a)	23. (c)	24. (b)	25. (a)	26. (d)	27. (b)	28. (c)
29. (a)	30. (d)	31. (b)	32. (a)	33. (d)	34. (c)	35. (a)
36. (c)	37. (b)	38. (d)	39. (b)	40. (a)	41. (d)	42. (a)
43. (c)	44. (b)	45. (d)	46. (c)	47. (a)	48. (c)	49. (a)
50. (d)	51. (b)	52. (c)	53. (a)	54. (b)	55. (d)	56. (c)
57. (a)	58. (c)	59. (b)	60. (d)	61. (b)	62. (a)	63. (d)
64. (c)	65. (b)	66. (d)	67. (d)	68. (b)	69. (a)	70. (a)
71. (a)	72. (a)	73. (a)	74. (a)	75. (a)	76. (a)	77. (a)
78. (a)	79. (a)	80. (d)	81. (a)	82. (a)	83. (d)	84. (a)
85. (c)	86. (c)	87. (a)	88. (d)	89. (a)	90. (c)	91. (b)
92. (c)						

II. ANSWERS

1. universal	2. pH	3. 0.1N HCl
4. buffers	5. diffusion	6. gravity
7. $Cu_2Fe(CN)_6$	8. lower	9. turgid

10. surface tension 11. digestion 12. hydrotropy
13. Viscosity 14. Bile Salts 15. Dialysis
16. artificial kidney 17. dehydration 18. isotopes
19. 1^{131}-Oleic acid 20. Cr^{51} 21. Fe^{59}
22. I^{131} 23. P^{32} 24. Hg^{198}
25. uniports, symports 26. ATP, ADP 27. erythrocytes
28. 1,25-dihydroxycholecalciferol, Ca^{++} 29. coagulation
30. interfere 31. tetany 32. phytic acid, phytin
33. ionized, calcitonin 34. low molecular weight, iron chelate.
35. ferric 36. familial hypophosphataemic rickets
37. vitamin D-resistant rickets. 38. phosphorylation 39. decrease
40. parathyroid hormone 41. cation 42. acidosis
43. low weakness 44. low 45. sodium
46. aldosterone 47. heat cramps, skeletal muscle
48. adrenocortical steroids 49. uraemia 50. K^+
51. sulphate 52. aldosterone 53. erythrocytes
54. decreases, insulin 55. digitalis 56. amylase
57. edema 58. aldosterone

III. ANSWERS

1. True	2. True	3. False	4. False	5. False
6. True	7. True	8. False	9. False	10. True
11. True	12. False	13. False	14. False	15. True
16. True	17. True	18. False	19. True	20. True
21. True	22. False	23. False	24. False	25. False
26. False	27. True	28. True	29. False	30. True
31. True	32. False	33. False	34. True	35. True
36. False	37. True	38. False	39. False	40. True
41. True	42. True	43. True	44. False	45. True
46. False	47. False	48. True	49. True	

2

THE CELL

CHOOSE THE CORRECT ANSWER

1. Who among the following proposed the hypothesis, "the bodies of animals and plants are composed of cells and products of cells"?
   (a) Robert Hooke (b) Rudolf Virchow
   (c) Theodore Schwann (d) none of these
2. All living cells
   (a) arise from pre-existing cells
   (b) are basically alike in chemical composition and metabolic activities
   (c) both (a) and (b)
   (d) none of these
3. During cell division the hereditary characters are transmitted in the form of
   (a) citric acids (b) nucleic acids (c) amino acids (d) none of these
4. Sections through the nuclear pore show the presence of an electron-dense ring or cylinder within the pore which has been called
   (a) perinuclear space (b) central granule (c) annulus (d) fibrous lamina
5. The plasma membrane is covered by a cell wall in cells of
   (a) plants (b) bacteria (c) fungi and algae (d) all of these
6. The principal functions of the cell wall include
   (a) protection (b) mechanical support
   (c) maintenance (d) all of these

7. The layer common to two adjacent cells and called middle lamella is composed of
   (a) calcium phosphate (b) calcium sulphate
   (c) calcium carbonate (d) calcium pectate
8. The cell wall is interrupted by narrow pores carrying fine strands of cytoplasm, which interlink the contents of the cells. These are called
   (a) plasmahole (b) microvilli (c) plasmodesmata (d) plasmalemma
9. The movement of cytoplasm within the eukaryotic cells is referred to as
   (a) exocytosis (b) pinocytosis (c) cyclosis (d) none of these
10. Many folds of the inner membrane of mitochondria project into the matrix as incomplete partitions which are called
    (a) cristae (b) grana
    (c) inner membrane particles (d) plastids
11. Thylakoids in a plastid are placed one above the other like a stack of coins to form a
    (a) stroma (b) matrix (c) granum (d) crista
12. Colourless plastids are
    (a) chloroplasts (b) chromoplasts (c) leucoplasts (d) all of these
13. Which of the following statements regarding Golgi complex is true?
    (a) The lumen of ER cisternae and tubules and Golgi cisternae communicate with each other forming passage for transport of secretory proteins, lipids and sterols.
    (b) Secretion is the main function of the Golgi complex.
    (c) The Golgi complex participates in recycling of plasma membrane.
    (d) All of these.
14. Each ribosome consists of two unequal subunits composed of
    (a) RNA and proteins (b) RNA and carbohydrates
    (c) only RNA (d) proteins and DNA
15. The cytoplasm of eukaryotic cells contains an extensive network of minute tubules and filaments, appropriately called
    (a) transitional vesicles (b) microbodies
    (c) microtubules and microfilaments (d) zymogen granules
16. Which of the following increases the free surface of the cell to enhance absorption?
    (a) microfilaments (b) microtubules
    (c) microbodies (d) microvilli

17. Which of the following statements regarding microtubules is true?
    (a) Microtubule wall is made of 13 longitudinal filaments, each a polymer chain to tubulin.
    (b) Microtubules help in the maintenance of cell form, cell mobility and movement of chromosomes.
    (c) both (a) and (b)
    (d) none of these
18. Which of these is a non-membranous organelle?
    (a) mitochondrion (b) ribosome
    (c) plastid (d) endoplasmic reticulum
19. Vacuoles are sap-filled vesicles in the cytoplasm, covered by a membrane called
    (a) protoplast (b) leucoplast (c) tonoplast (d) chromoplast
20. Which of the following statements is true?
    (a) Starch grains occur in plant cells only, particularly in storage organs such as rhizomes and tubers.
    (b) Aleurone grains contain stored proteins in plant cells.
    (c) Crystals of various chemical compounds occur in plant cells.
    (d) All of these.
21. The first part of interphase is called
    (a) synthetic phase (b) G_1 phase (c) G_2 phase (d) M phase
22. Consider the following statements regarding cell cycle and choose the correct answer using the codes given below:
    I. During G_1 phase the cell grows in size and there is an active synthesis between the RNA and proteins.
    II. The third stage of the cell cycle is termed G_2 phase.
    III. In the mitotic phase, the DNA molecule of each chromosome replicates by the synthesis of a new DNA molecule on the template of the existing DNA.

    Codes
    (a) Only I and II are correct (b) Only II and III are correct
    (c) Only III is correct (d) Only I and III are correct
23. The nucleoli and nuclear envelope disappear in which phase of mitosis?
    (a) metaphase (b) prophase (c) anaphase (d) telophase
24. The chromosomes move towards the equator of the spindle in
    (a) prophase (b) prometaphase (c) metaphase (d) interphase

25. Cytokinesis of a plant cell is achieved by the formation of

    (a) cleavage furrow (b) groove (c) cell plate (d) all of these

26. Regarding prophase

    (a) progressive condensation and coiling of chromatin fibres occur at Leptonema stage
    (b) two homologous chromosomes each derived form one parent during sexual reproduction come together and form pairs during Zygonema state
    (c) the nucleolus and nuclear envelope disappear and a spindle begins to be formed at the end of Diakinesis stage
    (d) all of these

27. Match List I with List II and choose the correct answer using the codes given below:

List I	**List II**
A. Mitochondria	I. Suicidal bags
B. Lysosomes	II. Breakdown of fats
C. Ribosomes	III. Powerhouses of cell
D. Glyoxysomes	IV. Sites of protein synthesis

Codes

	A	B	C	D
(a)	I	II	III	IV
(b)	III	IV	II	I
(c)	II	I	IV	III
(d)	III	I	IV	II

28. Which of the following is a nonliving structure which is formed by the living protoplast?

    (a) cell membrane (b) cell wall
    (c) plasmalemma (d) none of the above

29. Which of the following is formed between adjacent cell walls during cell division?

    (a) primary cell wall (b) secondary cell wall
    (c) tertiary cell wall (d) middle lamella

30. The tertiary cell wall formed on the inner surface of the secondary cell wall is composed mainly of

    (a) xylan (b) xylene (c) cellulose (d) glucose

31. The primary cell wall, in dicots, contains complex polysaccharides linked to each other and to the cellulose microfibrils by covalent linkages which include

    (a) xyloglucans (b) arabinogalactans
    (c) rhamnogalacturonans (d) all of the above

32. The part of the pit membrane surrounding the torus is called
    (a) primary pit field (b) pit chamber (c) matrix (d) pit aperture
33. A membrane which permits one substance to pass through more easily than another is called
    (a) permeable (b) semipermeable
    (c) selectively permeable (d) none of these
34. According to the Singer–Nicolson fluid mosaic model, the pores may not be stable, but may be constantly appearing and disappearing. This concept is known as
    (a) membrane pore concept (b) statistical pore concept
    (c) random pore concept (d) none of these
35. The Golgi complex is formed from which of the following cell organelles?
    (a) plasmalemma (b) nuclear envelope
    (c) endoplasmic reticulum (d) all of these
36. The cell walls of plants are made up of fibrils which predominantly contain
    (a) glucose (b) proteins (c) phospholipids (d) polysaccharides
37. The rough endoplasmic reticulum is also known as
    (a) protoplasm (b) tonoplast (c) ergastoplasm (d) all of these
38. Which of the following statements regarding endoplasmic reticulum are correct?
    (a) The endoplasmic reticulum appears to have a combination of triple-layered and micellar structures.
    (b) Geometric symmetry is lacking in ER membranes, which are asymmetrical.
    (c) The smooth endoplasmic reticulum (SER) membranes appear to be involved in the formation of lipoprotein complexes.
    (d) All of these.
39. The outer membrane of the nuclear envelope forms finger-like processes which are pinched off into the cytoplasm. This process has been called
    (a) budding (b) blebbing (c) detoxification (d) none of these
40. Consider the following statements and choose the correct answer using the codes given below:
    I. The outer membrane of the nuclear envelope forms vesicles by 'blebbing'.
    II. The Golgi membranes have been variously described as originating from the plasmalemma, nuclear envelope, annulate lamellae and ER.
    III. Vesicles arising from the Golgi complex fuse to form the plasmalemma.
    IV. The Golgi complex cisternae form primary lysosomes by blebbing.

    Codes

    (a) Only I, II and III are correct (b) Only II and III are correct
    (c) Only I and III are correct (d) I, II, III and IV are correct

41. Eukaryotic ribosomes are called
    (a) 80S ribosomes (b) 70S ribosomes (c) 60S ribosomes (d) 55S ribosomes
42. Which of the following statements regarding eukaryotic ribosomes are correct?
    (a) They contain larger number of proteins.
    (b) They have four molecules of RNA.
    (c) The RNA–protein ratio is nearly 1 : 1 in eukaryotic ribosomes.
    (d) All of these
43. Protein synthesis takes place on groups or clusters of ribosomes which are called
    (a) polyribosomes (b) polysomes (c) ergosomes (d) all of these
44. In eukaryotes, the site of synthesis of most of the ribosomal RNA (rRNA) is the
    (a) nucleus (b) nucleolus (c) mitochondrion (d) lysosome
45. In prokaryotic cells, the subunit RNA precursors are trimmed to form
    (a) 16S and 22S RNA (b) 16S and 23S RNA
    (c) 16S and 24S RNA (d) 5S and 24S RNA
46. In prokaryotes, the nuclear material is not bounded by a definite nucleus which is referred to as the
    (a) nucleoloid (b) nucleoid (c) mononucleate (d) polynucleate
47. On the inner side of the nuclear envelope of many cell types is present fibrous material which has been called the
    (a) fibrous lamina (b) middle lamella (c) epiblema (d) fibrous lamella
48. Which of the following statements regarding the nuclear pores are correct?
    (a) At the margin of each pore the outer and inner membranes are continuous.
    (b) The pores are randomly distributed in many nuclei.
    (c) Both (a) and (b) above
    (d) None of the above
49. Match List I with List II and choose the correct answer using the codes given below:

List I		**List II**	
A	Nuclear envelope	I.	Karyotheca
B.	False nucleoli	II.	Plasmasomes
C.	True nucleoli	III.	Heterosomes
D.	Sex chromosomes	IV.	Karyosomes

Codes

	A	B	C	D
(a)	I	II	IV	III
(b)	I	IV	III	II
(c)	I	IV	II	III
(d)	I	III	II	IV

50. Macromolecules can cross the nuclear envelope
    (a) through the nuclear pores and endoplasmic reticulum
    (b) by active transport
    (c) by blebbing
    (d) all of the above
51. The structural elements of cytoplasm include
    (a) myofilaments (b) microtubules (c) microfilaments (d) all of the above
52. Nuclear membranes appear to have electron transport systems similar to those of the
    (a) mitochondrion (b) chloroplast
    (c) endoplasmic reticulum (d) none of these
53. The nucleolus was first recognized by
    (a) Robert Hooke (b) Fontana (c) Stelly (d) Feldherr
54. Match List I (Conditions of chromosome) with List II (Chromosome numbers) and choose the correct answer using the codes given below:

List I	**List II**
A Monosomics	I. 2n + 2
B. Trisomics	II. 2n – 2
C. Nullisomics	III. 2n – 1 – 1
D. Double monosomics	IV. 2n + 1
E. Tetrasomics	V. 2n – 1

Codes

	A	B	C	D	E
(a)	II	III	IV	I	V
(b)	III	IV	II	I	V
(c)	V	IV	II	III	I
(d)	V	IV	I	III	II

55. The structures present in chromosome include
   (a) matrix, grains, thylakoids and cristae
   (b) centromere, secondary constrictions, telomeres and satellites
   (c) ribosome, cisternae, centromere and cristae
   (d) secondary constrictions, telomeres, satellites and grana

56. Which of the following statements regarding centromere are correct?
   (a) The position of the centromere is constant for a particular chromosome.
   (b) The structure and function of the centromere are different from that of the rest of the chromosome.
   (c) During division the centromere is functional, while the rest of the chromosome is genetically inactive.
   (d) All of these.

57. Match List 1 with List II and choose the correct answer using the codes given below.

List I		**List II**	
A	Metacentric	I.	Centromere near the end of the chromosome
B.	Submetacentric	II.	Centromere near the middle of the chromosome
C.	Acrocentric	III.	Centromere truly terminal
D.	Telocentric	IV.	Centromere situated some distance away from the middle

Codes

	A	B	C	D
(a)	II	III	IV	I
(b)	III	II	I	IV
(c)	II	IV	I	III
(d)	IV	II	I	III

58. The arm ratio is high in which of the following types of chromosomes?
   (a) telocentric chromosomes (b) acrocentric chromosomes
   (c) metacentric chromosomes (d) acentric chromosomes

59. In each diploid set of chromosomes, two homologous chromosomes have additional 'constrictions' called
   (a) secondary constrictions-II (b) nucleolar organizers
   (c) satellites (d) none of these

60. Chromosomes bearing satellites are called SAT chromosomes, where SAT stands for
   (a) sine acid thymonucleinico (b) sine acid thyamine
   (c) satellite acquired thyamine (d) sine acid thyroxide

61. Homologous pairs of identified chromosomes can be arranged in a series of decreasing lengths and such an arrangement is called

 (a) hologram (b) monogram (c) idiogram (d) polygram

62. Which of the following statements regarding supernumerary chromosomes are correct?

 (a) They form synapses among themselves.
 (b) They are generally smaller than normal chromosomes.
 (c) They are found more commonly in plants than in animals.
 (d) None of the above.

63. Five supernumerary chromosomes are found in

 (a) *Tradescantia edwardsiana* (b) *Zea mays*
 (c) *Datura stramonium* (d) none of these

64. L-chromosomes diff2er from supernumerary chromosomes in that

 (a) they are constant in all individuals of the species having them
 (b) they are smaller in size
 (c) they do not possess centromere
 (d) all of the above

65. Certain segments of the chromosomes, or the entire chromosomes, are more condensed than the rest of the karyotype during various stages of the cell cycle. Such difference in thickening has been called

 (a) pinocytosis (b) homopycnosis (c) heteropycnosis (d) polytenic

66. The non-condensed chromosomes which extend during interphase are called

 (a) L chromosomes (b) heterochromosomes
 (c) M chromosomes (d) euchromosomes

67. Which of the following differences between heterochromatin and euchromatin are correct?

 (a) Heterochromatin stains deeply while euchromatin stains less deeply.
 (b) Heterochromatin is late replicating while euchromatin replicates during the early stage of the S phase.
 (c) Heterochromatin does not become acetylated but euchromatin takes up acetic acid on its histone during interphase.
 (d) All of these.

68. A mass of heterochromatin at one end of the chromosome represents the so-called

 (a) nucleolar organizer (b) nucleolus (c) satellite (d) centromere

69. Match List I (Chromosome models) with List II (Discoverers) and choose the correct answer using the codes given below:

List I	**List II**
A. Simple multi-stranded model	I. DuPraw
B. Centipede model	II. Nebel
C. Coiled coil model	III. Steffensen
D. Folded fibre model	IV. Taylor

Codes

	A	B	C	D
(a)	IV	II	III	I
(b)	IV	I	II	III
(c)	IV	III	I	II
(d)	III	IV	II	I

70. Woodcock (1973) showed that chromatin appears to have a 'string of beads' structure and is made up of a number of repeating units which are called

(a) nucleosomes (b) nu bodies (c) B fibres (d) protamines

71. When chromatin is in a partly extended state, the diameter of the beads is reduced to about 70Å which are called

(a) protamines (b) A fibres (c) nucleohistones (d) nu bodies

72. Mitotic chromosomes are made up by the coiling of a large cylindrical fibre called

(a) type A fibre (b) type B fibre (c) unit fibre (d) none of these

73. The string of nucleosomes is coiled into a 300Å diameter structure, called

(a) super solenoid (b) solenoid (c) chromonema (d) seminucleosome

74. The solenoidal model for the superstructure in chromatin was first proposed by

(a) Kihlman (b) DuPraw (c) Finch and Clug (d) Freese and Taylor

75. The processes which are preparatory to cell division during interphase are

(a) replication of DNA
(b) division of centrioles, forming a pair of new centrioles
(c) synthesis of energy-rich compounds which provide energy for mitosis
(d) all of these

76. Match List I (Mitotic phases) with List II (Actions) and choose the correct answer using the codes given below :

List I	List II
A. Prophase	I. The chromosomes are lined up in one place to form the equatorial plate
B. Metaphase	II. Cell becomes spheroid and there is increase in viscosity and refractivity
C. Anaphase	III. The two sets of chromosomes migrate towards the poles
D. Telophase	IV. Nuclear membrane and nucleolus reappear

Codes

	A	B	C	D
(a)	II	III	IV	I
(b)	II	I	IV	III
(c)	II	III	IV	I
(d)	II	I	III	IV

77. Which of the following statements regarding the significance of mitosis are correct?
   (a) The constant number of chromosomes in all the cells of the body is because of mitosis.
   (b) Mitosis restores the surface–volume ratio of the cell.
   (c) Repair of the body take place because of addition of cells by mitosis.
   (d) All of these

78. The spindle without centrioles and blepharoplasts, typically found in higher plants is known as
   (a) anastral spindle
   (b) astral spindle
   (c) nonsymmetrical spindle
   (d) interzonal spindle

79. The fibres extending from the kinetochore (centromere) to the pole of the spindle are called
   (a) continuous fibres
   (b) chromosomal fibres
   (c) interzonal fibres
   (d) unit fibres

80. Fertilization in lower plants is immediately followed by which of the following types of meiosis, giving rise to cells with the haploid chromosome number?
   (a) gametic meiosis
   (b) sporic meiosis
   (c) zygotic meiosis
   (d) none of these

81. During the first meiotic division, the primary oocyte divides into a large cell, the secondary oocyte, and a smaller cell called
   (a) spermatocyte
   (b) first polar body
   (c) spermatozoan
   (d) second polar body

82. Sporic or intermediate meiosis is a characteristic of
    (a) higher plants and some thallophytes (b) algae and fungi
    (c) algae only (d) fungi only

83. Replication of DNA is confined to which of the following phases?
    (a) synthetic phase (b) post-mitotic gap phase
    (c) pre-mitotic gap phase (d) prophase-I

84. DNA synthesis takes place simultaneously on both strands under the action of the enzyme
    (a) endonuclease (b) ligase (c) DNA polymerase (d) exonuclease

85. The synthesis of two strands of DNA takes place in the
    (a) 5′ to 3′ direction (b) 3′ to 5′ direction (c) 3′ to 3′ direction (d) 5′ to 5′ direction

86. Leptotene chromosomes may be irregularly arranged or may be polarized towards the centrioles forming
    (a) matting threads (b) a bouquet
    (c) protein fibrils (d) a synaptonemal complex

87. In plant cells, the chromosomes may sometimes form a tangle of threads, called
    (a) procentric pair (b) proterminal pair (c) synizetic knot (d) matting threads

88. When pairing of homologous chromosomes begins at the ends and proceeds towards the centromere, the phenomenon is called
    (a) intermediate pairing (b) procentric pairing
    (c) proterminal pairing (d) none of these

89. Which of the following statements regarding the pachytene substage of prophase I during meiosis are correct?
    (a) Pachytene is said to begin when synapsis is completed.
    (b) During pachytene the chromosomes are associated in bivalents or tetrads.
    (c) Shortening and thickening of the chromosomes takes place due to longitudinal contraction.
    (d) All of these.

90. The interchange and fusion of broken ends between two homologous chromosomes is known as
    (a) linkage (b) crossing over (c) intermediate pairing (d) synapsis

91. The zygotene and pachytene bivalents are organized into structures called
    (a) synaptonemal complexes (b) synizetic knots
    (c) axial filaments (d) transverse units

92. The separating chromosomes during diplotene are held together at one or more points, where breaks and fusion had occurred, called as
    (a) centromere (b) chiasmata (c) syndesis (d) polar bodies

93. Chiasmata formed along the lengths of chromosomes are called
    (a) terminal chiasmata (b) interzonal chiasmata
    (c) intertwined chiasmata (d) interstitial chiasmata

94. The process by which the terminal chiasma slips off the ends of the chromosome, and its position is taken up by an interstitial chiasma is called
    (a) relational coiling (b) terminalization
    (c) achiasmatic meiosis (d) none of these

95. When there is a single chiasma, the arms of the bivalent rotate through
    (a) 180° (b) 90° (c) 45° (d) 380°

96. The synaptonemal complexes mostly disappear during
    (a) pachytene (b) diplotene (c) zygotene (d) leptotene

97. The process by which cells gradually specialize and undergo change in shape and function is called
    (a) cell differentiation (b) cell division
    (c) parthenogenesis (d) none of these

98. In many cells, the protein released from the endoplasmic reticulum is combined with carbohydrate to produce complex carbohydrates like glycoproteins, mucopolysaccharides, glycogen and glycolipids. This process is called
    (a) glycolysis (b) glycosylation (c) glycolation (d) none of these

99. In the Golgi complex the proteins are concentrated and transformed into
    (a) zymogen granules (b) mucigen granules
    (c) secretory granules (d) secretory vesicles

100. The cisternae on the forming face of the Golgi complex are constantly being formed by fusion of vesicles derived from the
    (a) mitochondria (b) endoplasmic reticulum
    (c) nuclear membrane (d) lysosomes

101. Which of the following statements regarding the functions of Golgi complex are correct?
    (a) synthesizes and secretes polysaccharides
    (b) takes part in sulphate metabolism (sulphation)
    (c) takes part in the formation of primary lysosomes
    (d) all of these

102. Parts of the endoplasmic reticulum adjacent to the Golgi complex are often without ribosomes and are called
    (a) ergastoplasm (b) transitional elements
    (c) cisternae (d) none of these

103. In plant cells the microbodies are known as
   (a) glycosomes (b) glyoxysomes
   (c) glycogenomes (d) glycogen granules

104. Who among the following first observed mitochondria as granular structures?
   (a) Kolliker (1850) (b) Altmann (1892) (c) Michaelis (1900) (d) Clande (1944)

105. Who among the following referred to the mitochondria as bioplasts, and introduced the bioplast theory?
   (a) Kingssbury (1912) (b) Palade (1940)
   (c) Flemming (1882) (d) Altmann (1892)

106. Which four respiratory chain complexes are parts of the electron transport system (ETS)?
   (a) I, II, III and V (b) II, III, IV and V
   (c) I, II, III and IV (d) I, III, IV and V

107. Which of the following have been called as mobile electron carriers?
   (a) Coenzyme Q and cytochrome *c* (b) Coenzyme Q and cytochrome *a*
   (c) Coenzyme Q and cytochrome *b* (d) Cytochrome *a*, cytochrome *b* and cytochrome *c*

108. Match List I (Complexes) with List II (Alternative names) and choose the correct answer using the codes given below:

List I		**List II**	
A	Complex I	I.	Succinate: CoQ reductase
B.	Complex II	II.	Cytochrome *c* oxidase
C.	Complex III	III.	ATPase
D.	Complex IV	IV.	NADH/NADPH: CoQ reductase
E.	Complex V	V.	Reduced CoQ: cytochrome *c* reductase

Codes

	A	B	C	D	E
(a)	I	II	III	IV	V
(b)	II	III	IV	I	V
(c)	IV	I	V	II	III
(d)	III	I	V	IV	II

109. Ubiquinone or coenzyme Q can exist in which of the following states of oxidation?
   (a) quinone (b) semiquinone (c) hydroquinone (d) all of these

110. The fully reduced form of ubiquinone, with hydrogen atoms attached to both oxygens is called as
   (a) semiquinone (b) hydroquinone (c) quinone (d) none of these

111. During the Quinone cycle (Q cycle), a total of how many protons are transported outside the mitochondrion?

(a) one proton (b) two protons (c) three protons (d) four protons

112. Which of the following is a hydrogen carrier?

(a) cytochrome *a* (b) cytochrome *b* (c) cytochrome *c* (d) coenzyme Q

113. Which of the following are fully capable of converting pyruvic acid to carbon dioxide and water?

(a) endoplasmic reticulum (b) plastids
(c) mitochondria (d) lysosomes

114. Which of the following chromoplasts are photosynthetically inactive?

(a) phaeoplasts (b) carotenoids
(c) rhodoplasts (d) blue-green chromoplasts

115. The grana in the chloroplast are interconnected by a network of anastomosing tubules called

(a) stroma lamellae (b) middle lamellae
(c) thylakoid membranes (d) stroma thylakoids

116. The membranous sacs of the thylakoids consist of

(a) protein membranes (b) cellulose membranes
(c) lipoprotein membranes (d) lipid membranes

117. Substances causing instability of the lysosomal membrane, leading to release of enzymes from the lysosomes, are called

(a) inhibitors (b) labilizers (c) stabilizers (d) none of these

118. Which of the following statements are correct?

(a) Autophagy is the digestion of endogenous material.
(b) Heterophagy is the intake of exogenous material into the cell and the subsequent breakdown of this material by enzymes.
(c) Both pinocytosis and phagocytosis require energy.
(d) All of these.

119. Both autophagy and heterophagy may occur simultaneously in the same vesicle called as

(a) ambilysosomes (b) phagocytes
(c) phagocytic vesicles (d) oxylysosomes

120. Microtubules were first discovered by

(a) DeRobertis and Franchi (b) Stautterback
(c) Porter (d) Dovzansky

121. The most abundant glycerophospholipids in higher plants are
   (a) sphingomyelin and cephalin
   (b) lecithin and cephalin
   (c) phosphatidyl choline and lecithin
   (d) Cephalin and phosphatidyl inositol

122. The lecithin molecule of the lipid layer consists of two lipid chains of glycerol and a polar head containing
   (a) cholesterol and cephalin
   (b) cholesterol and phosphate
   (c) phosphate and choline
   (d) none of these

123. The common artificial membranes that have been used as analogues to the plasmalemma in biological study include
   (a) epidermis form the scale of an onion
   (b) outer membrane of the intestine
   (c) copper ferrocyanide membrane
   (d) none of these

124. The carrier protein molecules are presumed to move to and from across the membrane by
   (a) group translocation
   (b) simple diffusion
   (c) thermal diffusion
   (d) osmosis

125. Transport of a variety of sugars like glucose, fructose and mannitol takes place across bacterial membranes through the
   (a) electron transport system
   (b) phosphotransferase system
   (c) simple active transport system
   (d) fluid mosaic system

126. Which of the following energy forms does the phosphotransferase system (PTS) always use?
   (a) adenosine triphosphate (ATP)
   (b) nicotinamide dinucleotide phosphate (NADP)
   (c) adenosine monophosphate (AMP)
   (d) phosphoenol pyruvate (PEP)

127. The antigenic determinants on the surface of erythrocytes are mainly
   (a) glycoproteins
   (b) sialoglycoproteins
   (c) glycolipids
   (d) phospholipids

128. Which of the following enzyme is activated as a result of hormone-receptor binding in the cell membrane?
   (a) adenylate cyclase
   (b) plasmosomes
   (c) lipid dehydrogenase
   (d) isomerase

129. The bulk ingestion of solid food into the cell is known as
   (a) pinocytosis
   (b) exocytosis
   (c) phagocytosis
   (d) none of these

130. Cells, in living organisms, adhere to each other by means of junctions called
(a) dictyosomes (b) plasmosomes (c) glycosomes (d) desmosomes

131. Which of the following are synthesized by Golgi membranes?
(a) triglycerides (b) phosphatidylcholines
(c) glycoproteins (d) all of these

132. The process of glycosylation begins in the
(a) Golgi complex and is completed in the endoplasmic reticulum
(b) endoplasmic reticulum and is completed in the Golgi complex
(c) endoplasmic reticulum and is completed in the mitochondria
(d) Golgi complex and is completed in the mitochondria

133. When cells are ground or homogenized, the cell membranes break up into fragments, which round off to form bodies called
(a) allosomes (b) microbodies (c) microsomes (d) ribosomes

134. A nonliving structure of cell is
(a) cell wall (b) plasma membrane (c) cytoplasm (d) nucleus

135. The chemical present in the cell wall is
(a) pectin (b) lignin (c) cellulose (d) all of these

136. All plant cells possess
(a) middle lamella (b) primary wall (c) lysosomes (d) centrioles

137. Middle lamella occurs
(a) inner to primary wall (b) inner to secondary wall
(c) outer to secondary wall (d) outer to primary wall

138. Middle lamella contains
(a) cellulose (b) pectate (c) lignin (d) cutin

139. Matrix of cell wall is made of
(a) pectin (b) hemicellulose (c) glycoprotein (d) cellulose

140. Chemical absent from matrix of cell wall is
(a) lipid (b) water (c) glycoprotein (d) cellulose

141. Hydrophilic chemical of cell wall is
(a) pectin (b) suberin (c) fat (d) lignin

142. Structural element of cell wall is
(a) matrix (b) microfibrils (c) microtubules (d) arabinogalactans

143. Cellulose microfibrils get bound to pectins of matrix through

(a) hemicellulose (b) lignin
(c) peptidoglycan (d) glycoprotein

144. Different layers of cell wall are

(a) middle lamella and primary wall
(b) primary wall and secondary wall
(c) middle lamella, primary wall and secondary wall
(d) wall layers exclude middle lamella

145. Which is the outermost structure of cell wall?

(a) primary wall (b) secondary wall
(c) tertiary wall (d) middle lamella, if present

146. The first wall layer of cell is

(a) tertiary wall, if present (b) secondary wall
(c) primary wall (d) middle lamella, if present

147. The innermost layer of cell wall is

(a) tertiary wall, if present (b) secondary wall
(c) primary wall (d) middle lamella, if present

148. Which component of cell wall is normally in contact with plasmalemma?

(a) primary wall (b) secondary wall
(c) plasmodesmata (d) middle lamella

149. Primary wall grows by

(a) accretion (b) introgression (c) intussusception (d) none of these

150. Secondary wall grows in thickness by

(a) intercalation (b) introgression (c) accretion (d) epiboly

151. Secondary wall is commonly formed of

(a) single layer (b) many layers (c) two layers (d) three layers

152. Load bearing part of the plant cell is

(a) middle lamella (b) secondary wall (c) primary wall (d) tertiary wall

153. Primary wall has a thickness of

(a) 0 : 1–3.0 μm (b) 0.01–0.03 μm (c) 0.1–0.5 μm (d) 0.5–1.5 μm

154. Primary wall is generally elastic due to absence of

(a) lignin (b) suberin (c) cutin (d) silica

155. Plant cells are distinguishable from animal cells in containing
(a) mitochondria (b) ribosomes (c) ER (d) cell wall

156. The structural material of fungal cell wall is
(a) pectin (b) cellulose (c) peptidoglycan (d) chitin

157. Ripe fruits soften due to
(a) degeneration of cell walls
(b) partial solubilization of pectic compounds
(c) metabolism of tannins
(d) exosmosis

158. Hardness of woody tissue is due to
(a) silica (b) lignin (c) cellulose (d) suberin

159. Impermeability of cork is related to
(a) silica (b) lignin (c) cellulose (d) suberin

160. Cutin occurs in
(a) phloem tissue (b) xylem tissue (c) endodermis (d) epidermis

161. In primary wall, cellulose microfibrils are
(a) small, loose and wavy (b) long, loose and wavy
(c) small, compact and straight (d) long, compact and straight

162. Cellulose content is high in
(a) primary wall (b) secondary wall (c) tertiary wall (d) middle lamella

163. Tertiary wall is known from
(a) compression wood of dicots (b) tension wood of gymnosperms
(c) cotton fibres (d) all hard woods

164. Plasmodesmata were discovered by
(a) Hamtein (b) Kolliker (c) Strasburger (d) Garnier

165. Part of endoplasmic reticulum present in a plasmodesmata is called
(a) sesmotubule (b) cisterna (c) vesicle (d) myeloid body

166. Uncutinized and non-suberized cell wall is
(a) semipermeable (b) permeable
(c) impermeable (d) selectively permeable

167. Adjacent tracheids and vessels can transfer sap through thin areas in their wall called
(a) plasmodesmata (b) gap junctions (c) tight junctions (d) pits

168. Pits are
   (a) depressions in primary walls
   (b) depressions in secondary walls
   (c) both (a) and (b)
   (d) plasmodesmal connections

169. A complete pit is
   (a) depressions in secondary wall
   (b) pit chamber and primary wall
   (c) pit chambers of two adjacent cells and pit membrane
   (d) pit chamber, primary wall and middle lamella

170. Pit membrane consists of
   (a) primary wall
   (b) middle lamella
   (c) middle lamella + primary wall
   (d) primary wall + middle lamella + secondary wall

171. A pit present in the wall of cell lying adjacent to an intercellular space is
   (a) complete pit
   (b) blind pit
   (c) a pit without its partner
   (d) both (b) and (c)

172. A disc-shaped thickening present on the pit membrane is
   (a) torus
   (b) callus
   (c) tylosis
   (d) stoma

173. Apoplast is
   (a) cytoplasm
   (b) protoplast and plasmodesmata
   (c) cell walls
   (d) nonliving continuum made of free or outer spaces of cells

174. Cell coat consists of
   (a) glycocalyx
   (b) cellulose
   (c) cellulose + hemicellulose + pectin
   (d) protein

175. Glycocalyx is
   (a) glycoproteins and glycolipids
   (b) oligosaccharide part of glycolipids and glycoproteins
   (c) lipid and protein parts of glycolipids and glycoproteins
   (d) mucopolysaccharides attached to cell wall

176. Separated cells of two sponge species are mixed up. They will
   (a) remain separate
   (b) aggregate tissue-wise
   (c) aggregate and fuse to form hybrids
   (d) aggregate species-wise and reconstruct the sponges

177. Separated cells of two vertebrates are mixed up. They will
   (a) aggregate species-wise
   (b) aggregate tissue-wise
   (c) aggregate species-wise and then tissue-wise
   (d) aggregate species-wise and then reconstruct the animals

178. Glycocalyx is responsible for
   (a) antigens like those of blood groups ABO (b) immune reactions and histocompatibility
   (c) hormone receptors (d) all of the above

179. The term protoplasm was coined by
   (a) Huxley (b) Purkinje (c) Dujardin (d) Suhultze

180. A unit of protoplasm having a nucleus and covered by plasmalemma is called
   (a) ectoplast (b) cell (c) cytoplast (d) none of these

181. Protoplasm is
   (a) alveolar (b) granular (c) fibrillar (d) crystallo-colloidal

182. Who proposed crystallo-colloidal nature of protoplasm?
   (a) Kolliker (b) Fromann (c) Velton (d) Hanstein

183. Protoplast is
   (a) granular protoplasm (b) whole protoplasm of an organism
   (c) unit of protoplasm contained in a cell (d) all of these

184. The term protoplast was coined by
   (a) Strasburger (b) Hanstein (c) Butschli (d) Fischer

185. Who differentiated prokaryotic and eukaryotic cells?
   (a) Huxley (b) Linnaeus (c) Whittaker (d) Dogherty

186. Mesokaryotic condition was distinguished by
   (a) Whittaker (b) Dodge (c) Copeland (d) Haeckel

187. An akaryotic cell is
   (a) single-nucleated (b) prokaryotic
   (c) denucleated (d) both (b) and (c)

188. Protoplast excluding nucleus is called
   (a) cytoplasm (b) endoplasm (c) ectoplasm (d) protoplasm

189. Cell structure between plasmalemma and karyotheca is
   (a) vacuole (b) nucleoplasm (c) endoplasm (d) cytoplasm

190. Which one is an extracytoplasmic cell organelle?
(a) vacuole (b) ER (c) nucleus (d) Golgi apparatus

191. The term cytoplasm was coined by
(a) Sachs (b) Strasburger (c) Hanstein (d) Flemming

192. Jelly-like semifluid complex of cytoplasm is called
(a) endoplast (b) cytosol (c) cytoplasmic matrix (d) both (b) and (c)

193. Plasmagel or gel part of cytosol in contact with plasmalemma is
(a) ectoplast (b) hyaloplasm (c) hyalosome (d) both (a) and (b)

194. Plasmasol on sol part of cytosol is known as
(a) hyalosome (b) hyaloplasm (c) endoplast (d) both (b) and (c)

195. Which part of protoplast shows streaming or cyclosis?
(a) ectoplast (b) endoplast (c) endoplasmic matrix (d) nucleoplasm

196. Cytoplasmic streaming or cyclosis is absent in
(a) plant cells (b) animal cells (c) protozoan protists (d) prokaryotes

197. Cyclosis was first studied by
(a) Robert Brown (b) Dalton and Felix (c) Amici (d) Sachs

198. Amici (1818) studied cyclosis for the first time in
(a) hydrilla (b) amoeba (c) chara (d) acetabularia

199. Cyclosis is caused by activity of
(a) microtubules (b) microfilaments
(c) Intermediate filaments (d) none of these

200. Circulation type of protoplasmic streaming is studied in
(a) staminal hair cell of *Tradescantia* (b) *Hydrilla* leaf cells
(c) *Vallisneria* leaf cells (d) both (b) and (c)

201. In circulation streaming, protoplasm moves in
(a) one direction
(b) two opposite directions around a vacuole
(c) different directions around different vacuoles
(d) both (a) and (b)

202. In rotation type of cyclosis, the cytoplasmic matrix flows in
(a) one direction (b) two opposite directions
(c) different directions (d) sideways

203. The term organoid is used for
   (a) an ill-defined organ (b) a distinct tissue
   (c) idioblast (d) cell organelle
204. A membrane-lined system of channels present throughout the cytoplasm is
   (a) endoplasmic reticulum (b) Golgi apparatus
   (c) microtubules (d) both (b) and (c)
205. Percentage of cell membranes contained in ER is
   (a) 10–20% (b) 20–30% (c) 30–60% (d) 60–75%
206. ER is absent in
   (a) animal cells (b) prokaryotes (c) plant cells (d) protista and fungi
207. Eukaryotic cells devoid of ER are
   (a) liver cells (b) kidney cells
   (c) mature leukocytes (d) mature erythrocytes
208. An intracellular structure believed to be formed by inpushing of plasmalemma is
   (a) endoplasmic reticulum (b) nuclear envelope
   (c) mitochondrion (d) chloroplast
209. Eukaryotic cells which contain very little of ER are
   (a) early embryonic cells (b) ova
   (c) resting cells (d) none of these
210. ER is made of
   (a) cisternae (b) tubules (c) vesicles (d) none of these
211. Spermatocytes possess ER in the form of
   (a) a few vesicles (b) a few tubules
   (c) abundant tubules and vesicles (d) cisternae, tubules and vesicles
212. Sarcoplasmic reticulum is endoplasmic reticulum of
   (a) adipose cells (b) muscle cells (c) nerve cells (d) leukocytes
213. In nerve cells, ER forms
   (a) myeloid bodies (b) neurotubules (c) Nissl granules (d) neurofilaments
214. Myeloid bodies are granular structures formed of ER in
   (a) retinal cells (b) adipose cells (c) plasma cells (d) reticulocytes
215. In adipose cells, ER is represented by
   (a) cisternae and tubules (b) a few tubules
   (c) cisternae and vesicles (d) a few vesicles

216. ER was discovered by
   (a) Palade (b) Porter *et al.* (c) Thompson (d) both (b) and (c)
217. The term endoplasmic reticulum was coined by
   (a) Thompson (b) Palade (c) Porter (d) Garnier
218. The term ergastoplasm was given to basophilic membranous structures by
   (a) Palade (b) Garnier (c) Schimper (d) Flemming
219. ER was discovered from
   (a) liver cells (b) kidney cells (c) muscle cells (d) nerve cells
220. Membrane thickness of ER is
   (a) 75 Å (b) 90 Å (c) 50–60 Å (d) 30–40 Å
221. ER without association with ribosomes is called
   (a) transitional ER (b) SER (c) granal ER (d) both (b) and (c)
222. Rough endoplasmic reticulum is the one that contains
   (a) abundant tubules (b) association with ribosomes
   (c) fenestrations (d) both (b) and (c)
223. RER is
   (a) neutrophilic (b) acidophilic (c) basophilic (d) neurogenic
224. What is more abundant in SER?
   (a) cisternae and vesicles (b) cisternae and tubules
   (c) tubules and vesicles (d) cisternae
225. What is more abundant in RER?
   (a) cisternae (b) vesicles (c) tubules (d) both (a) and (b)
226. P_{450} and P_{448} occur over
   (a) SER (b) RER (c) annulate ER (d) transitional ER
227. SER takes part in synthesis of
   (a) lipids and steroids (b) vitamins (c) carbohydrates (d) none of these
228. RER is specialized for synthesis of
   (a) local proteins (b) local proenzymes
   (c) proteins and proenzymes for transport (d) hormones
229. Pollutants and carcinogens are detoxified by
   (a) P_{450} and P_{448} (b) SER in liver (c) RER in liver (d) both (a) and (b)

230. Ribophorins are required for
   (a) synthesis of ribosomes in nucleolus
   (b) attachment of ribosomes over RER
   (c) attachment of ribosome subunits
   (d) attachment of mRNA to ribosomes for protein synthesis

231. Which of the following provides mechanical support to cell?
   (a) ribosomes (b) Golgi bodies
   (c) lysosomes (d) endoplasmic reticulum

232. Endoplasmic reticulum occurs in the form of
   (a) vesicles (b) cisternae (c) tubules (d) none of these

233. Cells specialized in secretion of proteinaceous substances possess
   (a) bound ribosomes (b) free ribosomes
   (c) abundant amino acids (d) abundant mRNA

234. Besides proteins, ribosomes contain
   (a) DNA (b) RNA (c) both DNA and RNA (d) lipids

235. Sedimentation unit of ribosome is
   (a) μ (micron) (b) μm (millimicron) (c) Å (Angstrom) (d) S (Svedberg)

236. Cytoribosomes of eukaryotes are different from those of bacterial cells in having
   (a) smaller size (70S type) (b) larger size (80S type)
   (c) differential chemical structure (d) none of these

237. A ribosome is composed of
   (a) a single unit (b) two subunits (c) three subunits (d) four subunits

238. Ribosomes develop from
   (a) nucleus (b) nucleolus (c) endoplasmic reticulum (d) mitochondria

239. Polysome is a chain of
   (a) oxysomes (b) sphaerosomes (c) ribosomes (d) dictyosomes

240. The enzyme peptidyl transferase occurs in
   (a) larger subunit of ribosome (b) blue-green algae
   (c) endoplasmic reticulum (d) nucleus

241. Organelle ribosomes occur in
   (a) bacteria (b) blue-green algae
   (c) plastids and mitochondria (d) nucleus

242. Organelle ribosomes resemble
(a) organelle ribosomes of prokaryotes (b) cytoribosomes of prokaryotes
(c) cytoribosomes of eukaryotes (d) all of these

243. Ribosomes were first observed by
(a) Claude (b) Palade (c) George (d) DeLuve

244. Palade granules are
(a) glycoprotein (b) pigment granules (c) excretory vesicles (d) ribosomes

245. Ribosomes were discovered by
(a) Robinson and Brown (b) Thompson
(c) Perner (d) Schiraper

246. Organelle and prokaryotic ribosomes are generally
(a) 80S (b) 100S (c) 70S (d) 45S

247. Which of these ribosomes do eukaryotes possess?
(a) 60S (b) 70S (c) 70S (d) Both (b) and (c)

248. An organelle devoid of membrane covering is
(a) lysosome (b) peroxisome (c) ribosome (d) sap vacuole

249. The element required for bringing about the union of the ribosome subunits is
(a) Ca^{2+} (b) Mg^{2+} (c) Fe^{2+} (d) Cu^{+}

250. The size of 80S ribosome is
(a) 150–200 Å or 135–150 Å (b) 200–290 Å or 170–210 Å
(c) 250–300 Å or 180–220 Å (d) 300–340 Å or 200–240 Å

251. Dalton weight of 70S ribosome is
(a) 2.7–3.0 million (b) 3.1–3.5 million (c) 3.6–4.0 million (d) 4.0–4.5 million

252. rRNA–protein ratio of 80S ribosome is
(a) 40–44 : 56–60 (b) 45–50 : 50–55 (c) 50–55 : 45–50 (d) 60–65 : 35–40

253. rRNA present in 40S subunit of ribosome is
(a) 5S (b) 5.8S (c) 16S (d) 18S

254. rRNA present in 60S subunit of ribosome is
(a) 5S (b) 5.8S (c) 28S (d) all the above

255. rRNA present in 50S subunit of ribosome is
(a) 23S (b) 5S (c) both (a) and (b) (d) 23S, 5.8S and 5S

256. The number of proteins associated with 60S ribosome subunit is
(a) 40 (b) 34 (c) 30 (d) 21

257. Chloramphenicol prevents protein synthesis over
(a) prokaryotic ribosomes (b) organelle ribosomes
(c) both (a) and (b) (d) 80S ribosomes

258. Most abundant organelles of the cell are
(a) mitochondria (b) plastids (c) ribosomes (d) microbodies

259. Golgi apparatus was first seen by
(a) George (b) Golgi (c) Cajal (d) Robinson and Brown

260. Who studied Golgi apparatus for the first time?
(a) Golgi (b) George (c) Cajal (d) Koltzoff

261. Golgi was able to differentiate Golgi apparatus through
(a) phase contrast microscopy (b) metallic impregnation technique
(c) electron microscopy (d) special redox dye

262. Golgi studied Golgi apparatus in
(a) nerve cells of dog and fish (b) goblet cells of dog's stomach
(c) nerve cells of barn owl and cat (d) goblet cells of barn owl and cat

263. The metallic stain used by Golgi was
(a) lead acetate (b) osmium chloride and silver salts
(c) phosphotungstate (d) palladium

264. Golgi apparatus was first seen under electron microscope by
(a) Novikoff (b) Dalton and Felix
(c) Rhodin (d) De Robertis and Franchi

265. A cell organelle with a definite polarity is
(a) ribosome (b) mitochondrion (c) Golgi apparatus (d) chloroplast

266. Golgi apparatus is made of
(a) cisternae (b) tubules and vesicles
(c) golgian vacuoles (d) none of these

267. On which side of Golgi apparatus are the membranes thin?
(a) concave distal side (b) concave proximal side
(c) convex distal side (d) convex proximal side

268. Golgian vacuoles develop from
    (a) tubules (b) convex proximal cisternae
    (c) concave distal cisternae (d) transition vesicles

269. Golgi apparatus receives biochemicals with the help of transition vesicles formed by
    (a) ER (b) plasmalemma (c) lysosomes (d) nuclear blebs

270. Number of cisternae present in Golgi apparatus of an animal cell is
    (a) 4–8 (b) 8–13 (c) 13–20 (d) 20–30

271. Space between adjacent cisternae and Golgi apparatus is
    (a) 15 Å (b) 30 Å (c) 80–100 Å (d) 100–300 Å

272. Inter-cisternal space is occupied by
    (a) cytosol (b) cementing materials (c) fibrils (d) none of these

273. Golgi apparatus helps in
    (a) cell plate formation (b) matrix formation of connective tissue
    (c) secretion of tears (d) all of these

274. Golgi apparatus is concerned with
    (a) excretion (b) secretion (c) ATP synthesis (d) RNA synthesis

275. Golgi complex is not found in
    (a) nerve cells (b) RBCs (c) germ cells (d) none of these

276. Golgi apparatus is found in
    (a) cryptogams only (b) phanerogams only (c) prokaryotic cells (d) eukaryotic cells

277. Cell organelle specialized in forming acrosome part of sperm is
    (a) mitochondrion (b) centriole (c) peroxisome (d) Golgi apparatus

278. Amongst plants, Golgi apparatus is absent in
    (a) sieve tube cells (b) sperms of bryophytes
    (c) sperms of pteridophytes (d) none of these

279. Membrane flow occurs in
    (a) Golgi apparatus (b) ER
    (c) karyotheca (d) contractile vacuoles

280. Isolated units of Golgi apparatus found in plant cells are called
    (a) Golgisomes (b) dictyosomes (c) lipochondria (d) cisternae

281. Dictyosomes are unicisternal in
    (a) fungi (b) protistans (c) algae (d) bryophytes

282. Which cell organelle is involved in formation of saliva, tears and sweat?
(a) RER (b) SER (c) Golgi apparatus (d) both (b) and (c)

283. Golgi apparatus takes part in synthesis of
(a) carbohydrates (b) glycoproteins (c) hormones (d) none of these

284. Lysosomes were first seen and named by
(a) De Duve (b) Palade (c) Novikoff (d) Robertson

285. Lysosomes originate from
(a) plasmalemma (b) Golgi apparatus (c) both (a) and (b) (d) RER

286. Main function of lysosomes is
(a) secretion (b) respiration
(c) extracellular digestion (d) intracellular digestion

287. Lysosomes are
(a) All alike (b) of three types (c) of four types (d) of two types

288. De Duve discovered lysosomes from
(a) orchid root cells (b) rat liver cells (c) rat kidney cells (d) leaf cells

289. Lysosomes are absent in
(a) erythrocytes (b) plasma cells (c) nerve cells (d) muscle cells

290. pH of lysosome interior is
(a) 10–12 (b) 8–10 (c) 5–7 (d) 4–5

291. The number of hydrolases identified from different lysosomes is
(a) 30 (b) 35 (c) 50 (d) 70

292. An enzyme often present in lysosomes and having antiseptic property is
(a) lysozyme (b) plasminogen activator
(c) lipofuscin (d) both (b) and (c)

293. Secondary lysosomes are
(a) digestive vacuoles (b) autophagic vacuoles
(c) residual vacuoles (d) none of these

294. Lysosomes containing inactive enzymes are called
(a) primary lysosomes (b) secondary lysosomes
(c) autophagosomes (d) residual bodies

295. Lysosomes along with food content is known as
(a) primary lysosomes (b) secondary lysosomes
(c) microbodies (d) residual bodies

296. Autophagic vacuoles digest
   (a) cell organelles
   (b) solid particles of phagosomes
   (c) fluid droplets of pinosomes
   (d) all of these

297. Residual vacuoles throw their undigested materials by
   (a) pinocytosis (b) phagocytosis (c) ephagy (d) diffusion

298. Lipofusion granules present in nerve cells are actually
   (a) primary lysosomes
   (b) digestive vacuoles
   (c) residual bodies
   (d) newly digested lipids

299. Cartilage matrix is digested during its osteogenesis through
   (a) intracellular autophagic activity
   (b) extracellular lysosomal activity
   (c) intercellular heterophagic activity
   (d) both (b) and (c)

300. Disease caused by hyperactivity of lysosomes is
   (a) arthritis (b) gout (c) lung fibrosis (d) none of these

301. A disease caused by reduced ephagic activity of residual bodies is
   (a) hepatitis (b) polynephritis (c) hypertension (d) both (a) and (b)

302. Which of the following is a lysosomal activity?
   (a) reabsorption of tadpole tail
   (b) mobilization of stored substances
   (c) removal of obstructions
   (d) all of these

303. When are lysosomes extra-active?
   (a) seed maturation (b) seed germination (c) flowering (d) fruiting

304. Cell organelles getting stained with redox dye Janus Green is
   (a) lysosomes (b) mitochondrion (c) ribosome (d) Golgi apparatus

305. An early name of mitochondrion was
   (a) fila (b) sarcosome (c) bioplast (d) none of these

306. Mitochondria were discovered by
   (a) Michaelis (b) Benda (c) Kolliker (d) Krebs

307. The scientist who related mitochondria to aerobic respiration is
   (a) Kinsbury (b) Michaelis (c) Seekevitz (d) Fernandes-Moran

308. A eukaryotic aerobic cell that does not possess mitochondria is
   (a) liver cell (b) kidney cell (c) erythrocyte (d) leukocyte

309. Single mitochondrion is found in
   (a) *Microasterias* (b) *Chlorella* (c) *Micromonas* (d) none of these

310. The number of mitochondria present in a sperm is
   (a) 4–6 (b) 12–14 (c) 20–24 (d) 40–50
311. The number of mitochondria found in a kidney cell is
   (a) 100–150 (b) 300–400 (c) 500–600 (d) 1000–1200
312. The number of mitochondria found in a liver cell may be
   (a) 500–1000 (b) 1000–1200 (c) 1500–2000 (d) 2000–4000
313. Giant amoeba (*Chaos chaos*) has ______ mitochondria
   (a) 5000–1000 (b) 50,000 (c) 80,000 (d) 100,000
314. The number of mitochondria found in flight muscle cell can be
   (a) 500,000 (b) 200,000 (c) 100,000 (d) 50,000–100,000
315. In animal cell, a mitochondrion is the
   (a) largest organelle (b) second largest organelle
   (c) third largest organelle (d) none of these
316. Outer mitochondrial membrane resembles bacterial membrane and outer chloroplast membrane in having
   (a) selective permeability (b) single ion channels
   (c) porin (d) none of these
317. Cristae were discovered by
   (a) Palade (b) Fernandes-Moran
   (c) Nass and Afzelius (d) Luck and Rich
318. ______ are the powerhouses of the cell.
   (a) ATP (b) lysosomes (c) mitochondria (d) chloroplasts
319. The number of oxisomes present in a mitochondrion is believed to be
   (a) 10^3–10^4 (b) 10^4–10^5 (c) 10^5–10^6 (d) 10^6–10^7
320. In the inner mitochondrial membrane, proton channel is constituted by
   (a) F_0 (b) F_1 (c) NADH (H^+) (d) cytochrome oxidase
321. Mitochondrial matrix contains
   (a) enzymes (b) DNA and RNA (c) ribosomes (d) none of these
322. Small particles attached to inner mitochondrial membrane are
   (a) ergosomes (b) cristae (c) elementary particles (d) quantasomes
323. Mitochondria are concerned with
   (a) oxidative phosphorylation (b) intermediate metabolism
   (c) Krebs cycle (d) none of these

324. In prokaryotes oxidative phosphorylation is carried out by
(a) chondriome (b) plasma membrane (c) lomasome (d) chromatophore

325. Mitochondria are semi-autonomous due to
(a) presence of functional naked DNA (b) presence of ribosomes
(c) ability to divide (d) all of these

326. Pigment-free plastids are
(a) chloroplasts (b) chromoplasts (c) lysosomes (d) leucoplasts

327. The term plastid was given by
(a) Schimper (b) Haeckel (c) Hanstein (d) Strasburger

328. All types of plastids are formed from
(a) proplastids (b) leucoplasts (c) amyloplasts (d) aleuroplasts

329. Proplastids are found in
(a) root cells (b) storage cells (c) meristematic cells (d) cortical cells

330. The plastid which can form all other types of plastids is
(a) leucoplast (b) amyloplast (c) chloroplast (d) chromoplast

331. Chromoplasts are formed from chloroplasts during
(a) ripening of tomato (b) ripening of chilli
(c) development of carrot (d) both (a) and (b)

332. Chromoplasts are formed from leucoplasts in
(a) rose petals (b) carrot (c) damia florets (d) none of these

333. The plastids with irregular shape are
(a) leucoplasts (b) chloroplasts (c) chromoplasts (d) amyloplasts

334. Irregular shape of chromoplasts is due to
(a) genetic differences (b) crystallization of carotenoids
(c) formation of lipids (d) destruction of lamellae

335. Leucoplasts are present in
(a) green cells (b) pigmented cells other than green
(c) nonpigmented cells (d) both (a) and (b)

336. Organelle covered by double membrane is
(a) nucleus (b) mitochondrion (c) plastid (d) none of these

337. Plastids with 3 or 4 membrane coverings are found in
(a) gymnosperms (b) pteridophytes
(c) euglenoids and brown algae (d) bryophytes

338. Plastids contain

(a) double membrane covering (b) DNA, RNA and ribosomes
(c) lamellae (d) none of these

339. Starch is stored in

(a) chromoplasts (b) amyloplasts (c) chloroplasts (d) both (b) and (c)

340. Plastids storing proteins are called

(a) elaioplasts (b) oleosomes (c) aleuroplasts (d) phaeoplasts

341. Plastids storing fat are called

(a) elaioplasts (b) sphaerosomes (c) adipocytes (d) pyrenoids

342. The three types of plastids were named by

(a) Meyer (b) Schimper (c) Hanstein (d) Flemming

343. _______ are the autoplasts of Meyer.

(a) leucoplasts (b) proplastids (c) chloroplasts (d) both (a) and (b)

344. Structural elements of chloroplasts are

(a) plastoglobuli (b) photosynthetic pigments
(c) thylakoids (d) quantasomes

345. Thylakoids of chloroplasts are also called

(a) lamellae (b) fret membranes (c) loculi (d) grana

346. In chloroplasts, the photosynthetic pigments occur

(a) in matrix (b) grana
(c) membranes of thylakoids (d) loculi and fret channels of thylakoids

347. Grana is present in

(a) mitochondria (b) chloroplasts (c) Golgi bodies (d) ribosomes

348. Grana are

(a) protein-storing plastids (b) coloured plastids
(c) stacks of thylakoids (d) individual thylakoids present in stroma

349. The number of grana present in a chloroplast is

(a) 10–20 (b) 20–30 (c) 30–40 (d) 40–60

350. The number of thylakoids in a granum is

(a) 10–100 (b) 5–10 (c) 100–200 (d) 200–500

351. Particles of thylakoid membranes involved in ATP synthesis are called

(a) quantasomes (b) $CF_0 - CF_1$ (c) photosystems (d) pyrenoids

352. Structure associated with chloroplast of green algae is
   (a) pyrenoid (b) stigma
   (c) both (a) and (b) (d) endoplasmic reticulum
353. Plastids which provide bright colour to flowers and fruits are
   (a) chloroplasts (b) leucoplasts (c) chromoplasts (d) proplastids
354. Peroxisomes and glyoxisomes are
   (a) energy-transforming organelles (b) membrane-less organelles
   (c) macrobodies (d) microbodies
355. Which one is a microbody?
   (a) sphaerosome (b) lysosome (c) peroxisome (d) none of these
356. The organelle covered by a single membrane is
   (a) sphaerosome (b) peroxisome (c) glyoxisome (d) none of these
357. The microbody present only in plants is
   (a) sphaerosome (b) peroxisome (c) glyoxisome (d) both (b) and (c)
358. Microbodies resemble mitochondria in
   (a) using oxygen (b) producing reducing power
   (c) having catalase (d) formation of ATP
359. Microbodies differ from mitochondria in
   (a) single membrane (b) absence of DNA (c) direct oxidation (d) none of these
360. New sphaerosomes develop from
   (a) old sphaerosomes (b) ER (c) Golgi apparatus (d) prosphaerosomes
361. Sphaerosomes are involved in
   (a) utilization of alcohol (b) storage of fat
   (c) synthesis and storage of fat (d) synthesis and storage of carbohydrates
362. Sphaerosome was discovered by
   (a) Rhodin (b) Perner (c) Koltzoff (d) Claude
363. An organelle reported to have lysosomal activity in plants is
   (a) sphaerosome (b) glyoxysome (c) peroxisome (d) Centriole
364. A cell organelle called uricosome is
   (a) lysosome (b) sphaerosome (c) peroxisome (d) glyoxysome
365. Microbodies possess
   (a) hydrolases (b) oxidases (c) isomerases (d) none of these

366. The enzyme catalase occurs in
 (a) lysosome (b) sphaerosome
 (c) peroxisome (d) peroxisome and glyox

367. Peroxisome was discovered by
 (a) De Duve (b) Rhodin
 (c) De Robertis and Franchi (d) Beevers

368. The cell organelle having the enzyme uricase or urate oxidase is
 (a) lysosome (b) glyoxisome (c) peroxisome (d) sphaerosome

369. Major function of peroxisomes is oxidation of
 (a) excess purine (b) surplus amino acids
 (c) alcohol and drugs (d) none of these

370. Glyoxisomes occur in
 (a) leaf cells (b) fatty seeds (c) roots (d) meristematic cells

371. Glyoxysomes are useful in
 (a) converting sugars to fat
 (b) converting fats to sugars
 (c) deamination and converting amino acids to fatty acids
 (d) amination and changing fatty acids to amino acids

372. The term cytoskeleton was given by
 (a) Koltzoff (b) Rhodin (c) Menke (d) Park

373. The term microtubule was coined by
 (a) De Robertis and Franchi (b) Mayer
 (c) Palade (d) Slautterback

374. A microtubule has a diameter of
 (a) 100 Å (b) 150 Å (c) 250 Å (d) 100 nm

375. A microtubule is made of
 (a) protofilaments (b) microfilaments (c) microfibrils (d) elementary fibrils

376. Number of protofilaments in a microtubule is
 (a) 10 (b) 13 (c) 16 (d) 18

377. Microtubules are formed of a protein called
 (a) tubulin (b) actin (c) flagellin (d) myosin

378. Microtubules are present in
 (a) bacteria (b) viruses (c) eukaryotic cells (d) mycoplasma

379. The protein tubulin occurs in microtubules as
   (a) monomers (b) homodimers (c) heterodimers (d) heteropolymers
380. Microtubule assembly is inhibited by
   (a) GTP (b) Ca^{2+} (c) Mg^{2+} (d) colchicine
381. Microtubule assembly is promoted by
   (a) calmodulin (b) GTP (c) Ca^{2+} and Mg^{2+} (d) None of these
382. Microfilaments were discovered by
   (a) Slautterback (b) Paleviz *et al.* (c) Altman (d) Ledbetter and Porter
383. The diameter of a microfilament is
   (a) 5–6 nm (b) 10 nm (c) 15 nm (d) 25 nm
384. Microfilaments are required for
   (a) movement of flagella and cilia (b) cell polarity
   (c) sol-gel changes (d) all of these
385. Cell polarity is determined by
   (a) intermediate filaments (b) microtubules
   (c) protofilaments (d) centrioles
386. Cytoskeletal components that determine orientation of cellulose microfibrils are
   (a) microfilaments (b) microtubules
   (c) intermediate filaments (d) basal bodies
387. Which of the following provides support to microvilli, membrane rufflings and pseudopodia?
   (a) microfilaments (b) desmin and vimentin filaments
   (c) microtubules (d) ER
388. The contractile constituent of cytoskeleton is
   (a) microtubules (b) intermediate filaments
   (c) microfilaments (d) microfibrils
389. The diameter of intermediate filament is
   (a) 1 nm (b) 5 nm (c) 7–8 nm (d) 10 nm
390. Tonofibrils are
   (a) intermediate filaments (b) microfilaments
   (c) cross-linked microtubules (d) both (a) and (b)
391. In nerve fibres, intermediate filaments form
   (a) neurotubules (b) neurofilaments (c) neurofibrils (d) dendrites

392. In muscle cells, intermediate filaments produce
(a) Z-line (b) M-line (c) H-line (d) both (a) and (b)

393. Centrioles are found
(a) singly (b) in pairs (c) in triplets (d) in quadruplets

394. The centriole pair occurs in a complex called
(a) centrosome (b) centromere (c) kinetochore (d) basal plate

395. The two centrioles of a pair occur
(a) parallel to each other (b) at right angles to each other
(c) at an angle other than 90° (d) end to end

396. Each centriole has on its periphery, MTOCs called
(a) spokes (b) pericentriolar thickenings
(c) massules (d) both (a) and (b)

397. Centrosome was discovered by
(a) Boveri (b) Porter (c) Massules (d) Schimper

398. Cell organelle having a cartwheel constituent is
(a) centriole and basal body (b) microtubule
(c) microfilament (d) basal plate

399. The pattern of organization in centrioles is
(a) 9 + 2 (b) 9 + 3 (c) 9 + 0 (d) 9 + 1

400. In ultrastructure, sphaeroplasts resemble
(a) centrioles (b) flagella (c) cilia (d) none of these

401. Subfibres of cilia and flagella are made of
(a) tubulin (b) elastin (c) myosin (d) actin

402. Pattern of organization of eukaryotic cilia flagella is
(a) 9 + 0 (b) 9 + 1 (c) 9 + 2 (d) 9 + 3

403. Bacterial flagellum is made of
(a) flagellin (b) actin (c) elastin (d) myosin

404. Number of microtubules present in a centriole is
(a) 20 (b) 25 (c) 27 (d) 18

405. Number of microtubules found in a cilium/flagellum is
(a) 11 (b) 13 (c) 18 (d) 20

406. Doublet fibrils of a cilium or flagellum are tilted at an angle of

(a) 5° (b) 10° (c) 15° (d) 20°

407. Triplet fibrils of a centriole are tilted at angle of

(a) 40° (b) 30° (c) 20° (d) 10°

408. Which microtubule or subfibre of the triplet disappears while passing through basal plate?

(a) C (b) A (c) B (d) None of these

409. The number of dynein arms are

(a) four attached to subfibre B (b) two attached to subfibre B
(c) two attached to subfibre A (d) four attached to subfibre A

410. Cilium or flagellum is structurally bilateral due to the presence of

(a) central sheath (b) singlet fibrils (c) double bridge (d) both (b) and (c)

411. A cilium beats

(a) asymmetrically by sweeping action (b) symmetrically by sweeping action
(c) symmetrically by undulatory action (d) asymmetrically by undulatory action

412. The beat of a flagellum is

(a) independent, undulatory and asymmetrical
(b) independent, undulatory and symmetrical
(c) coordinated, pendular and symmetric
(d) coordinated, pendular and asymmetric

413. The coordinated beating of cilia is

(a) undulatory (b) metachronous
(c) metachronous or isochronous (d) both (a) and (b)

414. A typical vacuole possesses

(a) nucleoplasm (b) cytoplasm (c) hydroplasm (d) both (b) and (c)

415. Hydroplasm of a vacuole possesses

(a) water (b) minerals (c) water and minerals (d) air

416. A vacuole without a regular covering membrane is

(a) contractile vacuole (b) food vacuole
(c) sap vacoule (d) gas vacuole

417. Which one possesses only a protein membrane?

(a) feeding canals of contractile vacuole (b) vesicles of a gas vacuole
(c) contractile vacuole (d) gas vacuole

418. Gas vacuoles occur in

(a) prokaryotes (b) protistans (c) both (a) and (b) (d) fishes

419. Gas vacuole takes part in

(a) storage of metabolic gases
(b) protection from intense radiations
(c) buoyancy regulation
(d) all of the above

420. Food vacuole is formed from

(a) absorbed and digested food
(b) phagosome + lysosome
(c) feeding canals + lysosome
(d) feeding canals + phagosome

421. Vacuoles are

(a) cytoplasmic organelles
(b) noncytoplasmic organelles
(c) Noncytoplasmic sacs
(d) cytoplasmic sacs

422. Contractile vacuoles take part in

(a) storage of wastes
(b) osmoregulation
(c) excretion
(d) both (b) and (c)

423. Filling of contractile vacuole and its swelling is called

(a) diastole
(b) diapause
(c) systole
(d) both (a) and (b)

424. Bursting of contractile vacuole to throw its contents is

(a) ephagy
(b) systole
(c) diapause
(d) dehydration

425. The number of sap vacuoles present in an animal cell is

(a) one and large (b) many and large (c) many and small (d) none of these

426. Young plant cells possess sap vacuole(s) that is/are

(a) single, large and central
(b) many, large and dispersed
(c) many, small and dispersed
(d) none of these

427. A single large central vacuole occurs in

(a) mature animal cells
(b) mature plant cells
(c) developing animal cells
(d) developing plant cells

428. The liquid content of a vacuole is called

(a) cell sap (b) matrix (c) nucleoid (d) core

429. Protoplasm present inside the nucleus is called

(a) hyaloplasm (b) nucleoplasm (c) nuclear matrix (d) hydroplasm

430. Nucleus was discovered by Robert Brown in
(a) orchid root cells (b) bean root cells (c) maize stem cells (d) wheat stem cells

431. A well organized nucleus is present in
(a) bacteria (b) prokaryotes (c) blue green algae (d) eukaryotes

432. A plant (cell) having more than two nuclei is known as
(a) syncytial (b) coenocyte (c) polynucleate (d) plasmodium

433. An animal cell with numerous nuclei is called
(a) coenocyte (b) syncytial (c) plasmodium (d) both (a) and (b)

434. Nuclear envelope is composed of
(a) single membrane (b) two membrane
(c) proteinaceous strand (d) both (a) and (b)

435. A biological membrane containing true pores is
(a) nuclear envelope (b) plasma membrane
(c) vitelline membrane (d) plastid envelope

436. Karyotheca is derived from
(a) envelope remains (b) ER
(c) Golgi apparatus (d) Both (a) and (b)

437. Perinuclear space has a width of
(a) 100–150 Å (b) 100–300 Å (c) 400–700 Å (d) 300–1000 Å

438. Diameter of the nuclear pores is
(a) 50–150 Å (b) 100–200 Å (c) 300–500 Å (d) 300–1000 Å

439. Nucleoplasm is also called
(a) nuclear sap (b) karyolymph (c) both (a) and (b) (d) nuclear matrix

440. Nucleus controls cytoplasmic functioning by sending out
(a) cholesterol (b) protein (c) RNAs (d) DNA copies

441. In a nondividing cell, most of DNA is located in
(a) mitochondria (b) chloroplasts (c) chromosomes (d) chromatin

442. Chromatin material which remains condensed during interphase is called
(a) heterochromatin (b) euchromatin (c) chromonematin (d) megachromatin

443. Heterochromatin is
(a) genetically and metabolically very active (b) genetically and metabolically inert
(c) very active genetically (d) inert metabolically

444. Nucleolus is a
(a) distinct membrane-bound organelle (b) part of chromosome
(c) ribonucleoprotein entity (d) DNA component

445. Nucleolus was discovered by
(a) Robert Brown (b) Leeuwenhoek (c) Robert Hooke (d) Fontana

446. Size of nucleolus is large where
(a) protein synthesis is active (b) protein synthesis is small
(c) there is no protein synthesis machinery (d) lipid metabolism is rapid

447. Nucleolus contains
(a) genetic instructions (b) ribosome assembly line
(c) protein synthetic machinery (d) enzymes for polysaccharide formation

448. The organelle having a significant role in mitosis is
(a) lysosomes (b) mitochondria (c) Golgi bodies (d) nucleolus

449. Granules of nucleolus are precursors of
(a) chromosomes (b) ribosomes (c) RNA (d) none of these

450. Nucleolus is formed from
(a) nucleus (b) nuclear sap (c) SAT chromosome (d) giant chromosome

451. Temporary inactivation of one chromosome of a pair is
(a) facultative heterochromatin (b) constitutive heterochromatin
(c) inactivated chromatin (d) none of the above

452. Barr body is
(a) nucleolus (b) facultative heterochromatin
(c) constitutive heterochromatin (d) euchromatin

453. Components of nucleus are
(a) karyotheca, nucleolus, chromatin, nucleoplasm and nuclear matrix
(b) nuclear envelope, nucleolus, chromatin and nucleoplasm
(c) nuclear envelope, nucleoplasm and chromatin
(d) all of the above

454. Nuclear matrix is
(a) heterochromatin complex (b) crystallo-colloidal complex
(c) fibrous complex (d) nucleoplasm

455. Nuclear matrix is formed of
(a) histones (b) neutral proteins (c) acid proteins (d) nucleosomes

456. Dense fibrous complex lying adjacent to inner membrane of nuclear envelope is
(a) chromatin (b) fibrous lamina (c) nuclear lamina (d) both (b) and (a)

457. Structural element of chromatin is
(a) histone (b) acid protein and DNA
(c) nucleosome (d) nuclear matrix

458. Largest organelle of the cell is
(a) nucleus (b) chloroplast (c) mitochondrion (d) vacuole

459. The term chromatin was coined by
(a) Heitz (b) Flemming (c) Fontana (d) Bowman

460. The term nucleoplasm was given by
(a) Strasburger (b) Flemming (c) Harris and James (d) Bowman

461. Nucleolus contains
(a) no membrane covering (b) amorphous matrix and granular zone
(c) fibrillar zone and chromatin (d) all of these

462. The term nucleolus was coined by
(a) Bowman (b) Fontana (c) Hanstein (d) Strasburger

463. Constitutive heterochromatin is
(a) condensed chromatin (b) present in all cells
(c) made of repetitive bases (d) none of these

464. Nonliving cell inclusions are called
(a) ergastic substances (b) deuteroplasmic substances
(c) paraplasmic substances (d) none of these

465. Idioblast is
(a) plant cell different from others (b) plant cell having cell inclusions
(c) both (a) and (b) (d) animal cell different from others

466. Aleurone layer present in the outer part of maize grain is rich in
(a) stored protein (b) stored fat (c) stored starch (d) none of these

467. An aleurone grain is
(a) leucoplast (b) proteinoplast (c) aleuroplast (d) none of these

468. Crystalloid and globoid are components of
(a) sphaerosomes (b) elaioplasts (c) aleurone grains (d) starch grains

469. Sticky and water-soluble heteropolysaccharides formed during breakdown of plant cell walls are
   (a) resins (b) gums (c) alkaloids (d) mucus

470. Alkaloids are
   (a) sour acidic substances (b) bitter basic substances
   (c) sweet neutral substances (d) tasteless ingredients

471. A common alkaloid taken in daily is
   (a) atropine (b) nicotine (c) caffeine (d) coccaine

472. A colloidal substance soluble in alcohol which exudes from plain wounds is
   (a) resin (b) tannin (c) alkaloid (d) latex

473. Asafoetida is a/an
   (a) oleoresin (b) gum resin (c) hard resin (d) none of these

474. Resin got from pinus is
   (a) hard resin (b) mixture of resin and latex
   (c) oleoresin (d) gum resin

475. A resin obtained from animals is
   (a) damar (b) shellac (c) salai (d) dhaora

476. An alkaloid used as drug is
   (a) quinine (b) reserpine (c) atropine (d) none of these

477. Tannins are
   (a) sweet (b) sour (c) astringent (d) mucopolysaccharides

478. Tannins are used in
   (a) dyeing (b) inks (c) treating hides (d) none of these

479. Water-soluble yellowish pigment present in petals of Dahlia is
   (a) carotene (b) xanthophyll (c) anthoxanthin (d) anthocyanin

480. Silica incrustation is found on the leaf surface of
   (a) mango (b) grasses (c) castor (d) oleander

481. Calcium is deposited in plant cells as
   (a) calcium carbonate (b) calcium oxalate (c) calcium sulphate (d) none of these

482. A bunch of crystals called cytolith consists of
   (a) calcium oxalate (b) calcium carbonate
   (c) calcium carbonate over a cellulose core (d) calcium oxalate over a cellulose core

483. Dry scales of onion possess in their cells
   (a) raphides (b) prismatic crystals (c) sphaeraphides (d) crystal sand
484. Crystal sand found in Atropa cells is made of
   (a) calcium oxalate (b) calcium carbonate (c) silica (d) silica and iron
485. A star-shaped mass of crystals of calcium oxalate found in the cells of Colocasia is called
   (a) crystal sand (b) sphaeraphide (c) druse (d) both (b) and (c)
486. Cell theory states that
   (a) all cells are living
   (b) all cells have nucleus
   (c) cells are fundamental structural units of living organisms
   (d) cells reproduce by mitosis and meiosis
487. Which organisms do not have cellular structure?
   (a) PPLO (b) rickettsia (c) viruses (d) archaebacteria
488. Cell principle is not applicable to
   (a) bacteria (b) viruses (c) algae (d) fungi
489. Distinction of individual cells is absent in coenocytic organisms like
   (a) *Ulothrix* (b) *Volvox* (c) *Escherichia* (d) *Rhizopus*
490. Nucleus is absent in the mature
   (a) sieve tube cells (b) mammalian erythrocytes
   (c) monocytes (d) both (a) and (b)
491. The tissue having more nonliving material than living matter is the
   (a) epithelial tissue (b) parenchyma
   (c) connective tissue (d) nervous system
492. The theory proposing that body of an organism consists of incompletely divided cells is
   (a) organismal theory (b) protoplasmic theory
   (c) cell theory (d) theory of cell lineage
493. Cells are autonomous because
   (a) they synthesize components of living protoplasm from nonliving materials
   (b) they are able to grow and divide
   (c) each cell has its own lifespan
   (d) all of these

494. Prokaryote is characterized by
   (a) absence of nucleolus
   (b) absence of nuclear envelope
   (c) dispersed DNA and lack of membrane-bound organelles
   (d) all of above

495. Ageing is slow or absent in
   (a) plants (b) parrot (c) hydra (d) unicells

496. Which of these are less efficient?
   (a) multicellular animals (b) multicellular plants
   (c) colonial organisms (d) unicellular organisms

497. A multicellular organism possesses
   (a) differentiated cells (b) undifferentiated cells
   (c) dedifferentiated cells (d) none of these

498. Number of types of cells found in human body is
   (a) 20 (b) 30 (c) 200 (d) 300

499. Cells which lose their nucleus during differentiation are
   (a) nerve cells (b) muscle cells (c) erythrocytes (d) leukocytes

500. A nucleated differentiated cell that has lost the power to dedifferentiate is
   (a) nerve cell (b) kidney cell (c) liver cell (d) none of these

501. Steward *et al.* performed an experiment to prove cellular totipotency on
   (a) tomato (b) carrot (c) tobacco (d) potato

502. Explant is a
   (a) propagule (b) callus used for subculturing
   (c) part of plant used in tissue culture (d) part of tissue culture used for planting

503. What was done by Steward *et al.* in order to separate individual cells of carrot root?
   (a) shaking in liquid medium (b) homogenization
   (c) pressure sieving (d) microsurgery

504. Single cells in Steward's culture formed
   (a) cellular clumps (b) embryoids (c) plantlets (d) none of these

505. Embryoids are
   (a) somatic embryo-like structures
   (b) small embryos through fertilization in culture
   (c) early embryo stages used for propagation in tissue culture
   (d) all of the above

506. In animals, cellular totipotency has been restricted to
   (a) germinal cells (b) epithelial cells
   (c) zygote (d) zygote and early blastomeres

507. Animal cloning is carried out by
   (a) artificial fertilization of ovum (b) direct growth of ovum
   (c) ovum with somatic nucleus (d) all of the above

508. Cells capable of division are
   (a) stem cells (b) meristematic cells
   (c) undifferentiated cells (d) none of these

509. Differentiated cells are
   (a) premitotic specialized (b) post-mitotic specialized
   (c) premeiotic specialized (d) post-meiotic specialized

510. RBCs are
   (a) differentiated cells (b) undifferentiated cells
   (c) dedifferentiated cells (d) dead cells

511. During differentiation, RBCs stop
   (a) aerobic respiration (b) DNA replication
   (c) RNA synthesis (d) none of these

512. Dedifferentiated cells are formed in the region of
   (a) injury (b) regeneration
   (c) secondary growth (d) none of these

513. Functionally important dead cells are
   (a) cork cells (b) tracheal elements
   (c) both (a) and (b) (d) endothelial cells

514. Rapidly dividing unorganized mass of cells in tissue culture is
   (a) callose (b) callus (c) embryoid (d) plantlet

515. First successful tissue culture was that of
   (a) tomato root (b) carrot root (c) potato stem (d) tobacco callus

516. Tissue used by Steward *et al.* (1957) to prove cellular totipotency was
   (a) pith of root (b) pith of stem (c) phloem of root (d) phloem of stem

517. White performed successful tissue culture in
   (a) 1939 (b) 1932 (c) 1929 (d) 1922

518. Prokaryotes have
   (a) one envelope system (b) double envelope system
   (c) mitochondria (d) lysosomes

519. Respiratory enzymes of prokaryotes are
   (a) dispersed in cytoplasm (b) attached to cell membrane
   (c) located in plasmids (d) absent

520. Intracellular compartmentalization is absent in
   (a) plant cells (b) animal cells (c) protoplasts (d) prokaryotes

521. An uncovered fibrillar nuclear structure is present in the cells of
   (a) prokaryotes (b) higher animals (c) viruses (d) eukaryotes

522. True nucleus is absent in
   (a) bacteria (b) fungi (c) algae (d) lichens

523. Nucleoid is present in
   (a) prokaryotes (b) protozoan protists (c) green algae (d) none of these

524. Nuclear envelope is absent in
   (a) eukaryotic cells (b) bacteria
   (c) both eukaryotic and prokaryotic cells (d) none of the above

525. In which of the following is DNA found instead of chromosomes
   (a) *Volvox* (b) *Anabaena* (c) *Chlamydomonas* (d) *Plasmodiophora*

526. 9 + 2 arrangement is absent in the flagellum or cilium of
   (a) *Chlamydomonas* (b) *Euglena* (c) *Paramecium* (d) *Bacterium*

527. Cells which lack nucleus will also lack
   (a) ribosomes (b) polyribosome (c) prochromosome (d) lysosome

528. No cell is without
   (a) chloroplast (b) mitochondria (c) ribosomes (d) nucleus

529. Animal cells are without
   (a) ribosomes (b) chloroplasts (c) lysosomes (d) vacuoles

530. Plant cells are generally without
   (a) lysosomes (b) cell wall (c) plastids (d) vacuole

531. Which of the following is present in both plant and animal cells?
   (a) primary wall (b) secondary wall (c) plasma membrane (d) plastids

532. Animal cell differs from plant cell in not having
   (a) cell wall (b) glyoxysome (c) plastids (d) none of these
533. Nucleus is peripheral in
   (a) plant cell (b) animal cell (c) protozoan protests (d) none of these
534. Cells with sap vacuoles generally belong to
   (a) animals (b) protozoans (c) plants (d) prokaryotes
535. Histones are absent in
   (a) fungi (b) protista (c) metazoa (d) monera
536. Genetic system is with fewer introns in
   (a) bacteria (b) cyanobacteria (c) mycoplasma (d) none of these
537. Nucleoid of prokaryote has ______ DNA.
   (a) circular (b) linear (c) fragmented (d) dispersed
538. DNA of prokaryotes is folded by means of
   (a) proteins (b) RNA (c) polyamines (d) both (b) and (c)
539. Prokaryotic DNA is naked because
   (a) it does not form chromosome (b) it does not lie inside the nucleus
   (c) it is without association with histone (d) all of the above
540. Plasmids are common in
   (a) bacteria (b) protista (c) algae (d) viruses
541. Which one is absent in prokaryotes?
   (a) endocytosis (b) exocytosis (c) cyclosis (d) none of these
542. 70S ribosomes occur in
   (a) prokaryotes (b) mitochondria (c) plastids (d) none of these
543. Vacuoles are small but numerous in
   (a) mature plant cell (b) mature animal cell
   (c) young animal cell (d) prokaryotic cell
544. Flow of intrinsic information involves
   (a) transcription (b) translation
   (c) both (a) and (b) (d) replication of DNA
545. Prochromosome is the name of
   (a) prophasic chromosome (b) separated chromosome in anaphase
   (c) nucleoid of prokaryotes (d) plasmid of prokaryotes

546. DNA has few introns in

(a) prokaryotes (b) protistans (c) fungi (d) animals

547. Protoplasm is

(a) alveolar (b) fibrillar
(c) granular (d) polyphasic crystallo-colloidal

548. Protoplasm undergoes which of the following changes

(a) alveolar–granular (b) fibrillar–reticular
(c) sol–gel (d) none of these

549. In emulsion

(a) a solid is dispersed in solid (b) solid is dispersed in liquid
(c) liquid is dispersed in liquid (d) liquid is dispersed in solid

550. New protoplasm develops

(a) outside the older protoplasm (b) de novo
(c) inside the older protoplasm (d) none of the above

551. pH of protoplasm is

(a) 6.5–7.0 (b) 7.0–8.0 (c) 5.0–6.5 (d) 4.1–5.8

552. As compared to water, viscosity of protoplasm is

(a) less than one (b) 2–10 times (c) 10–15 times (d) 15–25 times

553. Protoplasm is

(a) miscible with water (b) colourless and transparent
(c) immune to external stimuli (d) impermeable to external chemicals

554. What is characteristic of protoplasm?

(a) 1–4 refractivity (b) low electrical conductivity
(c) coagulates at high temperature (d) all of these

555. Major component of protoplasm is

(a) water (b) proteins (c) lipids (d) carbohydrates

556. Protoplasm is

(a) incompressible (b) compressible (c) stretchable (d) both (b) and (c)

557. Which one is capable of flowing?

(a) gel (b) sol (c) emulsion (d) both (b) and (c)

558. Cell organelles found only in plants are

(a) Golgi complex (b) mitochondria (c) plastids (d) ribosomes

559. Prokaryotes are organisms that are characterized by
   (a) well-defined nucleus
   (b) absence of nuclear envelope, nucleolus and nucleoplasm
   (c) absence of cell wall
   (d) absence of chlorophyll

560. Which one is a prokaryote?
   (a) green algae (b) bacteriophage (c) salmonella (d) agaricus

561. Which of the following is a characteristic of prokaryote?
   (a) absence of nuclear envelope (b) absence of membrane-bound organelles
   (c) absence of nucleolus (d) all of the above

562. An example of cell devoid of nuclear membrane and mitochondria is
   (a) sperm (b) protise (c) bacterial cell (d) sponge cell

563. A prokaryotic cell lacks
   (a) cell wall (b) cell membrane (c) cytoplasm (d) nucleolus

564. Prokaryotic genetic system has
   (a) neither DNA nor histones (b) both DNA and histones
   (c) DNA but no histones (d) either DNA or histones

565. Main difference between plant and animal cells is
   (a) animal cell has large vacuoles (b) plant cell has small vacuoles
   (c) animal cell lacks rigid cell wall (d) plant cell lacks rigid cell wall

566. Figures of cork cells observed by Robert Hooke were published in
   (a) *Genera Plantarum* (b) *Species Plantarum*
   (c) *Origin of Species* (d) *Micrographia*

567. The cells discovered in thin sections of cork by Robert Hooke were actually
   (a) cell walls (b) cellulose (c) protoplasm (d) nuclei

568. Nucleus was discovered by
   (a) Robert Brown (b) Leeuwenhoek
   (c) Robert Hooke (d) Schleiden and Schwann

569. Physical basis of life is
   (a) cell (b) nucleus (c) DNA (d) protoplasm

570. Smaller cell is
   (a) less active metabolically (b) with smaller nucleus
   (c) with larger nucleus (d) more active metabolically

571. A prokaryote
   (a) lacks ribosomes
   (b) shows different biochemistry
   (c) lacks DNA
   (d) has dispersed DNA without distinct nucleus and membrane-bound organelles

572. The difference between plant and animal cells is
   (a) plant cells have one large central vacuole while animal cells have many small vacuoles
   (b) animal cells have thinner cell walls than the ones present in plant cells
   (c) plant cells lack chloroplasts which are abundant in animal cells
   (d) none of the above

573. Which is correct about cell theory in view of current status of our knowledge about cell structure?
   (a) It needs modification due to discovery of subcellular structures like chloroplasts and mitochondria.
   (b) Modified cell theory means that all living beings are composed of cells capable of reproducing.
   (c) Cell theory does not hold good because all living beings (e.g. viruses) do not have cellular organization.
   (d) Cell theory means that all living objects consist of cells whether or not capable of reproducing.

574. A genophore (nucleoid) consists of
   (a) a single-stranded DNA (b) a single double-stranded DNA
   (c) histone and RNA (d) histone and nonhistone

575. Minimum cell size seen under light microscope is
   (a) 1 μm (b) 0.1 μm (c) 0.25 μm (d) 0.5 μm

576. Nuclear matter without envelope occurs in
   (a) cyanobacteria and red algae (b) bacteria and cyanobacteria
   (c) bacteria and green algae (d) mycoplasma and green algae

577. Mitochondria do not occur in
   (a) bacteria (b) brown algae (c) green algae (d) red algae

578. An exception to cell theory is
   (a) mycoplasma (b) virus (c) protistans (d) algae

579. Animal cell differs from plant cell in possessing
   (a) Golgi body (b) vacuole (c) plastid (d) centrosome

580. Cellular totipotency means

(a) synthesis of new cells
(b) formation of new species
(c) formation of new plants
(d) capability of a plant cell to form complete plant

581. Distinction of prokaryotes and eukaryotes is based on

(a) proteins (b) plasma membrane (c) nucleus (d) DNA

582. Intracellular compartments do not occur in

(a) prokaryotes (b) eukaryotes (c) lower plants (d) higher plants

583. Prokaryotic ribosomes are

(a) 30S (b) 50S (c) 70S (d) 80S

584. The suffix S in ribosome unit indicates

(a) sedimentation coefficient
(b) solubility
(c) surface
(d) Svedberg

ANSWERS

1. (c)	2. (d)	3. (b)	4. (c)	5. (d)	6. (d)	7. (d)
8. (c)	9. (c)	10. (a)	11. (c)	12. (c)	13. (d)	14. (a)
15. (a)	16. (d)	17. (c)	18. (b)	19. (c)	20. (d)	21. (b)
22. (a)	23. (b)	24. (b)	25. (c)	26. (d)	27. (d)	28. (b)
29. (d)	30. (a)	31. (d)	32. (c)	33. (c)	34. (b)	35. (d)
36. (d)	37. (c)	38. (d)	39. (b)	40. (d)	41. (a)	42. (d)
43. (d)	44. (b)	45. (a)	46. (b)	47. (a)	48. (c)	49. (c)
50. (d)	51. (d)	52. (c)	53. (b)	54. (c)	55. (b)	56. (d)
57. (c)	58. (b)	59. (b)	60. (a)	61. (c)	62. (d)	63. (a)
64. (a)	65. (c)	66. (d)	67. (d)	68. (c)	69. (d)	70. (a)
71. (d)	72. (c)	73. (b)	74. (c)	75. (d)	76. (d)	77. (d)
78. (a)	79. (b)	80. (c)	81. (b)	82. (a)	83. (a)	84. (c)
85. (a)	86. (b)	87. (c)	88. (c)	89. (d)	90. (b)	91. (a)
92. (b)	93. (d)	94. (b)	95. (a)	96. (b)	97. (a)	98. (b)
99. (a)	100. (b)	101. (d)	102. (b)	103. (b)	104. (a)	105. (d)
106. (c)	107. (a)	108. (c)	109. (d)	110. (b)	111. (d)	112. (d)
113. (c)	114. (b)	115. (a)	116. (c)	117. (b)	118. (d)	119. (a)
120. (a)	121. (a)	122. (c)	123. (c)	124. (c)	125. (b)	126. (d)
127. (c)	128. (a)	129. (c)	130. (d)	131. (c)	132. (b)	133. (c)
134. (a)	135. (d)	136. (b)	137. (d)	138. (b)	139. (a)	140. (d)
141. (a)	142. (b)	143. (a)	144. (c)	145. (d)	146. (c)	147. (a)
148. (b)	149. (c)	150. (c)	151. (d)	152. (b)	153. (a)	154. (a)

155. (d)	156. (d)	157. (b)	158. (b)	159. (d)	160. (d)	161. (a)
162. (b)	163. (b)	164. (c)	165. (a)	166. (b)	167. (d)	168. (b)
169. (c)	170. (d)	171. (d)	172. (a)	173. (d)	174. (a)	175. (b)
176. (d)	177. (b)	178. (d)	179. (b)	180. (b)	181. (d)	182. (a)
183. (c)	184. (b)	185. (d)	186. (b)	187. (d)	188. (a)	189. (d)
190. (c)	191. (b)	192. (d)	193. (a)	194. (d)	195. (b)	196. (d)
197. (c)	198. (c)	199. (b)	200. (a)	201. (c)	202. (a)	203. (d)
204. (a)	205. (c)	206. (b)	207. (d)	208. (a)	209. (d)	210. (d)
211. (a)	212. (b)	213. (c)	214. (a)	215. (b)	216. (d)	217. (c)
218. (b)	219. (a)	220. (c)	221. (a)	222. (b)	223. (c)	224. (c)
225. (a)	226. (a)	227. (d)	228. (c)	229. (d)	230. (b)	231. (d)
232. (d)	233. (a)	234. (b)	235. (d)	236. (b)	237. (b)	238. (b)
239. (c)	240. (a)	241. (c)	242. (b)	243. (a)	244. (d)	245. (a)
246. (c)	247. (d)	248. (c)	249. (b)	250. (d)	251. (a)	252. (a)
253. (d)	254. (a)	255. (c)	256. (a)	257. (c)	258. (c)	259. (a)
260. (a)	261. (b)	262. (c)	263. (b)	264. (b)	265. (c)	266. (d)
267. (d)	268. (c)	269. (a)	270. (a)	271. (d)	272. (d)	273. (a)
274. (a)	275. (b)	276. (d)	277. (d)	278. (d)	279. (a)	280. (b)
281. (a)	282. (c)	283. (d)	284. (c)	285. (b)	286. (d)	287. (c)
288. (b)	289. (a)	290. (d)	291. (c)	292. (a)	293. (d)	294. (a)
295. (b)	296. (a)	297. (c)	298. (c)	299. (b)	300. (d)	301. (d)
302. (d)	303. (b)	304. (b)	305. (d)	306. (c)	307. (a)	308. (c)
309. (d)	310. (c)	311. (b)	312. (a)	313. (b)	314. (a)	315. (b)
316. (c)	317. (a)	318. (c)	319. (b)	320. (a)	321. (d)	322. (c)
323. (d)	324. (d)	325. (d)	326. (d)	327. (b)	328. (a)	329. (c)
330. (a)	331. (d)	332. (b)	333. (c)	334. (b)	335. (c)	336. (d)
337. (c)	338. (d)	339. (d)	340. (c)	341. (a)	342. (b)	343. (c)
344. (c)	345. (a)	346. (c)	347. (b)	348. (c)	349. (d)	350. (a)
351. (b)	352. (c)	353. (c)	354. (d)	355. (c)	356. (d)	357. (c)
358. (a)	359. (d)	360. (b)	361. (c)	362. (b)	363. (a)	364. (c)
365. (b)	366. (d)	367. (a)	368. (c)	369. (d)	370. (b)	371. (b)
372. (a)	373. (d)	374. (c)	375. (a)	376. (b)	377. (a)	378. (c)
379. (c)	380. (d)	381. (d)	382. (b)	383. (a)	384. (c)	385. (b)
386. (b)	387. (a)	388. (c)	389. (d)	390. (a)	391. (c)	392. (a)
393. (b)	394. (a)	395. (b)	396. (c)	397. (a)	398. (a)	399. (c)
400. (a)	401. (a)	402. (c)	403. (a)	404. (c)	405. (d)	406. (b)
407. (a)	408. (a)	409. (c)	410. (d)	411. (a)	412. (b)	413. (c)
414. (c)	415. (c)	416. (d)	417. (b)	418. (a)	419. (d)	420. (b)
421. (a)	422. (d)	423. (a)	424. (b)	425. (c)	426. (c)	427. (b)
428. (a)	429. (b)	430. (a)	431. (d)	432. (b)	433. (b)	434. (b)
435. (a)	436. (d)	437. (b)	438. (d)	439. (c)	440. (c)	441. (d)
442. (a)	443. (b)	444. (c)	445. (d)	446. (a)	447. (b)	448. (d)
449. (b)	450. (c)	451. (a)	452. (b)	453. (a)	454. (c)	455. (c)

456. (d)	457. (c)	458. (a)	459. (b)	460. (a)	461. (d)	462. (a)
463. (d)	464. (d)	465. (c)	466. (a)	467. (d)	468. (c)	469. (b)
470. (b)	471. (c)	472. (a)	473. (b)	474. (c)	475. (b)	476. (d)
477. (c)	478. (d)	479. (c)	480. (b)	481. (d)	482. (c)	483. (b)
484. (a)	485. (d)	486. (c)	487. (c)	488. (b)	489. (d)	490. (d)
491. (c)	492. (a)	493. (d)	494. (d)	495. (d)	496. (d)	497. (d)
498. (c)	499. (c)	500. (a)	501. (b)	502. (c)	503. (a)	504. (d)
505. (a)	506. (d)	507. (c)	508. (d)	509. (b)	510. (a)	511. (d)
512. (d)	513. (c)	514. (b)	515. (a)	516. (c)	517. (b)	518. (a)
519. (b)	520. (d)	521. (a)	522. (a)	523. (a)	524. (b)	525. (b)
526. (d)	527. (d)	528. (c)	529. (b)	530. (a)	531. (c)	532. (d)
533. (a)	534. (c)	535. (d)	536. (d)	537. (a)	538. (d)	539. (c)
540. (a)	541. (d)	542. (d)	543. (b)	544. (c)	545. (c)	546. (a)
547. (d)	548. (c)	549. (c)	550. (c)	551. (a)	552. (b)	553. (b)
554. (d)	555. (a)	556. (d)	557. (d)	558. (c)	559. (b)	560. (c)
561. (d)	562. (c)	563. (d)	564. (c)	565. (c)	566. (d)	567. (a)
568. (a)	569. (d)	570. (d)	571. (d)	572. (a)	573. (c)	574. (b)
575. (c)	576. (b)	577. (a)	578. (b)	579. (d)	580. (a)	581. (a)
582. (a)	583. (c)	584. (a)				

3

AMINO ACIDS

CHOOSE THE CORRECT ANSWER

1. Primary structure of protein is due to
   (a) hydrogen bonds (b) peptide bonds (c) —S—S-linkages (d) ionic bonds
2. How many amino acid molecules form a molecule of haemocyanin?
   (a) 68 (b) 72 (c) 4500 (d) 8200
3. Secondary structure of protein is
   (a) α-helix (b) β-pleated (c) β-helix (d) both (a) and (b)
4. Proteins can be denatured by
   (a) carbon dioxide (b) carbon monoxide (c) heat (d) oxygen
5. Quaternary structure is present in
   (a) haemoglobin (b) histione (c) globulin (d) elastin
6. The protein present in milk is
   (a) casein (b) glutelin (c) myosin (d) globulin
7. The most abundant molecule of protoplast is
   (a) proteins (b) lipids (c) carbohydrates (d) nucleic acids
8. Molecular weight of smallest protein molecule ACTH (adrenocortiocotropic hormone) is
   (a) 5600 (b) 4500 (c) 3100 (d) 6200
9. In α-helix secondary structure, hydrogen bonds lie between imide group of one amino acid and carbonyl group of
   (a) second amino acid (b) third amino acid
   (c) fourth amino acid (d) fifth amino acid

10. Which is a type of secondary protein structure?

(a) α-helix (b) β-pleated (c) collagen helix (d) all of these

11. In β-pleated secondary protein structure, there are

(a) two or more polypeptide chains
(b) hydrogen bonds between adjacent polypeptide chains
(c) parallel or antiparallel polypeptide
(d) all of these

12. Haemoglobin is

(a) a monomeric protein (b) an oligomeric protein
(c) a chromoprotein (d) both (b) and (c)

13. Primary structure of polypeptide is stabilized or secondary structure of polypeptide is maintained by

(a) hydrogen bonds (b) disulphide bonds
(c) ionic bonds (d) hydrophobic interactions

14. Protomers are

(a) primitive proteins (b) protein subunits
(c) protein aggregates (d) none of these

15. Final structure in fibrous proteins is

(a) secondary structure (b) tertiary structure
(c) quaternary structure (d) primary structure

16. Quaternary structure is found in

(a) simple monomeric proteins (b) conjugate monomeric proteins
(c) oligoproteins (d) both (b) and (c)

17. Most abundant protein in the human body is

(a) haemoglobin (b) keratin (c) collagen (d) immunoglobulin

18. Proteins may be dissolved in the following.

(a) benzene (b) carbon tetrachloride
(c) anhydrous acetone (d) aqueous alcohol
(e) anhydrous alcohol

19. Protamines are

(a) very basic proteins (b) very neutral proteins
(c) very acidic proteins (d) none of these

20. Histones are
    (a) acidic proteins (b) basic proteins (c) neutral proteins (d) none of these
21. Albumins are
    (a) acidic proteins (b) basic proteins (c) neutral proteins (d) none of these
22. Globulins are
    (a) acidic proteins (b) neutral proteins (c) basic proteins (d) none of these
23. Albumins are soluble in
    (a) water (b) dilute salt solutions
    (c) half-saturated $(NH_4)_2SO_4$ solution (d) saturated $(NH_4)_2SO_4$ solution
24. Proteins may be separated by
    (a) fractional precipitation (b) electrophoresis and chromatography
    (c) ultracentrifugation molecular sieves (d) all of these
25. Proteins are detected with the help of
    (a) denaturation (b) alkaloidal reagents
    (c) biuret test (d) all of these
26. Phosphoproteins contain
    (a) orthophosphate (b) pyrophosphate (c) phosphodiesters (d) none of these
27. Milk casein is a
    (a) nucleoprotein (b) chromoprotein (c) phosphoprotein (d) glycoprotein
28. How many glycoproteins does plasma contain?
    (a) one (b) two (c) three (d) four
29. Chondroitin sulphuric acid on hydrolysis yields
    (a) *N*-acetylgalactosamine (b) glucuronic acid
    (c) sulphate (d) *N*-acetylglycosamine
30. Mucoitin sulphuric acid on hydrolysis yields
    (a) *N*-acetylglycosamine (b) glucuronic acid
    (c) sulphate (d) all of these
31. Clinically, methaemoglobinaemia is most often observed in
    (a) workers who absorb through the skin or lungs aromatic nitro and amino compounds, such as nitrobenzene, nitrophenols, and aniline
    (b) workers who inhale asbestos
    (c) patients who are severe diabetics
    (d) patients suffering from colon cancer

32. Methaemoglobinaemia is observed in patients taking large amounts of drugs such as
    (a) anticancerous drugs
    (b) acetanilide and sulphonamides
    (c) antidiabetic drugs
    (d) diuretics
33. Drug which enhances the renal clearance of uric acid is
    (a) sorbitrate (b) probenecid (c) quinine sulphate (d) vitamin B_6
34. The basic amino acids which occupy a less definite role in protein nutrition is/are
    (a) alanine (b) arginine (c) histidine (d) both (b) and (c)
35. By calculation it may be shown that the rate of total body protein synthesis in the state of nitrogen equilibrium is as high as
    (a) 250 g/day (b) 500 g/day (c) 750 g/day (d) 1000 g/day
36. Beans are found to be deficient in which of the following amino acid(s)?
    (a) lysine (b) leucine (c) isoleucine (d) methionine
37. A situation where the dietary intake exceeds the amount of nitrogen lost from the body by excretion, exfoliation and so forth is called
    (a) negative nitrogen balance
    (b) positive nitrogen balance
    (c) a state of nitrogen equilibrium
    (d) Lesch–Nyhan syndrome
38. A drug which reduces uric acid formation by inhibiting xanthine oxidase activity is
    (a) allopurinol (b) aspirin (c) diuretics (d) β-blockers
39. Low dietary intake of which amino acid(s) leads to eczema, reduced growth in infants and interferes with haemoglobin synthesis?
    (a) leucine (b) isoleucine (c) phenylalanine (d) tryptophan
40. Cholesterol content is found to be high in
    (a) egg white (b) egg yolk (c) milk (d) coffee
41. Corn is found to be severely deficient in which amino acid?
    (a) glycine (b) tryptophan (c) lysine (d) both (b) and (c)
42. Essential amino acids may be synthesized in
    (a) human beings (b) plants (c) bacteria (d) both (b) and (c)
43. Normal body growth demands the adequate ingestion of which branched-chain amino acid?
    (a) leucine (b) isoleucine (c) valine (d) all of these
44. The aromatic amino acids are
    (a) phenylalanine (b) tyrosine (c) tryptophan (d) all of these

45. Less dietary intake of which amino acids may lead to defective spermatogenesis?
(a) alanine (b) tryptophan (c) arginine (d) proline

46. Children affected with kwashiorkor may be cured by supplementing their diet with
(a) low-quality animal protein
(b) high-quality animal protein, such as powdered milk
(c) balanced vegetable protein mixture
(d) both (b) and (c)

47. Pre-beta lipoprotein is also known as
(a) LDL (b) HDL (c) IDL (d) VLDL

48. In kwashiorkor, one may find
(a) swelling in spleen (b) osteomalacia
(c) osteoporosis (d) lowered resistance to infection

49. The non-essential amino acids are termed as
(a) dispensable amino acids (b) semi-indispensable amino acids
(c) indispensable amino acids (d) lipotropic factors

50. Good proteins for human beings are from
(a) plants (b) animals (c) none of these (d) both (a) and (b)

51. The requirement of nucleic acids in the diet of normal human adult is
(a) 1.5 g/day/kg body weight (b) 2.0 g/day/kg body weight
(c) more than 5.0 g/day/kg body weight (d) not required

52. Pregnant and lactating females require ______ protein.
(a) 10% (b) 20% (c) 30% (d) 40%

53. In kwashiorkor, the level of haemoglobin remains
(a) normal (b) below normal
(c) high (d) so much high that it can't be corrected

54. In kwashiorkor serum albumin level becomes low which leads to
(a) oedema (b) no effect on the body
(c) greyish hairs (d) white patches in the skin

55. In kwashiorkor the level of very low density lipoproteins (VLDL) is reduced which leads to
(a) itching in the genitals (b) fatty liver
(c) swollen eyelids (d) poor salivary digestion

56. In kwashiorkor the activity of pancreatic enzymes gets diminished which leads to
    (a) poor brain functioning (b) poor renal functioning
    (c) poor liver functioning (d) poor intestinal digestion
57. In kwashiorkor, cell protein synthesis
    (a) is reduced (b) remains unaffected
    (c) gets increased (d) none of the above
58. Simple oligopeptides include
    (a) tocopherols (b) vitamin choline (c) endorphins (d) heparin
59. The average, normal adult, for example, maintains a healthy state on approximately
    (a) 0.4 g protein/kg body weight (b) 0.7 g protein/day/kg body weight
    (c) 1.4 g protein/day/kg body weight (d) 1.8 g protein/day/kg body weight
60. The requirement of proteins for newborn infants is over
    (a) 2.0 g/day/kg body weight (b) partially utilizable proteins
    (c) non-utilizable proteins (d) first class proteins
61. Skin and hair proteins called keratins are
    (a) utilizable proteins (b) partially utilizable proteins
    (c) non-utilizable proteins (d) first class proteins
62. Out of 200 different amino acids found in nature, how many amino acids are present in proteins?
    (a) 20 (b) 25 (c) 30 (d) 35
63. Enzyme-catalysed hydrolysis of proteins produces amino acids of the form
    (a) D-, (b) DL-, (c) L-, (d) all of these
64. The ionizable groups of amino acids are at least
    (a) 1 (b) 2 (c) 3 (d) 4
65. The carboxyl groups of amino acids exist almost entirely as the conjugated base at pH
    (a) 6.6 (b) 6.8 (c) 7.2 (d) 7.4
66. The pH of arginine is
    (a) 10.5 (b) 10.6 (c) 10.8 (d) 11.0
67. The neutral amino acid is
    (a) leucine (b) lysine (c) proline (d) histidine
68. The amino acid containing hydroxyl group is
    (a) alanine (b) isoleucine (c) arginine (d) threonine

69. The sulphur-containing amino acid is
    (a) DOPA (b) homoserine (c) methionine (d) valine
70. The basic amino acid is
    (a) glycine (b) histidine (c) proline (d) serine
71. All amino acids are optically active except
    (a) glycine (b) serine (c) threonine (d) tryptophan
72. The amino acid which synthesizes many hormones is
    (a) valine (b) phenylalanine (c) alanine (d) histidine
73. Amino acids are insoluble in
    (a) acetic acid (b) chloroform (c) ethanol (d) benzene
74. The melting point of amino acid is above
    (a) 100°C (b) 180°C (c) 200°C (d) 220°C
75. From two amino acids, peptide bond formulation involves removal of one mole of
    (a) water (b) ammonia (c) carbon dioxide (d) carboxylic acid
76. Polymers of more than 100 amino acids are called
    (a) proteins (b) polypeptides
    (c) both of the above (d) none of the above
77. Insulin degradation or disulphide bond formation is effected by
    (a) pyruvate dehydrogenase (b) xylitol reductase
    (c) glutathione reductase (d) xanthine oxidase
78. The example of globulins is
    (a) leucosin (b) tuberin (c) oryzenin (d) legunelin
79. The example of scleroproteins is
    (a) glutenin (b) gliadin (c) salmine (d) elastin
80. The example of phosphoprotein is
    (a) mucin (b) ovovitellin (c) ovomucoid (d) tendomucoid
81. The example of metalloprotein is
    (a) siderophilin (b) osseomucoid (c) elastin (d) all of these
82. The example of chromoprotein is
    (a) salmine (b) catalase (c) zein (d) gliadin

83. Protein structures are confirmed by weak bonds, the example of which is
(a) hydrophobic (b) disulphide (c) peptide (d) all of these

84. Proteases produce polypeptides from proteins by
(a) oxidizing (b) reducing (c) hydrolysing (d) none of these

85. Proteins react with Biuret reagent which is suggestive of two or more
(a) hydrogen bonds (b) peptide bonds
(c) disulphide bonds (d) hydrophobic bonds

86. Insulin is oxidized to separate the protein molecule into its constituent polypeptide chains without affecting the other part of the molecule by the use of
(a) performic acid (b) oxalic acid (c) citric acid (d) malic acid

87. The disulphide bond is not broken under the usual conditions of
(a) filtration (b) reduction (c) oxidation (d) denaturation

88. Each hydrogen bond is quite
(a) strong (b) weak (c) both (a) and (b) (d) none of these

89. A coiled structure in which peptide bonds are folded in a regular manner is the
(a) globular protein (b) fibrous protein (c) both of the above (d) none of these

90. In many proteins, the hydrogen-bonding produces a regular coiled arrangement called
(a) β-helix (b) α-helix (c) both (a) and (b) (d) none of these

91. Many globular proteins are stable in solution although they lack
(a) hydrogen bonds (b salt bonds
(c) non-polar bonds (d) disulphide bonds

92. How many amino acids does each turn of α-helix contain?
(a) 2.8 (b) 3.2 (c) 3.4 (d) 3.6

93. The distance travelled per turn of α-helix in nm is
(a) 0.34 (b) 0.44 (c) 0.54 (d) 0.64

94. The space covered by each amino acid residue of α-helix in nm is
(a) 0.09 (b) 0.12 (c) 0.15 (d) 0.18

95. α-helix is disrupted by certain amino acids like
(a) proline (b) arginine (c) histidine (d) lysine

96. α-helix is stabilized by
(a) hydrogen bonds (b) disulphide bonds (c) salt bonds (d) non-polar bonds

97. Glutamic dehydrogenase is a
    (a) monomer (b) tetramer (c) dimer (d) none of these
98. Aldolase molecule is a
    (a) dimer (b) trimer (c) both of (a) and (b) (d) none of these
99. Foetal haemoglobin contains
    (a) two α and two γ chains (b) two α and two β chains
    (c) both of (a) and (b) (d) none of these
100. There is change in structure when haemoglobin takes up O_2 due to closer movement of
    (a) β chains (b) α chains (c) γ chains (d) α and γ chains
101. The hydrogen bonds between peptide linkages are interfered by
    (a) guanidine (b) uric acid (c) salicylic acid (d) oxalic acid
102. The hydrogen bonds in the secondary and tertiary structure of proteins are directly attacked by
    (a) salts (b) alkalis (c) detergents (d) all of these
103. The digestibility of certain denatured proteins by proteolytic enzymes is
    (a) decreased (b) increased (c) normal (d) none of the above
104. The antigenic or antibody functions of proteins by denaturation are frequently
    (a) not changed (b) changed (c) both of (a) and (b) (d) none of these
105. In case of severe denaturation of protein, there is
    (a) reversible denaturation (b) moderate reversible denaturation
    (c) irreversible denaturation (d) none of these
106. How many disulphide bonds does bovine ribonuclease of single polypeptide chain of 124 amino acid residues with small molecular weight contain?
    (a) 2 (b) 3 (c) 4 (d) 6
107. When egg albumin is heated till it is coagulated, the secondary and tertiary structures of the proteins are completely lost resulting in a mixture of randomly arranged
    (a) dipeptide chains (b) tripeptide chains
    (c) polypeptide chains (d) all of these
108. In glycoproteins the carbohydrate is in the form of disaccharide units; the number of units are
    (a) 50–100 (b) 200–300 (c) 400–500 (d) 600–700
109. The disaccharide units of glycoproteins are attached to the peptide chain, one per
    (a) 3.4 amino acid residues (b) 4.4 amino acid residues
    (c) 5.4 amino acid residues (d) 6.4 amino acid residues

110. The milk protein in the stomach of infants is digested by

(a) pepsin (b) trypsin (c) chymotrypsin (d) rennin

111. Achylia gastrica is said to be the condition due to the absence of

(a) pepsin only (b) both pepsin and HCl
(c) HCl only (d) all of these

112. The pH of gastric juice become low in

(a) haemolytic anaemia (b) pernicious anaemia
(c) both (a) and (b) (d) none of these

113. In the small intestine, trypsin hydrolyses peptide linkages containing

(a) arginine (b) histidine (c) serine (d) aspartate

114. Chymotrypsin in the small intestine hydrolyses peptide linkages containing

(a) phenylalanine (b) alanine (c) methionine (d) valine

115. Carboxypeptidase B in the small intestine hydrolyses peptides containing

(a) leucine (b) isoleucine (c) arginine (d) cysteine

116. The transport of amino acids is regulated by active processes of different numbers

(a) 1 (b) 2 (c) 3 (d) 4

117. The third active process for amino acid transport involves

(a) basic amino acids (b) neutral amino acids
(c) acidic amino acids (d) sulphur-containing amino acids

118. The neutral amino acids for absorption need

(a) TPP (b) B_6-PO_4 (c) NAD^+ (d) $NADP^+$

119. If one amino acid is fed in excess, the absorption of another is

(a) slightly accelerated (b) moderately accelerated
(c) highly accelerated (d) retarded

120. Under normal conditions, the percentage of food proteins that are generally readily digested is

(a) 67 to 73 % (b) 74 to 81% (c) 82 to 89% (d) 90 to 97%

121. By overheating, the nutritional value of cereal proteins is

(a) increased (b) lowered (c) unchanged (d) none of these

122. More than half of the protein of the liver and intestinal mucosa are broken down and resynthesized in

(a) 10 days (b) 12 days (c) 15 days (d) 18 days

123. The half-life of antibody protein is about
   (a) 4 weeks (b) 3 weeks (c) 2 weeks (d) 1 week

124. Protein anabolism is stimulated by
   (a) ACTH (b) testosterone (c) glucagon (d) epinephrine

125. The metabolism of protein is integrated with that of carbohydrate and fat through
   (a) oxaloacetate (b) citrate (c) isocitrate (d) minerals

126. The building up and breaking down of protoplasm are concerned with the metabolism of
   (a) carbohydrates (b) fat (c) protein (d) minerals

127. The amino acids abstracted from the liver are not utilized for repair or special synthesis but are broken down to
   (a) keto acids (b) sulphur dioxide (c) water (d) ammonia

128. The unwanted amino acids abstracted from the tissues are either used by the tissue or in the liver converted into
   (a) ammonia (b) urea (c) ammonium salts (d) uric acid

129. Amino acids provide the nitrogen for the synthesis of
   (a) the bases of the phospholipids (b) exogenous metabolism
   (c) both of (a) and (b) (d) none of these

130. The metabolism of all proteins ingested over and above the essential requirements is called
   (a) endogenous metabolism (b) exogenous metabolism
   (c) both of (a) and (b) (d) none of these

131. Sulphur-containing amino acids after catabolism produce a substance which is excreted as
   (a) SO_2 (b) HNO_3 (c) H_2SO_4 (d) H_2PO_4

132. Ethereal sulphate synthesized from the amino acid is
   (a) neutral (b) acidic (c) basic (d) sulphur-containing

133. Keratin, the protein of hair, is synthesized from the amino acid
   (a) glycine (b) serine (c) proline (d) methionine

134. In human and other ureotelic organisms, the end product of amino acid nitrogen metabolism is
   (a) bile acids (b) ketone bodies (c) urea (d) uric acid

135. The end product of amino acid nitrogen metabolism in uricotelic organisms (e.g. reptiles and birds) is
   (a) bilirubin (b) urea (c) uric acid (d) biliverdin

136. The transaminase activity needs the coenzyme
   (a) ATP (b) B_6-FO_4 (c) FAD^+ (d) NAD^+

137. Transamination is
   (a) an irreversible process (b) a reversible process
   (c) both (a) and (b) (d) none of these

138. Most amino acids are substrates for transamination except
   (a) alanine (b) threonine (c) serine (d) valine

139. In mammals oxidative conversion of many amino acids to their corresponding α-keto acids occurs in the
   (a) liver and kidney (b) adipose tissue (c) pancreas (d) intestine

140. The α -keto acid is decarboxylated by H_2O_2 forming a carboxylic acid with one carbon atom less in the absence of the enzyme
   (a) catalase (b) decarboxylase (c) deaminase (d) phosphatase

141. The activity of mammalian L-amino acid oxidase, an FMN-flavoprotein, is quite
   (a) slow (b) rapid (c) both (a) and (b) (d) none of these

142. From dietary protein as well as from the urea present in fluids secreted into the gastrointestinal tract, intestinal bacteria produce
   (a) carbon dioxide (b) ammonia (c) ammonium sulphate (d) creatine

143. The symptom of ammonia intoxication includes
   (a) blurring of vision (b) constipation
   (c) mental confusion (d) diarrhoea

144. Ammonia intoxication symptoms occur when brain ammonia levels are
   (a) slightly diminished (b) highly diminished
   (c) increased (d) all of the above

145. Ammonia production by the kidney is depressed in
   (a) acidosis (b) alkalosis
   (c) both of (a) and (b) (d) none of these

146. Ammonia is excreted as ammonium salts during metabolic acidosis but the majority is excreted as
   (a) phosphates (b) creatine (c) uric acid (d) urea

147. Synthesis of glutamine is accompanied by the hydrolysis of
   (a) ATP (b) ADP (c) TPP (d) creatine phosphate

148. In brain, the major mechanism for removal of ammonia is the formation of
    (a) glutamate (b) aspartate (c) asparagine (d) glutamine

149. Carbamoyl phosphate synthetase structure is markedly changed in the presence of
    (a) *N*-acetyl glutamate (b) *N*-acetyl aspartate
    (c) neuraminic acid (d) oxalate

150. In bacteria, the synthesis of carbomoyl phosphate takes place from the substance
    (a) ammonium salts (b) ammonia (c) glutamine (d) aspartate

151. The competitive inhibitor of arginine is
    (a) citrulline (b) malate (c) lysine (d) serine

152. The biosynthesis of urea occurs mainly in the liver
    (a) cytosol (b) mitochondria (c) microsomes (d) nuclei

153. One mol. of urea is synthesized at the expense of how many mols. of ATP?
    (a) 2 (b) 3 (c) 4 (d) 5

154. Urea biosynthesis occurs mainly in the liver, how many amino acids does it involve?
    (a) 3 (b) 4 (c) 5 (d) 6

155. The normal daily output of urea through urine in moles/l is
    (a) 10 to 20 (b) 15 to 25 (c) 20 to 30 (d) 25 to 35

156. In severe acidosis, the output of urea is
    (a) decreased (b) slightly increased
    (c) highly increased (d) moderately increased

157. Uraemia occurs in
    (a) cirrhosis of the liver (b) nephritis
    (c) diabetes mellitus (d) coronary thrombosis

158. The clinical symptom of urea cycle disorder is
    (a) mental retardation (b) drowsiness
    (c) diarrhoea (d) oedema

159. The sparing action of methionine is by
    (a) tyrosine (b) cystine (c) arginine (d) tryptophan

160. NH_4^+ aminates glutamate to form glutamine requiring ATP and
    (a) Mg^{++} (b) Ca^{++} (c) Na^+ (d) K^+

ANSWERS

1. (b)	2. (d)	3. (d)	4. (c)	5. (a)	6. (a)	7. (a)
8. (b)	9. (c)	10. (d)	11. (d)	12. (d)	13. (a)	14. (b)
15. (a)	16. (c)	17. (c)	18. (d)	19. (a)	20. (b)	21. (c)
22. (b)	23. (a,b,c)	24. (d)	25. (d)	26. (a,b,c)	27. (c)	28. (b)
29. (a,b,c)	30. (d)	31. (a)	32. (b)	33. (b)	34. (d)	35. (b)
36. (d)	37. (b)	38. (a)	39. (d)	40. (b)	41. (d)	42. (d)
43. (d)	44. (d)	45. (c)	46. (d)	47. (d)	48. (d)	49. (a)
50. (b)	51. (d)	52. (b)	53. (b)	54. (a)	55. (b)	56. (d)
57. (a)	58. (c)	59. (b)	60. (a)	61. (c)	62. (a)	63. (c)
64. (b)	65. (d)	66. (c)	67. (a)	68. (d)	69. (c)	70. (b)
71. (a)	72. (b)	73. (d)	74. (c)	75. (a)	76. (a)	77. (c)
78. (b)	79. (d)	80. (b)	81. (a)	82. (b)	83. (a)	84. (c)
85. (b)	86. (a)	87. (d)	88. (b)	89. (a)	90. (b)	91. (d)
92. (d)	93. (c)	94. (c)	95. (a)	96. (a)	97. (b)	98. (b)
99. (a)	100. (a)	101. (a)	102. (b)	103. (b)	104. (b)	105. (c)
106. (c)	107. (c)	108. (d)	109. (d)	110. (d	111. (b)	112. (b)
113. (a)	114. (a)	115. (c)	116. (c)	117. (b)	118. (b)	119. (d)
120. (d)	121. (b)	122. (a)	123. (c)	124. (b)	125. (a)	126. (c)
127. (d)	128. (b)	129. (a)	130. (b)	131. (c)	132. (d)	133. (d)
134. (c)	135. (c)	136. (b)	137. (b)	138. (b)	139. (a)	140. (a)
141. (a)	142. (b)	143. (a)	144. (c)	145. (b)	146. (d)	147. (a)
148. (d)	149. (a)	150. (c)	151. (c)	152. (b)	153. (b)	154. (d)
155. (c)	156. (a)	157. (b)	158. (a)	159. (b)	160. (a)	

4

CARBOHYDRATES

Choose the Correct Answer

1. Which of the following sugars is a tetrose?
   (a) galactose (b) fructose (c) iodose (d) erythrose
2. Which of the following sugars is a pentose?
   (a) ribose (b) threose (c) mannose (d) gulose
3. Which of the following sugars is found in RNA?
   (a) xylose (b) threose (c) ribose (d) glucose
4. Which of the following sugars is found in DNA?
   (a) 2-deoxy-D-ribose (b) D-ribose (c) D-ribulose (d) D-xylulose
5. Which of the following is a sugar alcohol?
   (a) gluconic acid (b) maltose
   (c) mannitol (d) chondroitin sulphate
6. Which of the following sugars is a monosaccharide?
   (a) sucrose (b) fructose (c) raffinose (d) maltose
7. Which of the following sugars is a disaccharide?
   (a) starch (b) lactose (c) galactose (d) glucose
8. The minimum carbon number in a carbohydrate is
   (a) 6 (b) 4 (c) 3 (d) 1
9. The ratio between hydrogen and oxygen in a carbohydrate is
   (a) 5 : 1 (b) 4 : 3 (c) 3 : 1 (d) 2 : 1

10. $C_nH_{2n}O_n$ is the formula of
    (a) fatty acid (b) fat (c) glycerol (d) carbohydrate
11. Deoxyribose is
    (a) $C_5H_{10}O_5$ (b) $C_5H_{10}O_4$ (c) $C_6H_{12}O_6$ (d) $C_6H_{12}O_5$
12. The simplest form of carbohydrate is
    (a) carbon (b) starch (c) monosaccharide (d) cane sugar
13. Which of these is a monosaccharide?
    (a) lactose (b) sucrose (c) ribose (d) maltose
14. Monosaccharides have ______ carbons.
    (a) 20 (b) 10–15 (c) 2–10 (d) 3–7
15. The most common carbohydrate monomer is
    (a) maltose (b) lactose (c) glucose (d) galactose
16. The sweetest sugar is
    (a) fructose (b) glucose (c) sucrose (d) maltose
17. The sweetest chemical is
    (a) fructose (b) saccharin (c) monellin (d) sorbitol
18. A nonreducing sugar is
    (a) glucose (b) sucrose (c) fructose (d) maltose
19. The sugar present in milk is
    (a) casein (b) maltose (c) lactose (d) heparin
20. Xylitol is a/an
    (a) artificial sweetener (b) nutritive sweetener
    (c) nonnutritive sweetener (d) both (a) and (b)
21. Triose carbohydrate is
    (a) glycerol (b) glyceraldehyde
    (c) dihydroxyacetone (d) both (b) and (c)
22. Reducing sugars have
    (a) free aldehyde (b) bound aldehyde
    (c) free aldehyde or ketone (d) bound ketone
23. Galactose is
    (a) pentose sugar (b) hexose sugar (c) heptose sugar (d) triose carbohydrate

24. Deoxyribose is a
    (a) pentose sugar (b) tetrose carbohydrate
    (c) hexose sugar (d) triose carbohydrate
25. A ketose sugar is
    (a) glucose (b) fructose (c) glyceraldehyde (d) both (a) and (b)
26. Laevulose present in honey is a
    (a) disaccharide (b) glucose (c) fructose (d) pentose
27. L-fructose is found in
    (a) grapes (b) honey (c) most fruits (d) pentose
28. In Fehling's or Benedict's solution, a reducing sugar causes conversion of
    (a) ferric to ferrous state (b) ferrous to ferric state
    (c) cuprous to cupric state (d) cupric to cuprous state
29. Use of artificial sweetener saccharin has been discontinued because it is
    (a) liable to decompose in hot weather of tropics
    (b) carcinogenic
    (c) bitter in the beginning
    (d) all of these
30. The commonly used artificial sweetener is
    (a) acesulfame K (b) cyclamate (c) aspartame (d) saccharin
31. Mannose is a
    (a) disaccharide (b) hexosan monosaccharide
    (c) pentosan monosaccharide (d) tetrose
32. Mannitol is a/an
    (a) amino acid (b) amino alcohol (c) sugar alcohol (d) sugar acid
33. Which one is a sugar acid?
    (a) vitamin C (b) vitamin K (c) vitamin D (d) vitamin E
34. Cellobiose, the hydrolytic breakdown product of cellulose is a
    (a) monosaccharide (b) disaccharide (c) tetrasaccharide (d) trisaccharide
35. Inulin is a polymer of
    (a) glucose (b) galactose (c) fructose (d) arabinose
36. Raffinose is a
    (a) monosaccharide (b) disaccharide (c) trisaccharide (d) saccharin

37. The heptose ketose sugar formed as a result of chemical reaction in HMP shunt is
    (a) glucoheptose (b) galactoheptose (c) sedoheptulose (d) mannoheptose
38. The general formula for polysaccharide is
    (a) $(C_6H_{10}O_5)_N$ (b) $(C_6H_{12}O_6)_N$ (c) $(C_6H_{12}O_5)_N$ (d) $(C_6H_{10}O_6)_N$
39. The number of isomers of glucose is
    (a) 4 (b) 8 (c) 12 (d) 16
40. The epimers of glucose is
    (a) fructose (b) galactose (c) ribose (d) deoxyribose
41. Human heart muscle contains
    (a) D-arabinose (b) D-ribose (c) D-lyxose (d) D-xylose
42. The intermediate in hexose monophosphate shunt is
    (a) D-ribulose (b) D-arabinose (c) D-xylose (d) D-lyxose
43. Honey contains the hydrolytic product of
    (a) lactose (b) maltose (c) inulin (d) starch
44. On boiling, Benedict's solution is not produced by
    (a) sucrose (b) lactose (c) maltose (d) fructose
45. Glycosides are found in many
    (a) vitamins (b) drugs (c) minerals (d) nucleoproteins
46. Erythromycin contains
    (a) diethylamino sugars (b) trimethylamino sugars
    (c) dimethylamino sugars (d) triethylamino sugar
47. Galactose on oxidation with concentrated HNO_3 produces
    (a) gluconic acid (b) saccharic acid (c) saccharolactone (d) mucic acid
48. The distinguishing test between monosaccharides and disaccharides is
    (a) glucose (b) saccharic acid (c) saccharolactone (d) mucic acid
49. Barfoed's solution is not reduced by
    (a) glucose (b) mannose (c) sucrose (d) ribose
50. Cellulose is made up of the molecules of
    (a) α-glucose (b) β-glucose
    (c) both (a) and (b) (d) none of these

51. Iodine solution produces no colour with
(a) cellulose (b) starch (c) dextrin (d) glycogen

52. In glycogen structure, where is the branching present between α-glucose units?
(a) 4–101 (b) 6–12 (c) 8–14 (d) 12–18

53. Amylose contains ______ glucose units.
(a) 100–200 (b) 200–300 (c) 300–400 (d) 500–600

54. Each branch of amylopectin is at an interval of ______ glucose units.
(a) 14–20 (b) 24–30 (c) 34–40 (d) 44–50

55. *N*-acetylneuraminic acid is an example of
(a) sialic acid (b) mucic acid (c) glucuronic acid (d) hippuric acid

56. The molecular weight (in millions) of hyaluronic acid ranges from
(a) 1–2 (b) 1–4 (c) 1–6 (d) 1–8

57. In place of glucuronic acid, chondroitin sulphate B contains
(a) gluconic acid (b) gulonic acid (c) iduronic acid (d) sulphonic acid

58. Heparin has a molecular weight of about
(a) 14,000 (b) 15,000 (c) 16,000 (d) 17,000

59. Blood group substances consist of
(a) lactose (b) maltose (c) fucose (d) mucose

60. The component of cartilage and cornea is
(a) keratosulphate (b) chondroitin sulphate
(c) cadmium sulphate (d) antimony sulphate

61. Benedict's test is less likely to give weakly positive results with concentrated urine due to the action of
(a) urea (b) uric acid (c) ammonium salts (d) phosphates

62. Salivary amylase is activated by
(a) Na^+ (b) K^+ (c) HCO^-_3 (d) Cl^-

63. Active transport of sugar is depressed by
(a) oxaloacetate (b) fumarate (c) malonate (d) succinate

64. The absorption of glucose is interfered by the deficiency of
(a) vitamin A (b) thiamine
(c) magnesium sulphate (d) ferrous sulphate

65. Glucose absorptions may be decreased in
    (a) oedema (b) nephritis (c) rickets (d) osteomyelitis
66. Glycogen synthetase activity is depressed by
    (a) glucose (b) insulin (c) cyclic AMP (d) fructokinase
67. Glucose is removed from the blood following a meal by
    (a) hexokinase (b) glucokinase
    (c) both (a) and (b) (d) none of these
68. The branching enzyme acts on the glycogen when the glycogen chain has been lengthened to between ______ glucose units.
    (a) 1 and 6 (b) 2 and 7 (c) 3 and 9 (d) 6 and 11
69. Cyclic AMP is formed from ATP by the enzyme adenylate cyclase which is activated by the hormone
    (a) insulin (b) epinephrine (c) testosterone (d) progesterone
70. The synthesis of adenylate cyclase is increased by
    (a) thyroid hormones (b) growth hormones
    (c) ACTH (d) FSH
71. UDPG is essential for the synthesis of
    (a) lactose (b) maltose (c) sucrose (d) starch
72. Phosphorylase α in muscle is a tetramer containing 4 mol. of
    (a) ATP (b) NAD^+
    (c) pyridoxal phosphate (d) CoA
73. Inactive muscle phosphorylase β is activated by active protein kinase which is being stimulated by
    (a) ATP (b) NAD^+ (c) cyclic GMP (d) cyclic AMP
74. Hexokinase has a high affinity for glucose than
    (a) fructokinase (b) galactokinase
    (c) glucokinase (d) all of these
75. Fructose 1,6-diphosphate is converted to fructose 6-phosphate by the enzyme fructose 1.6-disphosphatase which is stimulated by
    (a) glucagon (b) insulin (c) ACTH (d) none of these
76. Dihydroxyacetone phosphate and glyceraldehyde 3-phosphate are interconverted by
    (a) triose isomerase (b) phosphotriose isomerase
    (c) diphosphotriose isomerase (d) dihydroxyacetone phosphorylase

77. In the liver, glyceraldehyde 3-phosphate is converted to
    (a) glycol (b) formaldehyde (c) formic acid (d) glycerol
78. Phosphoglycerate kinase responsible for the conversion for 1,3-diphosphoglycerate to 3-phosphoglycerate is inhibited by
    (a) arsenate (b) fumarate (c) citrate (d) cyanate
79. The reduced lipoate is reoxidized by
    (a) NAD^+ (b) $NADP^+$ (c) FAD^+ (d) FMN
80. The approximate number of mols. of pyruvate dehydrogenase in pyruvate dehydrogenase complex is
    (a) 19 (b) 28 (c) 39 (d) 49
81. Pyruvate is accumulated by the dietary deficiency of
    (a) vitamin B_6 (b) folic acid (c) vitamin B_{12} (d) thiamine
82. Citrate is converted to isocitrate by aconitase which contains
    (a) Ca^{++} (b) Fe^{++} (c) Zn^{++} (d) Mg^{++}
83. The reaction involving the conversion of succinyl-CoA to succinate requires
    (a) CDP (b) ADP (c) GDP (d) $NADP^+$
84. The carrier of the citric acid cycle is
    (a) succinate (b) fumarate (c) malate (d) oxaloacetate
85. Glucose 6-phosphatase is absent in
    (a) adipose tissue (b) intestine (c) kidney (d) heart
86. 6-phosphogluconate is converted to 3-keto-6-phosphogluconate in the presence of the coenzyme
    (a) FAD^+ (b) NAD^+ (c) $NADP^+$ (d) ATP
87. UDPG is oxidized to UDP glucuronic acid by UDP dehydrogenase in the presence of
    (a) FAD^+ (b) NAD^+ (c) $NADP^+$ (d) ADP
88. Fructokinase is present in
    (a) intestine (b) adipose tissue (c) heart (d) brain
89. Fructose 1-phosphate is split into glyceraldehyde and dihydroxyacetone phosphate by the enzyme
    (a) enolase (b) aldolase A (c) aldolase A and B (d) aldolase B
90. Galactose is phosphorylated by galactokinase to form
    (a) galactose 6-phosphate (b) galactose 1,6-diphosphate
    (c) galactose 1-phosphate (d) all of these

91. In galactosemic individuals UDP galactose is formed by epimerization from
(a) glucose (b) UDP glucose (c) CDP glucose (d) ITP glucose

ANSWERS

1. (d)	2. (a)	3. (c)	4. (a)	5. (c)	6. (b)	7. (b)
8. (c)	9. (d)	10. (d)	11. (b)	12. (c)	13. (c)	14. (d)
15. (c)	16. (a)	17. (d)	18. (b)	19. (c)	20. (c)	21. (d)
22. (c)	23. (b)	24. (a)	25. (b)	26. (c)	27. (b)	28. (d)
29. (a)	30. (c)	31. (b)	32. (c)	33. (a)	34. (b)	35. (c)
36. (c)	37. (c)	38. (a)	39. (d)	40. (b)	41. (c)	42. (a)
43. (c)	44. (a)	45. (b)	46. (a)	47. (d)	48. (b)	49. (c)
50. (b)	51. (a)	52. (d)	53. (c)	54. (b)	55. (a)	56. (b)
57. (c)	58. (d)	59. (c)	60. (a)	61. (b)	62. (d)	63. (c)
64. (b)	65. (a)	66. (c)	67. (b)	68. (d)	69. (b)	70. (a)
71. (a)	72. (c)	73. (d)	74. (c)	75. (a)	76. (b)	77. (d)
78. (a)	79. (c)	80. (b)	81. (d)	82. (b)	83. (c)	84. (d)
85. (a)	86. (c)	87. (b)	88. (a)	89. (d)	90. (c)	91. (b)

5

LIPIDS

CHOOSE THE CORRECT ANSWER

1. Range of concentration of cholesterol (total) in serum in normal adults is
   (a) 75–300 mg% (b) 250–400 mg% (c) 150–250 mg% (d) 200–350 mg%
2. Range of concentration of free cholesterol in plasma or serum in normal adults is
   (a) 60–100 mg/100 ml (b) 30–60 mg/100 ml
   (c) 60–150 mg/100 ml (d) 75–180 mg/100 ml
3. Range of concentration of phospholipids (total) in plasma/serum in normal adults is
   (a) 150–250 mg% (b) 110–280 mg% (c) 80–150 mg% (d) 110–250 mg%
4. Concentration of fatty acids (total) in serum in normal adults is
   (a) 250–800 mg% (b) 180–360 mg% (c) 250–550 mg% (d) 200–425 mg%
5. Concentration of non-esterified fatty acids (NEFA) in plasma/serum in normal adults is
   (a) 10–17 mg% (b) 30–30 mg% (c) 45–100 mg% (d) 60–90 mg%
6. Concentration of bile acids in plasma/serum in normal adults is
   (a) 0.2–3.0 mg% (b) 0.2–1.5 mg% (c) 0.2–1.0 mg% (d) 0.5–4.5 mg%
7. High-density lipoproteins are also called
   (a) α-lipoproteins (b) β-lipoproteins (c) γ-lipoproteins (d) α_1-lipoproteins
8. Low-density lipoproteins are also called
   (a) α-lipoproteins (b) β-lipoproteins (c) γ-lipoproteins (d) none of these
9. Concentration of very-low-density lipoproteins (VLDL) in plasma is
   (a) 150 mg% (b) 200 mg% (c) 50–100 mg% (d) 120–250 mg%

10. Concentration of low-density lipoproteins (LDL) in plasma is
    (a) 250–350 mg% (b) 350 mg% (c) 150–400 mg% (d) 120–150 mg%
11. Concentration of high-density lipoproteins (HDL) in plasma is
    (a) 300 mg% (b) 250–500 mg% (c) 150–300 mg% (d) 120–240 mg%
12. In 'Tangier disease' the HDL is
    (a) absent from plasma (b) sometimes absent from plasma
    (c) present in plasma (d) sometimes present in plasma
13. In cholesterol biosynthesis, acetoacetyl-CoA is converted to
    (a) β-hydroxy-β-methyl glutaryl-CoA (b) mevalonic acid
    (c) acetyl-CoA (d) squalene
14. In cholesterol biosynthesis, squalene is converted to
    (a) lanosterol (b) zymosterol
    (c) farnesyl pyrophosphate (d) desmosterol
15. In cholesterol biosynthesis the starting material is
    (a) acetoacetyl-CoA (b) oxaloacetic acid
    (c) acetyl-CoA (d) succinyl-CoA
16. In the biosynthesis of cholic acid, cholesterol gets firstly converted into
    (a) 7-hydroxycholesterol (b) chenodeoxycholic acid
    (c) 7-dehydrocholesterol (d) none of these
17. In the cholesterol biosynthesis, farnesyl pyrophosphate is converted to
    (a) geranyl pyrophosphate (b) isopentyl pyrophosphate
    (c) squalene (d) none of these
18. In cholesterol biosynthesis, desmosterol is converted to
    (a) cholesterol (b) zymosterol (c) lanosterol (d) none of these
19. In the absence of insulin, fat synthesis takes place
    (a) more (b) as usual (c) very little (d) uncertain
20. When ketosis is evident, fat synthesis is
    (a) increased (b) suppressed (c) as usual (d) none of these
21. If carbohydrate is ingested, fat oxidation is
    (a) dramatically reduced (b) dramatically accelerated (c) as usual (d) none of these
22. NEFA and ketone bodies affect the utilization and oxidation of glucose by
    (a) increasing it (b) inhibiting it
    (c) first increasing and then inhibiting it (d) uncertain

23. Most important site(s) of fat metabolism are
    (a) adipose tissue (b) liver (c) intestine (d) all of these
24. Fatty liver occurs in many conditions such as
    (a) poisoning by CCl_4 (b) many kinds of malnutrition
    (c) induced choline deficiency (d) all of these
25. Acyl-carrier protein, i.e., ACP, is
    (a) heat-stable (b) heat-labile (c) both (a) and (b) (d) none of these
26. Acyl-carrier proteins have been isolated from a number of
    (a) plant sources (b) animal sources (c) microbial sources (d) all of these
27. MW of acyl carrier proteins is in the range of
    (a) 9,000 to 11,000 g per mole of protein (b) 11,000 to 19,000 g per mole of protein
    (c) 17,000 to 26,000 g per mole of protein (d) none of these
28. Most of the enzymes responsible for fatty acid synthesis are contained in the
    (a) soluble portion of the cell cytoplasm (b) microsomal fraction
    (c) mitochondrial fraction (d) lysosomal fraction
29. Fatty acid synthesis requires
    (a) NAD (b) NADPH (c) TPP (d) FAD
30. Essential fatty acid(s) for human beings is/are
    (a) palmitic acid (b) stearic acid (c) linoleic acid (d) arachidonic acid
31. A source of essential fatty acids is
    (a) butter (b) mustard oil (c) desi ghee (d) sunflower oil
32. Arachidonic acid is a precursor of
    (a) vitamin A (b) prostaglandins (c) vitamin K (d) sex hormones
33. Chemically, prostaglandins are
    (a) saturated (b) unsaturated (c) both (a) and (b) (d) none of these
34. Prostaglandins are fatty acids containing ______ carbon atoms.
    (a) 19 (b) 20 (c) 21 (d) 22
35. In their structures, prostaglandins possess
    (a) cyclopentane ring (b) benzene ring
    (c) cyclopentanoperhydrophenanthrene ring (d) none of these
36. The cholesterol which is present in plasma originates largely from
    (a) hepatic synthesis (b) renal synthesis
    (c) brain synthesis (d) intestinal mucosa synthesis

37. The fatty acid which predominates in cholesterol esters is
    (a) stearic acid (b) oleic acid (c) linoleic acid (d) caproic acid
38. Ketosis does not develop on
    (a) high protein–high fat diets (b) high protein–low fat diets
    (c) low protein–high fat diets (d) low protein–low fat diets
39. Even in most severe diabetics not being treated with insulin, the excretion of ketone bodies per day does not exceed
    (a) 25 g (b) 50 g (c) 75 g (d) 100 g
40. In uncontrolled juvenile-type diabetes, the ketone body concentration in blood may be
    (a) 25–50 mg/100 ml (b) 50–100 mg/100 ml
    (c) 100–150 mg/100 ml (d) 150–200 mg/100 ml
41. In severe ketosis, reduction in plasma bicarbonate and the loss of cations in urine are
    (a) dangerous (b) not dangerous
    (c) sometime dangerous (d) having absolutely no effect
42. 50 g of ketone bodies is equivalent to ______ of stearic acid.
    (a) 100 g (b) 75 g (c) 50 g (d) 25 g
43. 1 g of ketone bodies is equivalent to ______ of energy.
    (a) 300 kJ (b) 600 kJ (c) 600 kJ (d) 1200 kJ
44. The clinical problem of ketoneuria is the loss of
    (a) energy (b) Na^+ (c) Ca^{++} (d) HCO_3^-
45. Hydroxymethylglutaryl-CoA synthetase is inhibited by
    (a) cholesterol (b) excess glucose concentrates
    (c) α-keto acids (d) methionine
46. If glucose is readily available, i.e., after a meal containing carbohydrate, fat synthesis
    (a) will occur (b) will not occur (c) will be uncertain (d) may occur
47. In the absence of insulin, ______ fat synthesis takes place.
    (a) little (b) very little (c) more (d) much
48. When carbohydrate is ingested, fat oxidation is
    (a) dramatically reduced (b) having no effect
    (c) increased (d) partially increased
49. Effect of insulin on the rate of fatty acid synthesis de novo is that
    (a) it increases the rate (b) it decreases the rate
    (c) it has no effect (d) it firstly decreases and then increases the rate

50. As a prosthetic group, transaminases require vitamin
    (a) B_1 (b) B_2 (c) B_6 (d) B_{12}
51. The enzyme glutamate dehydrogenase requires a coenzyme
    (a) vitamin B_1 (b) vitamin B_2 (c) NAD (d) TPP
52. Ammonium ion to the brain
    (a) is beneficial (b) is toxic (c) is non-toxic (d) has no effect
53. Urea is ______ in water.
    (a) soluble (b) partially soluble (c) very soluble (d) insoluble
54. To the tissues, urea in higher concentration
    (a) is useful (b) is partially useful
    (c) is completely harmless (d) has no effect
55. During menstruation in women, there is breakdown of the smooth muscle cells of the endometrium which causes excretion of creatinine in quantity that is
    (a) more (b) less than usual (c) no effect (d) very less
56. Creatinine is also found in urine in other conditions such as
    (a) thyrotoxicosis (b) diabetes mellitus (c) encephalitis (d) nephritis
57. The non-essential amino acids are
    (a) alanine (b) serine (c) tyrosine (d) isoleucine
58. The essential amino acids are
    (a) proline (b) aspartic acid (c) glycine (d) methionine
59. Methylation of guanidinoacetic acid leads to the formation of
    (a) DNA (b) RNA (c) creatine (d) creatinine
60. Oxidation of choline leads to the formation of
    (a) betaine (b) ethanolamine (c) cysteine (d) homocysteine
61. How much choline does daily normal diet contain?
    (a) 1–1.5 g (b) 0–1 g (c) 0.5–1 g (d) 1.5–2 g
62. How much choline gets synthesized in the body each day during the synthesis of lecithin?
    (a) 2.0–2.5 g (b) 0.5–1 g (c) 1.0–1.5 g (d) 0.2–0.8 g
63. Hypervalinaemia is a disease that is
    (a) common (b) common in south Indians
    (c) congenital (d) non-congenital
64. Hypervalinaemia is a condition in which valine accumulates in the
    (a) liver (b) optic lens (c) blood (d) spleen

65. An important derivative of histidine is the local hormone named as
    (a) adrenaline (b) noradrenaline (c) histamine (d) epinephrine
66. Histidine is catabolized by opening the ring to produce
    (a) formiminoglutamic acid (b) ornithine
    (c) γ-aminobutyric acid (d) π-aminobenzoic acid
67. For the synthesis of collagen, the amino acids required are
    (a) alanine (b) glycine (c) proline (d) hydoxyproline
68. The concentration of tryptophan in tissues is
    (a) low (b) very low (c) high (d) very high
69. Concentration of tryptophan in plasma is
    (a) undetectable (b) 5–10 mg% (c) 10–20 mg% (d) 30–50 mg%
70. A major pathway of tryptophan metabolism begins by opening the 5-membered ring to form
    (a) formylkynurenine (b) nicotinic acid
    (c) 5-hydroxytryptamine (d) kynurenic acid
71. How much of the vitamin, nicotinamide, does a man get from tryptophan?
    (a) a fraction (b) 20–40 mg (c) 40–60 mg (d) more than 100 mg
72. The conversion of 3-hydroxykynurenine into 3-hydroxyanthranilic acid requires vitamin
    (a) B_1 (b) B_2 (c) B_6 (d) B_{12}
73. In phenylketonuria, how much of phenylpyruvic acid gets excreted in urine daily?
    (a) 0.5–1.0 g (b) 1.0–1.5 g (c) 1.5–2.0 g (d) 2.0–2.5 g
74. Phenylketonuria is characterized by
    (a) gastritis (b) pain in the bones (c) nephrosis (d) mental deficiency
75. Tyrosinosis is a
    (a) rare disease (b) popular disease
    (c) disease of western countries (d) disease of south-east Asian countries
76. How much time do protein deficiency diseases take to develop?
    (a) 15–20 days (b) 7–15 days (c) 20–30 days (d) months
77. The end product of protein metabolism is
    (a) urea (b) creatinine (c) uric acid (d) creatine
78. The end product of nitrogen metabolism is
    (a) urea (b) uric acid (c) creatinine (d) creatine

79. Can adults live in a negative nitrogen balance for years?
    (a) yes (b) no (c) sometimes (d) not possible
80. Do inborn errors of tricarboxylic acid cycle metabolism exist?
    (a) yes (b) no (c) sometimes (d) not possible
81. Ornithine in the presence of carbamyl phosphate forms
    (a) citrulline (b) argininosuccinate (c) arginine (d) none of these
82. Who is the father of modern genetics?
    (a) Darwin (b) Watson (c) Mendel (d) Avogadro
83. Who crystallized urease and proved that it was a protein in nature?
    (a) Sumner (b) Mitchell (c) Davidson (d) Chargaff
84. Chemical mutagens include
    (a) 5-bromouracil (b) DNFB (c) PCA (d) EDTA
85. Inhibitor(s) of protein synthesis include
    (a) benzene hexachloride (b) 2, 6-dichlorophenol indophenol
    (c) 5-methyltryptophan (d) phosphoadenosine phosphosulphate (PAPS)
86. Inhibitor(s) of protein synthesis include
    (a) tetracycline (b) chlorotetracycline (aureomycin)
    (c) oxytetracycline (terramycin) (d) chloramphenicol
    (e) cycloheximide (f) streptomycin
    (g) puromycin (h) cardiac glycosides
    (i) cardiolipins (j) carnosine
    (k) barbiturates (l) caprylic acid
87. The two subunits of ribosomes can be dissociated *in vitro* by lowering the concentration of
    (a) Ca^{2+} (b) Mg^{2+} (c) Mn^{2+} (d) Zn^{2+} in the suspending medium
88. In protein synthesis, termination codons include
    (a) UAA (b) UAG (c) UGA (d) all of these
89. Serotonin, which is derived from tryptophan, functions in the central pathways of the brain as a
    (a) neurotransmitter (b) accelerator of glycolysis
    (c) inhibitor of TCA cycle (d) inhibitor of gluconeogenesis
90. Another name of serotonin is
    (a) 5-hydroxytryptophan (b) 5-hydroxytryptamine
    (c) 5-hydroxyindole acetic acid (d) indole acetic acid

91. In addition to its localization in the brain, serotonin also exists in high concentrations in
    (a) liver and pancreas
    (b) kidney and spleen
    (c) gastrointestinal tract
    (d) adipose tissue

92. Serotonin is converted to 5-hydroxyindole acetic acid by the enzyme
    (a) phosphofructokinase
    (b) monoamine oxidase
    (c) phenylalanine hydroxylase
    (d) phospholipase 'a'

93. The end product of serotonin catabolism is
    (a) tyrosine
    (b) norepinephrine
    (c) epinephrine
    (d) 5-hydroxyindoleacetic acid

94. The decarboxylation product of tyrosine is
    (a) glycine
    (b) alanine
    (c) tyramine
    (d) citrulline
    (e) arginine
    (f) serine

95. The enkephalins are
    (a) dipeptides
    (b) tripeptides
    (c) tetrapeptides
    (d) pentapeptides
    (e) hexapeptides
    (f) heptapeptides

96. Endorphins and enkephalins are
    (a) neuropeptides
    (b) recently discovered vitamins
    (c) higher lipids
    (d) gluconeogenic substances

97. Functional proteins include
    (a) myosin
    (b) haemoglobin
    (c) cytochromes
    (d) lipoproteins

98. Nucleotides include
    (a) ATP
    (b) UTP
    (c) CTP
    (d) GTP

99. Important nitrogen compounds include
    (a) catecholamines
    (b) thyroid hormones
    (c) creatine
    (d) haem–porphyrin ring

100. Which of the following nitrogenous metabolites present in excess in body fluid may exert toxic effects?
    (a) ammonia
    (b) amino acids
    (c) porphyrin derivatives
    (d) cholesterol

101. The most important metal requirements for human beings are
    (a) sodium
    (b) chromium
    (c) selenium
    (d) copper

102. The most important inorganic requirements for human beings are
   (a) potassium and calcium (b) magnesium
   (c) iodine (d) all of these

103. Which of the following should be added regularly in traces of drinking water so as to avoid dental caries?
   (a) chlorine (b) fluorine (c) bromine (d) iodine

104. Which of the following elements is needed in quantities greater than 1 mg for human beings?
   (a) calcium (b) copper (c) zinc (d) nickel

105. The element which is found in most abundance in human body is
   (a) sodium (b) potassium (c) calcium (d) phosphorus

106. Calcium present in a 70 kg man is approximately
   (a) 584 g (b) 884 g (c) 1184 g (d) 1484 g

107. What percentage of the total calcium of the body is found in the bones?
   (a) 39% (b) 59% (c) 79% (d) 99%

108. The chief constituent of bone is
   (a) phosphorus (b) calcium (c) magnesium (d) none of these

109. Absorption of calcium is enhanced in the body by the presence of vitamin
   (a) A (b) B_1 (c) B_2 (d) D

110. Which is the main inorganic cation present in intracellular fluid?
   (a) sodium (Na^+) (b) potassium (K^+) (c) calcium (Ca^{++}) (d) magnesium (Mg^{++})

111. Concentration of potassium per kg of muscle is about
   (a) 3.2 g (b) 4.2 g (c) 5.2 g (d) 6.2 g

112. Foods of vegetable origin are richer in
   (a) sodium (b) potassium (c) calcium (d) magnesium

113. The metabolism of chlorine cannot be separated from the metabolism of
   (a) calcium (b) magnesium (c) potassium (d) sodium

114. How much iodine does the body of a man weighing 70 kg contain?
   (a) approximately 100 mg (b) approximately 75 mg
   (c) approximately 25 mg (d) approximately 50 mg

115. Major portion of the iodine remains confined to the
   (a) thyroid gland (b) prostate gland (c) lungs (d) thymus

116. What is the percentage of fluorine present in a normal bone?
(a) 0.01–0.03% (b) 0.03–0.08% (c) 0.08–0.12% (d) 0.12–0.18%

117. What is the percentage of fluorine present in dental enamel?
(a) 0.04–0.08% (b) 0.01–0.02% (c) 0.08 –0.12% (d) 0.10 – 0.15%

118. Rich sources of iron are
(a) beans (dried) (b) egg yolk, cheese, wheat (entire grain)
(c) apples, spinach (d) all of these

119. Apples contain ______ iron per 100 g of fresh substance.
(a) 0.2 mg (b) 0.3 mg (c) 0.8 mg (d) 1.0 mg

120. The two richest sources of iron in the following list is
(a) beans (dried) (b) egg yolk
(c) wheat (entire grain) (d) all of these

121. The two richest sources of copper are
(a) liver (b) nuts (c) legumes (d) fish

122. Zinc is a component of active
(a) insulin (b) progesterone (c) prolactin (d) prostaglandins

123. The copper content of which of the following foods per kg is called essential level?
(a) liver 44.1 mg (b) nuts 11.6 mg
(c) legumes 1.2 mg (d) green legumes 1.7 mg
(e) all of these

124. Zinc is a part of the structure of the enzyme
(a) carboxypeptidase (b) pepsin
(c) trypsin (d) carbonic anhydrase

125. Manganese is involved in activating
(a) peptidases (b) carboxylase (c) cholinesterase (d) phosphatases
(e) all of these

126. How much manganese is sufficient for human beings per kg of body weight?
(a) 0.01–0.03 g (b) 0.02–0.03 g (c) 0.03–0.04 g (d) 0.04–0.05 g

127. The efficiency of iron absorption can increase greatly in
(a) pregnancy (b) iron deficiency
(c) most kinds of anaemias (d) all of these

128. How much percentage of the total iron of the human body does remain in haemoglobin?
(a) 25% (b) 50% (c) 75% (d) more than 75%

129. Good sources of calcium are
(a) milk and milk products (b) mango and guava
(c) gram and spinach (d) all of these

130. Molybdenum is rapidly and effectively absorbed in human beings by
(a) liver (b) kidney (c) intestinal tract (d) pancreas

131. Excretion of molybdenum in humans is primarily by way of the
(a) urine (b) faeces (c) sweat (d) tears

132. Which of the following is predominantly an intracellular ion?
(a) sodium (b) potassium (c) magnesium (d) calcium

133. Which of the following is predominantly an extracellular ion?
(a) sodium (b) potassium (c) magnesium (d) calcium

134. In heart failure, which myocardial ions are depleted?
(a) sodium and manganese (b) potassium
(c) calcium and magnesium (d) copper

135. In most cases, hyperkalaemia results on account of
(a) liver failure (b) pancreas failure (c) renal failure (d) none of these

136. Cushing's syndrome (over-activity of the adrenal cortex) may be diagnosed by
(a) hypokalaemia (b) hyperkalaemia (c) hyponatraemia (d) hypernatraemia

137. Loss of mineral mass is seen in all except
(a) glucocorticoids therapy (b) hypoparathyroidism
(c) immobilization (d) post-menopausal women

138. The term lipid was given by
(a) Sutherland (b) Bloor (c) Altmann (d) Berzelius

139. $C_nH_{2n}O_2$ is the formula of
(a) carbohydrate (b) fatty acid (c) fat (d) nucleic acid

140. Each molecule of fat has
(a) one glycerol molecule and one fatty acid
(b) one glycerol molecule and three fatty acid molecules
(c) three glycerol molecules and one fatty acid molecule
(d) three glycerol molecules and three fatty acid molecules

141. A saturated fatty acid is with
    (a) no double bond (b) high melting point
    (c) low melting point (d) both (a) and (b)

142. The melting point of unsaturated fatty acids
    (a) increases with increase in double bonds
    (b) decreases with increase in double bonds
    (c) rises in some and falls in others
    (d) shows that there exists no relationship between unsaturation and melting point

143. Essential fatty acids were discovered by
    (a) Evans and Burr (b) Bloor (c) Sutherland (d) Beevers

144. Number of essential fatty acids is
    (a) 6 (b) 4 (c) 3 (d) 2

145. Essential fatty acids are
    (a) saturated (b) unsaturated
    (c) both saturated and unsaturated (d) cyclic

146. Which of these is an essential fatty acid?
    (a) linoleic acid (b) linolenic acid (c) arachidonic acid (d) all of these

147. Essential fatty acids occur in
    (a) animal fat (b) plant oils (c) fish and fowl (d) both (a) and (b)

148. Deficiency of EFA causes
    (a) follicular keratosis (b) kidney failure (c) sterility (d) all of these

149. Number of double bonds present in arachidonic acid is
    (a) 4 (b) 3 (c) 2 (d) 1

150. Arachidonic acid gives rise to
    (a) terpenes (b) prostaglandins (c) gangliosides (d) cerebrosides

151. Hydrogenation
    (a) increases percentage of EFA (b) destroys EFA
    (c) converts waxes into neutral fats (d) forms conjugated linoleic acid

152. Major function of PUFA is
    (a) supply of essential fatty acids (b) lowering of cholesterol level
    (c) to prevent atherosclerosis (d) all of these

153. Which one yields twice as many calories as carbohydrate per unit weight?
    (a) protein (b) fat (c) amino acids (d) vitamins

154. Number of fatty acids present in a molecule of phospholipid is
(a) 2 (b) 3 (c) 1 (d) none of these

155. Cholesterol is a
(a) monosaccharide (b) protein (c) sterol (d) wax

156. In contact with water, fatty acids produce
(a) monolayer (b) bilayer (c) trilayer (d) membrane

157. In contact with water, phospholipids produce
(a) monolayer (b) bilayer (c) trilayer (d) emulsion

158. A steroid that prevents fertility is
(a) campesterol (b) cholesterol (c) diosgenin (d) testosterone

159. Wax present in human blood is
(a) mericyl palmitate (b) cetyl palmitate
(c) cholesterol palmitate (d) ceryl palmitate

160. Wax is an ester of fatty acid with
(a) long-chain dihydric alcohol (b) trihydric alcohol
(c) long-chain monohydric alcohol (d) short-chain monohydric alcohol

161. Beeswax consists of
(a) mericyl palmitate (b) mericyl stearate
(c) cetyl palmitate (d) cetyl stearate

162. Commonly, candles are prepared from
(a) beeswax (b) paraffin wax (c) stearic acid (d) both (b) and (c)

163. Lanolin or wool fat is a an
(a) hard fat (b) oil (c) wax (d) sterol

164. Phosphatidic acid is a
(a) fatty acid (b) phospholipid (c) ganglioside (d) wax

165. Cephalin is a
(a) phospholipid (b) terpenoid (c) sterol (d) phosphoprotein

166. Lycopene, the colouring agent of tomato, is a
(a) sterol (b) terpenoid (c) sphingolipid (d) phospholipid

167. Rubber is a
(a) monoterpene (b) diterpene (c) tetraterpene (d) polyterpene

168. Fats are solids at
   (a) 10°C (b) 20°C (c) 30°C (d) 40°C
169. Esters of fatty acids with higher alcohols other than glycerol are called
   (a) waxes (b) fats (c) both (a) and (b) (d) none of these
170. The combination of an amino-alcohol, fatty acid and sialic acid form
   (a) phospholipid (b) sulpholipids (c) glycolipids (d) aminolipids
171. Hydrolysis of fat by alkali is called
   (a) saponification number (b) saponification
   (c) both (a) and (b) (d) none of these
172. The number of millilitres of 0.1N KOH required to neutralize the insoluble fatty acids from 5 grams of fat is called
   (a) acid number (b) acetyl number
   (c) halogenation (d) Polenske number
173. The rate of fatty acid oxidation is increased by
   (a) phospholipids (b) glycolipids (c) aminolipids (d) all of these
174. Cardiolipin found in mitochondria is formed from
   (a) lipositol (b) phosphatidyl ethanolamine
   (c) phosphatidyl glycerol (d) none of these
175. Lecithin contains a nitrogenous base called
   (a) ethanolamine (b) choline (c) inositol (d) all of these
176. Lecithins contain an unsaturated fatty acid at position
   (a) α- (b) α- and β- (c) β- (d) none of these
177. Lecithins are soluble in ordinary fat solvents except
   (a) benzene (b) ethyl alcohol (c) methyl alcohol (d) acetone
178. When lecithins are exposed to air, they become
   (a) black (b) brown (c) red (d) yellow
179. Lecithins combine with protein to form
   (a) phosphoprotein (b) mucoprotein (c) lipoprotein (d) glycoprotein
180. Phosphatidyl inositol is found in
   (a) cabbages (b) soyabeans (c) cauliflowers (d) apples
181. Instead of an ester link, plasmalogens possess an ether link in position
   (a) α- (b) β- (c) γ- (d) none of these

182. The alkyl radical in plasmalogen is a/an ______ alcohol.
   (a) saturated
   (b) unsaturated
   (c) both of (a) and (b)
   (d) none of these

183. The concentrations of sphingomyelins are increased in
   (a) Gaucher's disease
   (b) Fabry's disease
   (c) febrile disease
   (d) Niemann–Pick disease

184. Sphingomyelins contain a complex amino alcohol called
   (a) serine (b) lysolecithin (c) sphingosine (d) glycol

185. The types of sphingomyelins are
   (a) 2 (b) 3 (c) 4 (d) 5

186. Glycolipids contain an amino alcohol
   (a) sphingosine
   (b) isosphingosine
   (c) both (a) and (b) above
   (d) none of these

187. Cerebrosides may also be classified as
   (a) sphingolipids (b) sulpholipids (c) aminolipids (d) glycolipids

188. Kerasin contains
   (a) nervonic acid
   (b) hydroxynervonic acid
   (c) cerebronic acid
   (d) lignoceric acid

189. Oxynervon contains
   (a) nervonic acid
   (b) hydroxynervonic acid
   (c) lignoceric acid
   (d) hydroxylignoceric acid

190. Gaucher's disease is characterized specially by the increase in
   (a) lignoceric acid
   (b) nervonic acid
   (c) cerebronic acid
   (d) hydroxynervonic acid

191. Gangliosides are the glycolipids occurring in
   (a) liver (b) brain (c) kidney (d) muscle

192. Some gangliosides also contain
   (a) sphingosine
   (b) dihydrosphingosine
   (c) both (a) and (b)
   (d) none of these

193. Up to how many molecules of sialic acid do most gangliosides contain?
   (a) 1 (b) 2 (c) 3 (d) 4

194. The percentage of triacylglycerol of lipoproteins is
(a) 30 (b) 35 (c) 40 (d) 45

195. The percentage of cholesterol and cholesteryl esters present in lipoproteins is
(a) 15 (b) 20 (c) 25 (d) 28

196. Lipoprotein present in cell membrane is by nature,
(a) hydrophobic (b) hydrophilic
(c) both (a) and (b) (d) none of these

197. The density of lipoproteins increases as the protein content
(a) rises (b) decreases
(c) highly decreases (d) slightly and promptly decreases

198. Lipoproteins may be identified more accurately by means of
(a) electrophoresis (b) centrifugation
(c) immunoelectrophoresis (d) ultracentrifugation

199. Very-low-density lipoproteins are also known as
(a) β-lipoproteins (b) pre-β-lipoproteins (c) α-lipoproteins (d) none of these

200. The protein moiety of lipoprotein is known as
(a) apoprotein (b) pre-protein (c) post-protein (d) pseudoprotein

201. The per cent of apoprotein in chylomicrons is
(a) 0.4 (b) 0.6 (c) 0.8 (d) 1.0

202. The β-lipoprotein fraction increases in severe
(a) diabetes mellitus (b) uraemia
(c) nephritis (d) muscular dystrophy

203. Sulpholipids have been isolated from
(a) heart (b) liver (c) brain (d) intestine

204. Δ^9 indicates a double bond between carbon atoms _______ of the fatty acids.
(a) 8 and 9 (b) 9 and 10 (c) 10 and 11 (d) 11 and 12

205. The number of carbon atoms in decanoic acid present in butter is
(a) 6 (b) 8 (c) 10 (d) 12

206. How many carbon atoms does lignoceric acid present in peanut oil contains?
(a) 18 (b) 20 (c) 22 (d) 24

207. How many double bonds does arachidonic acid contains?
(a) 2 (b) 3 (c) 4 (d) 5

208. The prostaglandins are synthesized from
(a) arachidonic acid (b) oleic acid (c) linoleic acid (d) linolenic acid

209. Chaulmoogric acid was used many years ago in the treatment of
(a) bronchitis (b) leprosy (c) nephritis (d) oedema

210. The iodine number of essential fatty acids of vegetable oils is
(a) high (b) very high (c) very low (d) low

211. The essential fatty acids retard
(a) atherosclerosis (b) diabetes mellitus (c) nephritis (d) oedema

212. Arachidonic acid is
(a) I-shaped (b) M-shaped (c) U-shaped (d) V-shaped

213. Waxes containing higher alcohols are called
(a) methyl (b) ethyl (c) phytyl (d) cetyl

214. An example of cardiac glycosides is
(a) digitonin (b) strophanthin (c) lycopyll (d) digitalis

215. The product of reduction of cholesterol by bacteria in the intestine occurs in faeces as
(a) ergosterol (b) demosterol (c) coprosterol (d) lanosterol

216. Lieberman–Burchard reaction is performed to detect
(a) cholesterol (b) glycerol (c) fatty acid (d) vitamin D

217. Lipase present in the stomach cannot hydrolyse fats owing to
(a) alkalinity (b) acidity (c) high acidity (d) neutrality

218. Before the action of lipase the fat is emulsified by
(a) lipoproteins (b) phospholipids (c) ergosterols (d) digitonin

219. Of the total amount of triacylglycerol in the intestinal lumen, the free glycerol present is about
(a) 16% (b) 18% (c) 20% (d) 22%

220. A great majority of absorbed fat appears in the form of
(a) HDL (b) chylomicrons (c) VLDL (d) LDL

221. More metabolic water is available on oxidation of
(a) fatty acids (b) glycerol (c) both (a) and (b) (d) none of these

222. Fatty acids are oxidized by
(a) α-oxidation (b) β-oxidation (c) ω-oxidation (d) all of these

223. The fatty acids containing even number and odd number of carbon atoms as well as the unsaturated fatty acids are oxidized by
(a) α-oxidation (b) β-oxidation (c) ω-oxidation (d) all of these

224. Long-chain fatty acids are first activated to acyl-CoA in the
(a) cytosol (b) mitochondria (c) microsomes (d) lysosomes

225. Activation of lower fatty acids occurs within the
(a) cytosol (b) mitochondria (c) ribosomes (d) microsomes

226. Long-chain acyl-CoA penetrates mitochondria in the presence of
(a) palmitate (b) carnitine (c) sorbitol (d) DNP

227. Acyl-CoA dehydrogenase converts acyl-CoA to α,β-unsatuated acyl-CoA in the presence of the coenzyme
(a) NAD^+ (b) $NADP^+$ (c) ATP (d) FAD^+

228. For the activation of long-chain fatty acids, the enzyme thiokinase requires the cofactor
(a) Mg^{++} (b) Ca^{++} (c) Mn^{++} (d) K^+

229. ω-oxidation takes place by the hydroxylase in microsomes involving
(a) Cytochrome *b* (b) Cytochrome *c*
(c) Cytochrome P-450 (d) Cytochrome a_3

230. Carboxylation of acetyl-CoA to malonyl-CoA takes place in the presence of
(a) FAD^+ (b) biotin (c) NAD^+ (d) $NADP^+$

231. Malonyl-CoA reacts with the central
(a) —SH group (b) —NH_2 group (c) —COOH group (d) —CH_2OH group

232. Fatty acid synthesis takes place in the presence of the coenzyme
(a) NAD^+ (b) reduced NAD (c) $NADP^+$ (d) reduced NADP

233. Fatty acids are activated to acyl-CoA by the enzyme thiokinase using
(a) NAD^+ (b) $NADP^+$ (c) CoA (d) FAD

234. Phospholipase A_1 attacks the ester bond of phospholipids in position
(a) 1 (b) 5 (c) 3 (d) all of these

235. Phospholipase C release, 1,2-diacylglycerol and phosphoryl base attacking the ester bond in position
(a) 1 (b) 2 (c) 3 (d) 4

236. Phospholipids help the oxidation of

(a) glycerol (b) fatty acids
(c) glycerophosphates (d) none of these

237. The desaturation and chain elongation system of polyunsaturated fatty acids are greatly diminished in the absence of

(a) insulin (b) glucagon (c) epinephrine (d) thyroxine

238. Prostaglandins are liberated in the circulation by the stimulation of

(a) anterior pituitary gland (b) posterior pituitary gland
(c) adrenal gland (d) thyroid gland

239. Prostaglandins have a common structure based on prostanoic acid which contains carbon atoms

(a) 12 (b) 16 (c) 18 (d) 20

240. The carbon chains of prostanoic acid are bonded at the middle of the chain by a

(a) 5-membered ring (b) 6-membered ring
(c) 8-membered ring (d) none of these

241. All active prostaglandins have at least one double bond between positions

(a) 7 and 8 (b) 9 and 10 (c) 11 and 12 (d) 13 and 14

242. The synthesis of prostaglandins is inhibited by

(a) aspirin (b) arsenite (c) fluoride (d) cyanide

243. The synthesis of prostaglandins requires the consumption of two molecules of oxygen and two molecules of reduced

(a) NAD (b) NADP (c) glutathione (d) lipoate

244. Prostaglandins lower cyclic AMP in

(a) thyroid (b) adipose tissue (c) platelets (d) lung

245. The inhibitory activity of prostaglandins in liver is effective on

(a) glycogen synthetase (b) hexokinase
(c) phosphorylase (d) glucose 6-phosphatase

246. In adipose tissue, prostaglandins decrease

(a) lipogenesis (b) ketogenesis (c) lipolysis (d) ketolysis

247. Prostaglandins increase intestinal motility and cause

(a) constipation (b) loose motion (c) diarrhoea (d) dysentery

248. The enzyme responsible for the metabolism of prostaglandins is blocked by the introduction of a methyl group at the

(a) C_9 position (b) C_{11} position (c) C_{13} position (d) C_{15} position

249. In adipose tissue, less glycerol 3-phosphate is formed in
(a) diabetes mellitus (b) nephritis
(c) coronary thrombosis (d) deafness

250. LDL contains the apoprotein
(a) C-I (b) C-II (c) C-III (d) B

251. The normal concentration of β-lipoproteins in mg per cent in plasma is
(a) 100 (b) 200 (c) 300 (d) 400

252. The concentration of triacylglyerol in pre-β-lipoproteins in per cent is
(a) 40 (b) 45 (c) 50 (d) 55

253. Both chylomicrons and VLDL are released from the intestine or hepatic cell by reverse
(a) pinocytosis (b) diffusion (c) osmosis (d) passive diffusion

254. HDL is synthesized and secreted from
(a) pancreas (b) liver (c) kidney (d) muscle

255. Serum LDL has been found to be increased in
(a) obstructive jaundice (b) hepatic jaundice
(c) haemolytic jaundice (d) septicaemia

256. The enzyme systems for lengthening and shortening, for saturating and desaturating of fatty acids occur in
(a) intestine (b) muscle (c) kidney (d) liver

257. The lowered glucokinase leading to diminished fatty acid synthesis in the liver is caused by the effect of
(a) feeding (b) overfeeding (c) starvation (d) diarrhoea

258. The increased levels of plasma fatty acids results from mobilization of fat from
(a) muscle (b) adipose tissue (c) kidney (d) intestine

259. Fatty appearance and enlargement of the liver are caused by
(a) pregnancy (b) nephritis
(c) abortion (d) toxaemia of pregnancy

260. Triacylglycerol is accumulated even in normal rate of fatty acid synthesis by the deficiency of
(a) lipotropic factor (b) vitamin A (c) calcium (d) vitamin C

261. Lipotropic activity is possessed by
(a) casein (b) cellulose (c) phospholipids (d) glycolipids

262. Fatty liver is caused by
(a) CH_3Cl (b) CCl_4 (c) $MgSO_4$ (d) CH_3COOH

263. Fatty liver results in the deficiency of
(a) stearic acid (b) caproic acid (c) vitamin A (d) pantothenic acid

264. The normal concentration of ketone bodies in mg per cent in blood does not exceed
(a) 1.0 (b) 1.5 (c) 2.0 (d) 2.5

265. Ketosis generally occurs in
(a) nephritis (b) oedema
(c) infective hepatic disease (d) coronary thrombosis

266. Ketone bodies are utilized in
(a) mitochondria (b) extrahepatic tissues
(c) nuclei (d) chromosomes

267. The excretion of ketone body in the urine involves the deficiency of
(a) Na^+ (b) Fe^{++} (c) Ca^{++} (d) Mg^{++}

268. The total amount of cholesterol in grams in the body of a man weighing 70 kg is about
(a) 200 (b) 180 (c) 160 (d) 140

269. The blood cholesterol level is increased due to the deficiency of
(a) vitamin D (b) vitamin B_2 (c) pyridoxine (d) inositol

270. Eicosanoids are formed from
(a) arachidonate (b) palmitate (c) stearate (d) butyrate

271. The three groups of eicosanoids synthesized from eicosanoic acid are
(a) PG, TX, LT (b) PG, TT, LT (c) TG, TX, LT (d) TG, TX, DT

272. Eicosanoic acids are derived from
(a) linoleic acid (b) linolenic acid (c) arachidonic acid (d) all of these

273. Prostaglandin endoperoxide synthase catalyses
(a) 4 moles of O_2 (b) 3 moles of O_2 (c) 2 moles of O_2 (d) 1 mole of O_2

274. Cyclo-oxygenase is termed as
(a) inhibiting enzyme (b) suicide enzyme
(c) oxidizing enzyme (d) reducing enzyme

275. PGH (nedoperoxide) is converted to prostaglandin
(a) D (b) E (c) F (d) all of these

276. Leukotrienes are formed in
(a) leukocytes (b) mastocytome (c) platelets (d) all of these

277. Leukotrienes are formed by only
(a) 5-lipoxygenase (b) lipase (c) phospholipase (d) none of these

278. Leukotriene A_4 is metabolized to leukotriene
(a) C_2 (b) C_3 (c) C_4 (d) D_4

279. Leukotriene C_4 is formed by the addition of
(a) ascorbic acid (b) glutathione (c) glutamate (d) aspartate

280. PG_3 and TX_3 inhibit the release of
(a) oleic acid (b) palmitoleic acid (c) palmitic acid (d) arachidonic acid

281. TXA_3 is a weaker aggregator than
(a) TXA_2 (b) TXA_4 (c) TXA_5 (d) TXA_6

282. Prostaglandins increase cAMP in
(a) platelets (b) thyroid (c) corpus luteum (d) all of these

283. Eskimos have low plasma concentration of
(a) cholesterol (b) triacylglycerol (c) LDL (d) all of these

284. The contraction of smooth muscle is caused by prostaglandins at a minimum concentration of
(a) 1 mg/ml (b) 10 mg/ml (c) 100 mg/ml (d) <10 mg/ml

285. Leukotrienes are important in
(a) allergic reaction (b) oxidation reaction
(c) reduction reaction (d) inhibitory reaction

ANSWERS

1. (c)	2. (b)	3. (a)	4. (d)	5. (a)	6. (a)	7. (a)
8. (b)	9. (a)	10. (b)	11. (a)	12. (a)	13. (a)	14. (a)
15. (c)	16. (a)	17. (c)	18. (a)	19. (c)	20. (b)	21. (a)
22. (b)	23. (d)	24. (d)	25. (a)	26. (d)	27. (a)	28. (a)
29. (b)	30. (c,d)	31. (d)	32. (b)	33. (b)	34. (b)	35. (a)
36. (a)	37. (b)	38. (a)	39. (b)	40. (b)	41. (a)	42. (d)
43. (c)	44. (b)	45. (a)	46. (a)	47. (b)	48. (a)	49. (a)
50. (c)	51. (c)	52. (b)	53. (c)	54. (c)	55. (a)	56. (a)
57. (a,b)	58. (d)	59. (c)	60. (a)	61. (a)	62. (b)	63. (c)

64. (c) 65. (c) 66. (a) 67. (c,d) 68. (b) 69. (a) 70. (a)
71. (a) 72. (c) 73. (a) 74. (d) 75. (a) 76. (d) 77. (a)
78. (a) 79. (a) 80. (b) 81. (a) 82. (c) 83. (a) 84. (a)
85. (c) 86. (a,b,c,d,e,f,g) 87. (b) 88. (d) 89. (a) 90. (b)
91. (c) 92. (b) 93. (d) 94. (c) 95. (d) 96. (a) 97. (a,d)
98. (a,d) 99. (a,d) 100. (a,c) 101. (a) 102. (e) 103. (b) 104. (a)
105. (c) 106. (c) 107. (d) 108. (b) 109. (d) 110. (b) 111. (a)
112. (b) 113. (d) 114. (c) 115. (a) 116. (a) 117. (b) 118. (d)
119. (b) 120. (a,b) 121. (a,b) 122. (a) 123. (e) 124. (d) 125. (e)
126. (b) 127. (d) 128. (c) 129. (d) 130. (c) 131. (a) 132. (b)
133. (a) 134. (b) 135. (c) 136. (a) 137. (b) 138. (b) 139. (b)
140. (b) 141. (d) 142. (b) 143. (a) 144. (c) 145. (b) 146. (d)
147. (d) 148. (d) 149. (a) 150. (b) 151. (b) 152. (d) 153. (b)
154. (a) 155. (c) 156. (a) 157. (b) 158. (e) 159. (c) 160. (c)
161. (a) 162. (d) 163. (c) 164. (b) 165. (a) 166. (b) 167. (d)
168. (b) 169. (a) 170. (c) 171. (b) 172. (d) 173. (a) 174. (c)
175. (b) 176. (c) 177. (d) 178. (b) 179. (c) 180. (b) 181. (a)
182. (b) 183. (d) 184. (c) 185. (b) 186. (c) 187. (c) 188. (d)
189. (b) 190. (c) 191. (b) 192. (b) 193. (c) 194. (d) 195. (a)
196. (b) 197. (a) 198. (c) 199. (b) 200. (a) 201. (d) 202. (a)
203. (c) 204. (b) 205. (c) 206. (d) 207. (c) 208. (a) 209. (b)
210. (d) 211. (a) 212. (c) 213. (d) 214. (b) 215. (c) 216. (a)
217. (c) 218. (b) 219. (d) 220. (b) 221. (a) 222. (d) 223. (b)
224. (a) 225. (b) 226. (b) 227. (d) 228. (a) 229. (c) 230. (b)
231. (a) 232. (d) 233. (c) 234. (a) 235. (c) 236. (b) 237. (a)
238. (c) 239. (d) 240. (a) 241. (d) 242. (a) 243. (c) 244. (b)
245. (a) 246. (a) 247. (b) 248. (d) 249. (a) 250. (d) 251. (c)
252. (c) 253. (a) 254. (b) 255. (a) 256. (d) 257. (c) 258. (b)
259. (d) 260. (a) 261. (a) 262. (b) 263. (d) 264. (a) 265. (c)
266. (b) 267. (a) 268. (d) 269. (c) 270. (a) 271. (a) 272. (d)
273. (c) 274. (b) 275. (d) 276. (d) 277. (a) 278. (c) 279. (b)
280. (d) 281. (a) 282. (d) 283. (d) 284. (d) 285. (a)

6

NUCLEIC ACIDS

Choose the Correct Answer

1. An alpha helix of a protein is most likely to be disrupted if a missense mutation introduces which one of the following amino acids within the α-helical structure?
   (a) alanine (b) aspartic acid (c) tyrosine (d) glycine
2. During gluconeogenesis reducing equivalents are transported from mitochondria to cytosol by
   (a) aspartate (b) oxaloacetate (c) malate (d) glutamate
3. Gluconeogenesis is the reversal of glycolytic pathway except for the presence of barriers, i.e.,
   (a) pyruvate and phosphoenolpyruvate
   (b) fructose 1,6-diphosphate and fructose 6-PO_4
   (c) glucose 6-phosphate and glucose
   (d) glucose 1-phosphate and glycogen
4. Which of the following separation techniques depends upon the molecular size of the protein?
   (a) chromatography on a carboxymethyl (CM) cellulose column
   (b) isoelectric focusing
   (c) gel filtration chromatography
   (d) chromatography on a diethylaminoethyl (DEAE) cellulose column
5. At the physiological pH, the DNA molecules are
   (a) positively charged (b) negatively charged
   (c) amphipathic (d) neutral
6. Which one of the following enzymes is not a protein?
   (a) DNase (b) abzyme (c) *Eco* RI (d) ribozyme

7. Which of the following hormones does not have any intracellular receptor?
   (a) vitamin D_3 (b) cortisone (c) adrenaline (d) thyroxine
8. Accumulation of sphingomyelin in phagocytes is a feature of
   (a) Gaucher's disease (b) Niemann–Pick disease
   (c) Tay–Sach's disease (d) Down syndrome
9. Accumulation of GM_2 gangliosides is a feature of
   (a) Gaucher's disease (b) Niemann–Pick disease
   (c) Tay–Sach's disease (d) Down syndrome
10. Accumulation of glucosyl ceramide is a feature of
    (a) Gaucher's disease (b) Nieman–Pick disease
    (c) Tay–Sach's disease (d) Down syndrome
11. In haemoglobin, the innate affinity of haem for CO_2 is diminished by the presence of
    (a) His E-S (b) His E-7 (c) Gly B-6 (d) Thr C-4
12. In which year was Flemming awarded the Nobel Prize?
    (a) 1945 (b) 1947 (c) 1942 (d) 1944
13. The most direct precursor of taurine is
    (a) glycine (b) cysteine (c) methionine (d) glutathione
14. The most important buffer in RBCs is
    (a) O_2Hb + Na haemoglobinate (b) O_2Hb + K haemoglobinate
    (c) H_2CO_3 + $KHCO_3$ (d) H_2CO_3 + $NaHCO_3$
15. Iron in haemoglobin is held by
    (a) polar bonds (b) non-polar bonds
    (c) co-ordinate bonds (d) covalent bonds
16. Rigor mortis is known as
    (a) relaxation of muscles (b) loss of NAD
    (c) deficiency of the enzyme hexokinase (d) none of these
17. Muscle relaxation means
    (a) more production of ATP
    (b) more production of CTP
    (c) dissociation of actin and myosin and a decrease in the cross-linking between these elements
    (d) loss of high-energy reserve, i.e., creatine PO_4 in muscles
18. Sarcoplasm contains large stores of
    (a) glucose (b) fructose (c) maltose (d) glycogen

19. The phenothiazines are widely used as
    (a) antitubercular drug (b) anticancer drug
    (c) antipsychotic drug (d) antidiabetic drug
20. The powerful antagonists of folic acid action are
    (a) aminopterin (b) dicumanol (c) dihydrofolic acid (d) warfarin
21. In 1869 who defined what we now know to be nucleic acids?
    (a) Chargaff (b) Davidson (c) Miescher (d) Watson
22. In an experiment, DNA was found to have 31% adenine and 19% guanine. The quantity of cytosine shall be
    (a) 38% (b) 31% (c) 19% (d) 62%
23. Helical structure of DNA was found out by
    (a) Wilkins and Franklin (b) Watson and Crick
    (c) Beadle and Tatum (d) Kornberg and Nirenberg
24. Double-helical structure of DNA was proposed by
    (a) Kornberg (b) Nirenberg
    (c) Watson and Crick (d) Wilkins and Franklin
25. Watson and Crick were awarded Nobel Prize for their finding that
    (a) RNA is single-stranded (b) DNA is double-stranded
    (c) DNA is the genetic material (d) DNA guides mRNA synthesis
26. DNA is made of
    (a) pentose sugar, phosphoric acid, purine and pyrimidines
    (b) pentose sugar, phosphoric acid and purines
    (c) phosphoric acid, pentose sugar and pyrimidines
    (d) pentose sugar, purines and pyrimidines
27. Purine bases of DNA are
    (a) A and G (b) A and C (c) C and T (d) A and U
28. Wilkin's X-ray diffraction showed DNA diameter to be
    (a) 200 Å (b) 100 Å (c) 20 Å (d) 50 Å
29. In DNA, guanine lies opposite
    (a) uracil (b) cytosine (c) adenine (d) thymine
30. Distance between two base pairs of DNA is
    (a) 34 nm (b) 3.4 nm (c) 0.68 nm (d) 0.34 nm

31. Base-pairing of DNA does not prove that
    (a) A/T is one
    (b) base sequence of RNA can be found out
    (c) A + T = C + G
    (d) C = G
32. Which of the following is not consistent with double helical structure of DNA?
    (a) A = T, C = G,
    (b) density of DNA decreases on heating
    (c) A + T/C + G is not constant
    (d) both (a) and (b)
33. tRNA is
    (a) messenger (b) soluble RNA (c) rRNA (d) microsomal RNA
34. A storage protein is
    (a) keratin (b) collagen (c) haemoglobin (d) glutelin/globulin
35. Uracil occurs in
    (a) RNA (b) DNA (c) ATP (d) protein
36. RNA takes part in synthesis of
    (a) DNA (b) carbohydrate (c) fat (d) protein
37. RNA responsible for directing proper sequence of amino acids in protein synthesis is
    (a) chromosomal RNA
    (b) ribosomal RNA
    (c) messenger RNA
    (d) transfer RNA
38. Inulin occurs in the root of
    (a) mango (b) dahlia (c) wheat (d) sugar cane
39. The bond present between two nucleotides of a polynucleotide is
    (a) covalent bond
    (b) hydrogen bond
    (c) phosphodiester bond
    (d) high-energy phosphate bond
40. A segment of DNA has 120 adenine and 120 cytosine bases. The total number of nucleotides present in the segment is
    (a) 120 (b) 240 (c) 60 (d) 480
41. mRNA is a polymer of
    (a) deoxyribonucleotides
    (b) ribonucleosides
    (c) ribonucleotides
    (d) deoxyribonucleosides
42. DNA strands are antiparallel because of
    (a) hydrogen bonds
    (b) phosphodiester bonds
    (c) disulphide bonds
    (d) glycosidic bonds

43. The smallest RNA is
    (a) tRNA (b) mRNA
    (c) rRNA (d) chromosomal RNA
44. DNA is composed of repeating units of
    (a) ribonucleosides (b) deoxyribonucleosides
    (c) ribonucleotides (d) deoxyribonucleotides
45. Hydrogen bonds present between cytosine and guanosine are
    (a) 2 (b) 3 (c) 1 (d) 4
46. The two strands of DNA are held together by bonds of
    (a) nitrogen (b) oxygen (c) hydrogen (d) carbon
47. Nucleic acids/nuclein were first discovered by
    (a) Altmann (b) Fleming (c) Miescher (d) Koch
48. In DNA, hydrogen bonds and base pairings occur between
    (a) A—U, C—G (b) A—C, G—T (c) A—G, C—T (d) A—T, C—G
49. DNA does not occur in
    (a) nucleus (b) ribosomes (c) mitochondria (d) plastids
50. In nucleic acids
    (a) DNA is single-stranded in some viruses (b) RNA is double-stranded occasionally
    (c) one turn of Z-DNA has 12 bases (d) all of these
51. In DNA, adenine pairs with
    (a) guanine (b) thymine (c) cytosine (d) uracil
52. Distance between two strands of DNA is
    (a) 34 Å (b) 20 Å (c) 3.4 Å (d) 340 Å
53. In double helix of DNA, the two DNA strands are
    (a) coiled around a common axis (b) coiled around each other
    (c) coiled differently (d) coiled over protein sheath
54. In RNA, adenine will pair with
    (a) thymine (b) guanine (c) uracil (d) cytosine
55. tRNA takes part in
    (a) formation of ribosomes (b) bringing mRNA in contact with ribosome
    (c) selection of amino acids (d) formation of mRNA

56. Iodine test can detect
    (a) carbohydrates (b) nucleic acids (c) lipids (d) proteins
57. RNA does not possess
    (a) uracil (b) thymine (c) adenine (d) cytosine
58. Most abundant RNA of the cell is
    (a) tRNA (b) rRNA (c) mRNA (d) tRNA threonine
59. Number of nitrogen bases found in RNA are
    (a) two (b) eight (c) six (d) four
60. Purines of RNA are
    (a) guanine and adenine (b) uracil and thymine
    (c) adenine and cytosine (d) uracil and guanine
61. Cytoplasmic DNA constitutes a total of
    (a) 95–99% (b) 45–50% (c) 65–70% (d) 1–5%
62. Anticodon occurs in
    (a) DNA (b) tRNA (c) mRNA (d) rRNA
63. Watson and Crick proposed the model of DNA in
    (a) 1955 (b) 1953 (c) 1943 (d) 1963
64. DNA and RNA show similarity in having
    (a) polymers of nucleotides (b) similar pyrimidines
    (c) double strands (d) similar sugars
65. *In vitro* synthesis of DNA was carried out by
    (a) Nirenberg (b) Watson and Crick (c) Khorana (d) Kornberg
66. Structure of DNA was given by
    (a) Kornberg (b) Nirenberg
    (c) Watson and Crick (d) Holley and Nirenberg
67. Which of the following does not have high-energy phosphate bond?
    (a) creatine phosphate (b) GTP
    (c) ATP (d) AMP
68. The principal energy-storing molecule is
    (a) NADP (b) FAD (c) ATP (d) ADP
69. A source of maximum energy is
    (a) carbohydrate (b) fat (c) protein (d) vitamins

70. A complex formed by deoxyribose sugar, phosphate and nitrogen base is
(a) amino acid (b) polypeptide (c) nucleoside (d) nucleotide

71. Which one is the hydrogen acceptor?
(a) CoA (b) $NADP^+$ (c) ATP (d) DNA

72. Adenine is a
(a) purine (b) pyrimidine (c) nucleoside (d) nucleotide

73. Which one is exclusively a ribose nucleotide?
(a) thymine–pentose sugar–phosphate
(b) guanine–pentose sugar–phosphate
(c) uracil–pentose sugar–phosphate
(d) cytosine–pentose sugar–phosphate

74. A nucleotide is formed of
(a) purine, pyrimidine and phosphate (b) purine, sugar and phosphate
(c) nitrogen base, sugar and phosphate (d) pyrimidine, sugar and phosphate

75. Pyrimidines of RNA are
(a) adenine and cytosine (b) adenine and guanine
(c) thymine and cytosine (d) uracil and cytosine

76. Nitrogen bases do not contain
(a) carbon (b) phosphorus (c) hydrogen (d) nitrogen

77. Which of the following are purines?
(a) adenine and guanine (b) adenine and thymine
(c) cytosine and thymine (d) cytosine and guanine

78. Which of the following is a dinucleotide?
(a) ADP (b) RNA (c) NAD (d) DNA

79. The correct sequence in protein synthesis is
(a) amino acid $\rightarrow$ DNA $\rightarrow$ RNA $\rightarrow$ protein
(b) RNA $\rightarrow$ DNA $\rightarrow$ amino acid $\rightarrow$ protein
(c) DNA $\rightarrow$ amino acid $\rightarrow$ RNA $\rightarrow$ protein
(d) DNA $\rightarrow$ RNA $\rightarrow$ amino acid $\rightarrow$ protein

80. Operon model for gene regulation was proposed by
(a) Benzer (b) Jacob and Monod
(c) Khorana (d) Beadle and Tatum

81. Bacteroids refer to
    (a) *Nitrosomonas* bacteria in soil
    (b) enlarged nonmobile cellular bacteria *Rhizobium leguminosarum* in root nodules of legumes
    (c) a bacterial cell infected with viruses
    (d) a mobile bacterium

82. Which level of protein structure is affected by DNA?
    (a) tertiary structure (b) primary structure
    (c) secondary structure (d) quaternary structure

83. The length of mRNA which carries the information necessary for the determination of a complete polypeptide chain of a protein molecule is called
    (a) codon (b) cistron (c) operon (d) muton

84. In an *E.coli* cell according to operon theory, an operator gene combines with
    (a) regulator protein to switch off structural gene transcription
    (b) regulator protein to switch on structural gene transcription
    (c) inducer gene to switch on structural gene transcription
    (d) regulator gene to switch off structural gene transcription

85. Attachment of amino acyl tRNA is catalysed by or transfer of specific amino acid to specific tRNA for participation in protein synthesis is mediated by
    (a) peptidases (b) peptidyl transferase
    (c) aminoacyl synthetase enzymes (d) hydrogen bonds

86. An antibiotic which inhibits translation in eukaryotes is
    (a) tetracyclin (b) flavicin (c) penicillin (d) chloromycetin

87. The types of RNA specifically responsible for directing the proper sequence of amino acids in protein synthesis is
    (a) mRNA (b) rRNA
    (c) chromosomal RNA (d) none of these

88. The codons of RNA are recognized by
    (a) rRNA (b) tRNA
    (c) chromosomal RNA (d) ribosomes

89. The activity of a repressor depends on whether
    (a) a key substance in the metabolic pathway is present
    (b) the repressor is positioned next to the operon
    (c) there is enough RNA polymerase present
    (d) the repressor is positioned next to promoter

90. An operon is a
    (a) cluster of structural genes with related functions
    (b) protein that accelerates gene expression
    (c) protein that suppresses gene expression
    (d) gene that switches other genes on or off

91. If a cell is treated with a chemical that blocked nucleic acid synthesis, which of the following processes would most likely be affected first?
    (a) tRNA synthesis (b) protein synthesis
    (c) DNA replication (d) mRNA synthesis

92. Transcription is
    (a) synthesis of mRNA from DNA and to carry over the information from DNA to mRNA
    (b) recognition of base sequence on mRNA by tRNA
    (c) recognition of amino acids by RNA synthetase enzyme
    (d) formation of polypeptide through the association of amino acids by mRNA

93. Protein synthesis involves
    (a) transcription and transformation (b) transcription and translation
    (c) translation and transformation (d) transcription and transduction

94. The production of repression protein is by
    (a) plasma gene (b) regulator gene (c) promoter gene (d) operator gene

95. Through which enzyme can RNA give rise to DNA?
    (a) RNA polymerase (b) DNA polymerase
    (c) reverse transcriptase (d) restriction enzyme

96. The best role of purine and pyrimidine nucleotides is to serve as the monomeric precursors of
    (a) RNA (b) DNA (c) both (a) and (b) (d) none of these

97. The purine nucleotides act as the components of
    (a) FAD^+ (b) NAD^+ (c) $NADP^+$ (d) all of these

98. The pyrimidine nucleotides act as the high-energy intermediates of
    (a) UDPG (b) ATP (c) ADP (d) AMP

99. The chemical name of thymine is
    (a) 2-oxy-4-aminopyrimidine (b) 2, 4-dioxy-5-methylpyrimidine
    (c) 2, 4-dioxypyrimidine (d) none of these

100. The chemical name 2-amino-6-oxypurine is that of
    (a) adenine (b) xanthine (c) guanine (d) hypoxanthine

101. The lactam form is the predominant automer of

(a) uracil (b) cytosine (c) adenine (d) xanthine

102. N_7-methylguanine has been found more recently in the nucleic acids of the cells of

(a) bacteria (b) yeast (c) plant (d) mammals

103. Hypoxanthine and ribose constitute

(a) adenosine (b) inosine (c) guanosine (d) cytidine

104. Thymine and deoxyribose form

(a) deoxycytidine (b) deoxyadenosine
(c) deoxythymidine (d) deoxyuridine

105. The most abundant intracellular free nucleotide is

(a) ATP (b) FAD^+ (c) NAD^+ (d) $NADP^+$

106. The concentration of ATP in living mammalian cells in mM is near

(a) 0.2 (b) 0.4 (c) 0.6 (d) 1.0

107. The intracellular CAMP concentration in μM is near

(a) 3.0 (b) 2.0 (c) 1.0 (d) 0.5

108. The biosynthesis of phosphoglycerides in animal tissues requires

(a) UTP (b) CTP (c) GTP (d) ATP

109. The chemical name 4-hydroxypyrazole pyrimidine is used for

(a) thioguanine (b) mercaptopurine (c) azathioprine (d) allopurinol

110. "Transforming factor" is used for the name of

(a) RNA (b) DNA (c) tRNA (d) none of these

111. By how many number of hydrogen bonds is guanosine nucleotide held by the cytosine nucleotide?

(a) 1 (b) 2 (c) 3 (d) 4

112. How many base pairs exist within the single turn of DNA?

(a) 4 (b) 6 (c) 8 (d) 10

113. Each turn of DNA structure has a pitch of

(a) 1.4 mm (b) 2.4 mm (c) 3.4 mm (d) 4.4 mm

114. The double-stranded DNA molecule loses its viscosity upon

(a) denaturation (b) filtration (c) sedimentation (d) concentration

115. DNA molecule contains

(a) 800 to 4000 nucleotides (b) 1000 to 6000 nucleotides
(c) 1200 to 8000 nucleotides (d) 1600 to 9000 nucleotides

116. DNA is denatured by

(a) heat (b) acid (c) alkali (d) all of these

117. How many repeating units in nm does chromatin contain

(a) 10 (b) 15 (c) 20 (d) 25

118. The guanine content in RNA does not necessarily equal its cytosine content and its adenine content does not necessarily equal its uracil content since it is a

(a) double-stranded molecule (b) single-stranded molecule
(c) stable molecule (d) unstable molecule 1

119. How many nucleotides does each RNA molecule contain?

(a) 40 to 4000 (b) 50 to 5000 (c) 60 to 6000 (d) 70 to 7000

120. Messenger RNA has a molecular weight of

(a) 15000 to 30000 (b) 20000 to 35000
(c) 25000 to 40000 (d) 30000 to 50000

121. Of the total cellular RNA molecules, the number of transfer RNA molecules amount to

(a) 5 to 10 (b) 8 to 16 (c) 10 to 20 (d) 15 to 30

122. How many nucleotides does each transfer RNA molecule contain?

(a) 70 (b) 75 (c) 80 (d) 85

123. In every cell, the number of tRNA molecules are at least

(a) 10 (b) 15 (c) 20 (d) 25

124. All tRNA molecules have a common CCA sequence at the

(a) 3′ terminus (b) 5′ terminus (c) 3′, 5′ termini (d) all of these

125. In nearly all tRNA molecules, there is a loop containing the nucleotides of

(a) uridine (b) thymidine
(c) ribothymine and pseudouridine (d) cytidine

126. In tRNA molecules, there is another loop containing the minor base

(a) uracil (b) dihydrouracil (c) cytosine (d) dihydrocytosine

127. The larger subunit of the ribosomal RNA is

(a) 60 (b) 70 (c) 80 (d) 90

128. The mammalian ribosome contains ______ major nucleoprotein subunits.
(a) 2 (b) 3 (c) 4 (d) 5

129. Gene is a segment of the DNA molecule containing base pairs of about
(a) 300 (b) 400 (c) 500 (d) 600

130. The fragments of DNA attached to an RNA initiator component were discovered by
(a) Watson and Crick (b) Okazaki (c) Peterson (d) Nelson

131. Synthesis of RNA molecule is terminated by a signal which is recognized by
(a) ρ-factor (b) δ-factor (c) α-factor (d) none of these

132. The promoter site is the site at which the synthesis of new RNA molecule begins with the help of another factor,
(a) σ (b) γ (c) β (d) α

133. The binding of prokaryotic DNA-dependent RNA polymerase to promoter sites of genes is inhibited by the antibiotic
(a) septron (b) streptomycin (c) aureomycin (d) rifampin

134. The N atoms at positions 3 and 9 of purine base are derived from the amide nitrogen of
(a) glutamate (b) glutamine (c) asparagine (d) aspartate

135. The carbon atoms at positions 4 and 5 and the N atom at position 7 of purine base are supplied from
(a) valine (b) alanine (c) glycine (d) serine

136. 5-phosphoribosylamine reacts with glycine to produce glycinamide ribosylphosphate by glycinamide kinosynthetase in the presence of
(a) ATP (b) GTP (c) UTP (d) CTP

137. The carbon atoms at position 4, 5 and 6 and the N atom at position 3 of pyrimidine base are derived from
(a) glutamic acid (b) aspartic acid (c) glycine (d) serine

138. The end product of purine catabolism in mammals, except man, is
(a) uric acid (b) allantoin (c) ammonia (d) creatinine

139. The net excretion of total uric acid in normal men in mg in 24 hours is
(a) 100–300 (b) 200–400 (c) 300–500 (d) 400–600

140. A portion of uric acid is converted to urea ammonia by intestinal
(a) urogenolysis (b) ureolysis (c) uricolysis (d) ureotolysis

ANSWERS

1.	(b)	2.	(c)	3.	(d)	4.	(c)	5.	(b,c)	6.	(d)	7.	(c)
8.	(b)	9.	(c)	10.	(a)	11.	(b)	12.	(a)	13.	(b)	14.	(b)
15.	(c)	16.	(d)	17.	(c)	18.	(d)	19.	(c)	20.	(a)	21.	(c)
22.	(c)	23.	(a)	24.	(c)	25.	(b)	26.	(a)	27.	(a)	28.	(c)
29.	(b)	30.	(d)	31.	(c)	32.	(c)	33.	(b)	34.	(d)	35.	(a)
36.	(d)	37.	(c)	38.	(b)	39.	(c)	40.	(d)	41.	(c)	42.	(b)
43.	(a)	44.	(d)	45.	(b)	46.	(c)	47.	(c)	48.	(d)	49.	(b)
50.	(d)	51.	(b)	52.	(b)	53.	(a)	54.	(c)	55.	(c)	56.	(a)
57.	(b)	58.	(d)	59.	(d)	60.	(a)	61.	(d)	62.	(b)	63.	(b)
64.	(a)	65.	(d)	66.	(c)	67.	(d)	68.	(c)	69.	(b)	70.	(d)
71.	(b)	72.	(a)	73.	(c)	74.	(c)	75.	(d)	76.	(b)	77.	(a)
78.	(c)	79.	(d)	80.	(b)	81.	(b)	82.	(b)	83.	(c)	84.	(a)
85.	(c)	86.	(d)	87.	(a)	88.	(b)	89.	(a)	90.	(a)	91.	(d)
92.	(a)	93.	(b)	94.	(b)	95.	(c)	96.	(c)	97.	(d)	98.	(a)
99.	(b)	100.	(c)	101.	(a)	102.	(d)	103.	(b)	104.	(c)	105.	(a)
106.	(d)	107.	(c)	108.	(a)	109.	(d)	110.	(b)	111.	(c)	112.	(d)
113.	(c)	114.	(a)	115.	(d)	116.	(d)	117.	(a)	118.	(b)	120.	(c)
120.	(d)	121.	(c)	122.	(b)	123.	(c)	124.	(a)	125.	(c)	126.	(b)
127.	(c)	128.	(a)	129.	(d)	130.	(b)	131.	(a)	132.	(d)	133.	(d)
134.	(b)	135.	(c)	136.	(a)	137.	(b)	138.	(b)	139.	(d)	140.	(c)

7

PROTEINS AND ENZYMES

CHOOSE THE CORRECT ANSWER

1. In competition inhibition
   (a) apparent K_m is unchanged
   (b) apparent K_m is decreased
   (c) V_{max} is decreased
   (d) V_{max} is unchanged
2. K_m is
   (a) dissociation constant for the enzyme–substrate complex
   (b) equal to half the substrate concentration required to achieve V_{max}
   (c) identical for all isozymes of an enzyme
   (d) substrate concentration that gives one-half V_{max}
3. Di-isopropyl fluorophosphate (DFP) reacts with serine proteases stoichiometrically and irreversibly and therefore is a
   (a) competitive inhibitor
   (b) non-competitive inhibitor
   (c) uncompetitive inhibitor
   (d) allosteric inhibitor
4. In non-competitive inhibition
   (a) concentration of active enzyme molecules is reduced
   (b) V_{max} is increased
   (c) concentration of active enzyme molecules is unchanged
   (d) apparent K_m is increased
5. In an enzyme assay in which substrate concentration is much lower than K_m, the rate
   (a) approaches V_{max}
   (b) shows zero-order kinetics
   (c) is proportional to substrate concentration
   (d) is independent of enzyme concentration

6. Isoenzymes
   (a) are enzymes that exist in more than one amino acid sequence in the same species
   (b) cannot be distinguished in a given species except immunologically
   (c) by definition must have the same amino acid composition
   (d) are single polypeptide chains that differ by an amino acid replacement

7. Which of the following is not a component of coenzyme A?
   (a) adenylic acid (b) pantothenic acid (c) cysteine (d) acetic acid

8. As a coenzyme, pyruvate decarboxylase requires
   (a) coenzyme A (b) NAD^+
   (c) FMN (d) thiamine pyrophosphate

9. Which of the following do dehydrogenases use as coenzymes?
   (a) NAD^+ (b) $NADP^+$ (c) FAD (d) all of these

10. Which of the following is an essential cofactor in carboxylation reactions?
    (a) Coenzyme A (b) CTP (c) lipoic acid (d) biotin

11. The enzyme that catalyses the reaction $2H_2O \longrightarrow 2H_2 + O_2$ is a
    (a) dehydrogenase (b) peroxidase (c) catalase (d) hydrolase

12. Which enzyme will cleave leucyl-glycyl-proline to leucine and glycyl-proline?
    (a) carboxypeptidase (b) glycyl peptidase
    (c) aminopeptidase (d) chymotrypsin

13. The enzyme of saliva that hydrolyses starch is
    (a) pepsin (b) β-amylase (c) lysozyme (d) α-amylase

14. The isoenzyme of lactate dehydrogenase is
    (a) histidine (b) tryptophan (c) aspartate (d) serine

15. The nerve gas, DFP, has been a useful reagent in enzyme chemistry. At the active site, which of the following amino acids of hydrolytic enzymes combines with DFP?
    (a) histidine (b) tryptophan (c) aspartate (d) serine

16. The isocitrate dehydrogenase reaction is analogous to (performs the same type of chemical reaction)
    (a) pyruvate dehydrogenase (b) α-ketoglutarate dehydrogenase
    (c) β-hydroxyacyl-CoA dehydrogenase (d) 6-phosphogluconate dehydrogenase

17. Antimetabolites act by
    (a) competitive inhibition when they combine irreversibly with an enzyme
    (b) competitive inhibition when they combine reversibly with an enzyme
    (c) noncompetitive inhibition if there is no relation between degree of inhibition and substrate concentration
    (d) noncompetitive inhibition by combining with the prosthetic group of the enzyme

18. K_m and V_{max} can be determined from the Lineweaver–Burk plot of Michaelis–Menten equation shown below. When V is the reaction velocity at substrate concentration S, the X-axis experimental data are expressed as

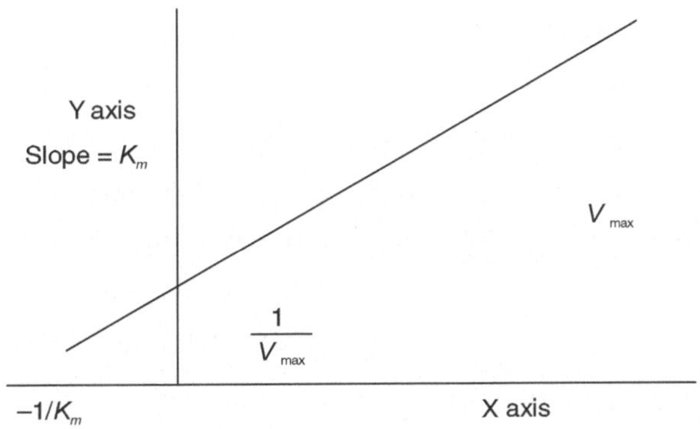

(a) $1/V$ (b) V (c) $1/S$ (d) S

19. Which of the following gastrointestinal enzymes are secreted as inactive zymogens (proenzymes)?
    (a) trypsin (b) chymotrypsin (c) carboxypeptidase (d) all of these

20. Which enzyme has the greatest specificity for peptide bonds on the carboxyl side of a cationic amino acid side chain?
    (a) carboxypeptidase (b) trypsin (c) rennin (d) chymotrypsin

21. Myosin is
    (a) a spherically symmetric molecule (b) low in α-helix content
    (c) a zinc-requiring enzyme (d) an actin-binding protein

22. If an enzyme behaves according to classic Michaelis–Menten kinetics, from a double-reciprocal plot of velocity versus substrate concentration, the value for the Michaelis-constant (K_m) of the substrate can be determined graphically as the
    (a) slope of curve
    (b) absolute value of the intercept of the curve with the X-axis
    (c) reciprocal of the absolute value of the intercept of the curve with the Y-axis
    (d) reciprocal of the absolute value of the intercept of the curve with the X-axis

23. Which of the following oxidation–reduction systems has highest redox potential?
    (a) fumarate/succinate
    (b) NAD^+/NADH
    (c) Fe^{+++} cytochrome *a*/Fe^{++}
    (d) Fe^{+++} cytochrome *b*/Fe^{++}

24. Dinitrophenol would most likely inhibit cell function by disrupting
    (a) TCA cycle
    (b) glycolysis
    (c) hepatic gluconeogenesis
    (d) oxidative phosphorylation

25. Polymyxin is unique among chemotherapeutic agents because it is bactericidal in the absence of cell growth. It exerts its effects by
    (a) binding to DNA as an insertion mutation
    (b) binding to polysome-bound mRNA
    (c) binding to DNA polymerase
    (d) detergent-like disruption of membranes

26. Cycloserine inhibits transpeptidation in the formation of the peptidoglycan cell wall network of gram-positive organisms. This action of cycloserine is competitively
    (a) L-lysine
    (b) D-alanine
    (c) D-glutamine
    (d) diaminopimelic acid

27. Some antibiotics act as ionophores, which means that they
    (a) interfere directly with bacterial cell-wall synthesis
    (b) inhibit only translation
    (c) have a detergent-like effect on cell membranes
    (d) increase cell membrane permeability to specific ions

28. De novo synthesis of an enzyme, promoted by the substrate on which it acts, is characterized by the term
    (a) induction (b) gratuity (c) activation (d) de-repression

29. Control of metabolic pathway may be exerted by enzyme repression or induction. In vertebrates, this form of enzyme control occurs primarily in the
    (a) heart (b) brain (c) liver (d) skeletal muscle

30. A hypothetical biosynthetic pathway is shown in the diagram below. A microbial organism defective in one enzyme of this path is grown in a medium containing X. Large amounts of M and L are found in the organism but none of Z. In which enzyme is the mutation expressed?

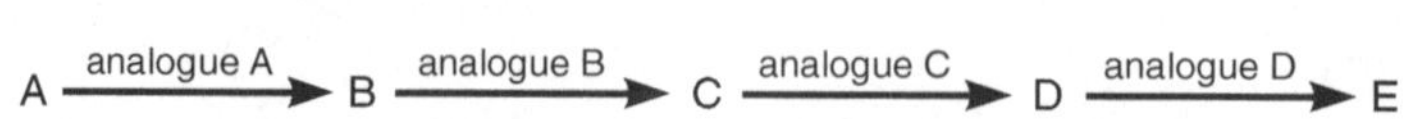

    (a) enzyme A (b) enzyme B (c) enzyme C (d) enzyme D

31. The activity of most single-polypeptide enzymes can be represented by the hyperbolic curve shown below. Which of the areas in the curve is homotropic?

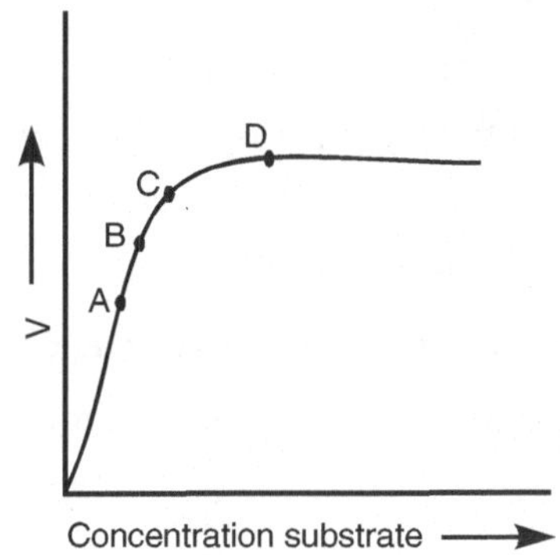

(a) region A
(b) region B
(c) region C
(d) region D

32. Trypsinogen and chymotrypsinogen
    (a) have considerable homologies in primary sequence
    (b) are secreted by exocrine cells in the pancreas
    (c) can be converted into active enzyme by limited digestion with trypsin
    (d) all of these

33. Most enzymes
    (a) increase the rapidity of the reaction they catalyse
    (b) are specific for the substrate as well as the reaction catalysed
    (c) are large polypeptides with molecular weight greater than 5,000 daltons
    (d) all of these

34. Proteins that contain a porphyrin ring include
    (a) catalase (b) cytochrome (c) haemoglobin (d) all of these

35. Which of the following proteins contain iron?
    (a) haemoglobin (b) cytochrome *c* (c) peroxidase (d) all of these

36. The immunoglobulin present in urine
    (a) contains the antibody-combining site
    (b) is Bence-Jones protein
    (c) contains the "hypervariable region" sequences
    (d) is the heavy chain

37. Papain cleaves an enzyme to form two A fragments and one B fragment. It is true that in fragment A
    (a) all allosteric enzymes are polymers
    (b) each subunit bears both a catalytic and an allosteric site
    (c) binding affinity of a subunit for a ligand may differ according to the conformational state of the subunit
    (d) transitions state involve changes in all subunits

38. Phosphorylation sites in the mitochondrial respiratory chain may occur between
    (a) coenzyme Q and cytochrome *b* (b) cytochrome *b* and cytochrome *c*
    (c) pyruvate and NAD (d) NAD and a flavoprotein

39. Sickle cell anaemia
    (a) is widely distributed in areas of high malaria fatality
    (b) may exhibit different K_m values for substrates or cofactors
    (c) results from a single amino acid change in haemoglobin
    (d) is seen in homozygous individuals only

40. Isozymes of a given enzyme
    (a) have different substrate specificities
    (b) may exhibit different K_m values for substrates or cofactors
    (c) are composed of distinct multimeric complexes
    (d) usually exhibit identical electrophoretic mobility

41. Heavy chains of IgG antibody may be separated from light chains with
    (a) ethanolamine (b) pepsin (c) papain (d) mercaptoethanol

42. How many gram atoms of iron per mole in the ferrous state does haemoglobin contains?
    (a) 1 (b) 2 (c) 3 (d) 4

43. The molecular weight of haemoglobin is
    (a) 44,450 (b) 54,450 (c) 64,450 (d) 74,450

44. The porphyrins are cyclic compounds formed through methylene bridges by the linkages of how many pyrrole rings?
    (a) 4 (b) 3 (c) 2 (d) 1

45. In porphyrins found in nature the various side chains are substituted for ______ hydrogen atoms.
    (a) 2 (b) 4 (c) 6 (d) 8

46. A porphyrin with a completely symmetrical arrangement of the substituents is classified as a porphyrin of
    (a) type I (b) type II (c) type III (d) all of these
47. The more abundant porphyrin found in nature is
    (a) type I series (b) type II series (c) type III series (d) none of these
48. In the biosynthesis of porphyrins the activation of glycine needs the coenzyme
    (a) NAD^+ (b) B_6-PO_4 (c) FAD^+ (d) ATP
49. Anaemia has been observed in the deficiency of the vitamin
    (a) biotin (b) inositol (c) niacin (d) pantothenic acid
50. Dipyrryl compounds are of two types,
    (a) A and B (b) B and C (c) A and C (d) B and D
51. A type III porphyrin results with the condensation of
    (a) two A (b) one A and one B
    (c) two B (d) one A and one C
52. The enzyme coproporphyrinogen oxidase is able to act on coproporphyrinogen
    (a) type I (b) type II (c) type III (d) all of these
53. In mammalian liver, the conversion of corproporphyrinogen to protoporphyrin requires
    (a) molecular oxygen (b) water (c) ATP (d) B_6-PO_4
54. Haem is synthesized by the incorporation of ferrous ion (Fe^{++}) into protoporphyrin III being catalysed by the enzyme
    (a) ferroxidase (b) ferroreductase (c) ferrochelatase (d) none of these
55. The globin of the haemoglobin is a protein composed of closely packed polypeptide chains of
    (a) 6 parallel layers (b) 4 parallel layers
    (c) 3 parallel layers (d) 2 parallel layers
56. Two α-chains of globin have identical amino acid composition of
    (a) 111 (b) 121 (c) 131 (d) 141
57. The total number of amino acids in globin is
    (a) 544 (b) 554 (c) 564 (d) 574
58. How many of molecules of oxygen does haemoglobin take up?
    (a) 1 (b) 2 (c) 4 (d) 6
59. Carboxyhaemoglobin is formed by
    (a) CO (b) CO_2 (c) HCO_3 (d) HCN

60. How many histidine residues does one mol. of haemoglobin contain?
(a) 5 (b) 15 (c) 25 (d) 35

61. Methaemoglobin is formed as a result of the oxidation of haemoglobin by
(a) oxygen of air (b) hydrogen peroxide
(c) potassium ferricyanide (d) potassium permanganate

62. Methaemoglobin can be reduced to haemoglobin by
(a) removal of hydrogen (b) vitamin C
(c) glutathione (d) creatinine

63. The colour of cyanomethaemoglobin is
(a) yellow (b) pink (c) brown (d) bright red

64. The iron of haem is coordinated in chains at positions
(a) 43 and 72 (b) 53 and 82 (c) 63 and 92 (d) 73 and 102

65. How many gram atoms of iron per mole does myoglobin contain?
(a) 1 (b) 2 (c) 3 (d) 4

66. How many gram atoms of iron per mole do catalases contain?
(a) 1 (b) 2 (c) 3 (d) 4

67. Homologous proteins have similar
(a) sequences (b) structures (c) functions (d) all of these

68. Most prevalent elements of protein secondary structure are
(a) α-helix (b) β-strand (c) sheet (d) all of these

69. A polypeptide chain is synthesized by a complex process called
(a) translation (b) folding (c) degradation (d) none of these

70. Folding proteins *in vivo* is promoted by
(a) proteasomes (b) chaperones
(c) messenger RNA (d) none of these

71. Chaperones
(a) are located in every cellular compartment
(b) bind a wide range of proteins
(c) function in the general protein-folding mechanism
(d) all of these

72. Molecular chaperones
    (a) directly facilitate folding protein
    (b) bind and stabilize in folded or partly folded proteins
    (c) both (a) and (b) above
    (d) none of these

73. In bacteria, the percentage of proteins released from their chaperones is
    (a) 70% (b) 80% (c) 85% (d) 95%

ANSWERS

1. (d)	2. (d)	3. (b)	4. (a)	5. (c)	6. (a)	7. (d)
8. (d)	9. (d)	10. (d)	11. (c)	12. (c)	13. (d)	14. (d)
15. (d)	16. (d)	17. (b)	18. (c)	19. (d)	20. (b)	21. (d)
22. (d)	23. (c)	24. (d)	25. (d)	26. (b)	27. (d)	28. (a)
29. (c)	30. (c)	31. (d)	32. (d)	33. (d)	34. (d)	35. (d)
36. (b)	37. (b)	38. (d)	39. (a,c,d)	40. (b,c)	41. (d)	42. (d)
43. (c)	44. (a)	45. (d)	46. (a)	47. (c)	48. (b)	49. (d)
50. (a)	51. (b)	52. (c)	53. (a)	54. (c)	55. (b)	56. (d)
57. (d)	58. (c)	59. (a)	60. (d)	61. (c)	62. (b)	63. (d)
64. (c)	65. (a)	66. (d)	67. (d)	68. (d)	69. (a,d)	70. (b)
71. (d)	72. (b)	73. (c)				

8

VITAMINS

CHOOSE THE CORRECT ANSWER

1. Fat-soluble vitamins include
   (a) vitamin K (b) ascorbic acid (c) riboflavin (d) both (a) and (b)
2. Vitamin C is
   (a) quite strongly acid (b) a mild acid
   (c) a base (d) neither acid nor a base
3. The disease resulting from vitamin C deficiency is
   (a) pernicious anaemia (b) beri beri
   (c) keratomalacia (d) any other disease
4. Ascorbic acid deficiency affects the formation of
   (a) collagen (b) myosin (c) keratin (d) histone
5. Ascorbic acid is found in
   (a) gur (b) sugar (c) potatoes (d) fish oil
6. The immediate precursor of vitamin D is
   (a) cholecalciferol (b) calciferol
   (c) 7-dehydrocholesterol (d) cholesterol
7. 7-dehydrocholesterol is found in
   (a) skin (b) sebum
   (c) both (a) and (b) (d) germinating seeds

8. Absence of vitamin D causes
   (a) osteomalacia (b) myocardial infarction
   (c) dermatosis (d) tuberculosis of bones
9. The form of vitamin 'D' that becomes active when taken by mouth is
   (a) ergosterol (b) calciferol (c) both (a) and (b) (d) cholic acid
10. 7-dehydrocholesterol gets changed into vitamin D in the presence of
    (a) ultraviolet light (b) dim light
    (c) infra red light (d) light of bulbs and tube lights
11. The biologically active substance for vitamin D *in vivo* is
    (a) 1, 25-dihydroxycholecalciferol (b) calciferol
    (c) 7-dehydrocholesterol (d) 25-hydroxycholecalciferol
12. The best source(s) of vitamin D is/are
    (a) margarine (b) vegetables (c) wheat (d) sugar cane
13. The daily requirement of vitamin D is about
    (a) 100 IU (b) 250 IU (c) 500 IU (d) 1000 IU
14. From many plant sources, the daily human diet contains about
    (a) 150 mg vitamin E (b) 250 mg vitamin E
    (c) 500 mg vitamin E (d) 750 mg vitamin E
15. Epimers of tocopherol include
    (a) α-tocophenol (b) γ-tocophenol (c) δ-tocophenol (d) all of these
16. Vitamin K_1 is found in
    (a) green leafy plants (b) meat (c) fish (d) milk
17. Vitamin K_2 has a different sidechain known as
    (a) farnesyl (b) phytyl (c) calciferol (d) none of these
18. Vitamin K_2 is found in
    (a) plants (b) corn (c) oils (d) none of these
19. In man, the deficiency diseases of vitamin A are almost limited to the
    (a) eyes (b) ears (c) stomach (d) intestine
20. Vitamin A deficiency in human being causes
    (a) xanthomatosis (b) sterility
    (c) mental retardation (d) keratomalacia

21. Keratomalacia means
    (a) irreversible corneal opacity
    (b) irreversible dermatitis
    (c) irreversible conjunctivitis
    (d) irreversible night blindness
22. Very good, indirect sources of vitamin A include
    (a) carrots (b) tomatoes (c) both (a) and (b) (d) peas
23. Sources of vitamin A include
    (a) milk (b) butter (c) eggs (d) all of these
24. Good sources of thiamine include
    (a) cereals (b) husk of rice (c) seeds (d) all of these
25. Other popular names of the vitamin thiamine are
    (a) Vitamin B_1 (b) aneurin (c) both (a) and (b) (d) Vitamin B_2
26. The most serious deficiency disease of vitamin B_1 is
    (a) rickets (b) sterility (c) beri beri (d) pellagra
27. Alcoholic drinks are a source of
    (a) energy
    (b) wastage of energy
    (c) kidney diseases
    (d) vitamins
28. Beri beri is still endemic in the
    (a) east (b) west (c) both (a) and (b) (d) none of these
29. Beri beri is characterized by
    (a) bradycardia
    (b) anorexia (loss of appetite)
    (c) both (a) and (b)
    (d) dementia
30. Less serious deficiency disease of vitamin B_1 is
    (a) generalized peripheral neuritis
    (b) hepatitis
    (c) encephalitis
    (d) nephritis
31. The flavin found in the eggs is known as
    (a) lactoflavin (b) ovoflavin (c) hepatoflavin (d) verdoflavin
32. The flavin found in the milk is known as
    (a) lactoflavin (b) ovoflavin (c) hepatoflavin (d) verdoflavin
33. The extracellular enzyme is
    (a) glucose 6-phosphatase
    (b) hexokinase
    (c) glucokinase
    (d) pancreatic amylase

34. Which of the following enzymes is usually named by adding the suffix -ase to the main part of the name of the substrate on which they act?

    (a) maltase (b) lactase (c) sucrase (d) all of these

35. Which of the following enzyme is named by their functions only?

    (a) ptyalin (b) pepsin (c) reductases (d) trypsin

36. Which of the following enzyme acting on the substrates are freely described by the hydrolytic activity?

    (a) lactate dehydrogenase (b) lipolytic
    (c) cytochrome oxidase (d) phosphatase

37. The enzyme which uses oxygen as hydrogen acceptor is

    (a) tyrosinase (b) succinate dehydrogenase
    (c) aconitase (d) carboxylase

38. The enzyme which uses H_2O_2 as substrate is

    (a) catalase (b) malate dehydrogenase
    (c) ascorbic oxidase (d) phosphorylase

39. The enzyme which acts on single hydrogen donor with incorporation of oxygen is

    (a) succinate thiokinase (b) glycogen synthetase
    (c) tryptophan oxygenase (d) uricase

40. The enzyme which acts on paired donors with incorporation of oxygen into one donor is

    (a) pyruvate kinase (b) enolase
    (c) steroid hydroxylase (d) glycerokinase

41. The transglycosidase is

    (a) hexokinase (b) phosphorylase
    (c) GPT (d) aminoacyl transacetylase

42. The enzyme acting on peptide bonds is

    (a) hexokinase (b) chymotrypsin
    (c) GOT (d) glucose 6-phosphatase

43. Peroxidase has the coenzyme

    (a) FAD^+ (b) NAD^+ (c) $NADP^+$ (d) None of these

44. Enzyme exergonic reaction means the system undergoes a loss of free energy. An example is

    (a) synthetase (b) phosphatase (c) hexokinase (d) urease

45. The optimum temperature for plant urease is

    (a) 40ºC (b) 37ºC (c) 45ºC (d) 60ºC

46. The maximum activity of most of the enzymes is at the optimum pH between
    (a) 2 and 3 (b) 4 and 5 (c) 5 and 9 (d) 7 and 12
47. Oxidases are generally inhibited by
    (a) cyanides (b) fluorides (c) salts of mercury (d) salts of silver
48. The group transferring coenzyme is
    (a) CoA (b) NAD^+ (c) $NADP^+$ (d) FAD^+
49. The coenzyme containing an aromatic hetero ring in the structure is
    (a) biotin (b) TPP (c) sugar phosphate (d) coenzyme Q
50. The hydrogen-transferring coenzyme is
    (a) B_6-PO_4 (b) $NADP^+$ (c) TPP (d) ATP
51. Retinal is reduced to retinol by retinene reductase in the presence of coenzyme
    (a) NAD^+ (b) $NADP^+$ (c) $NADH^+H^+$ (d) $NADP^+H^+$
52. Retinol exists as an ester with higher fatty acids in the
    (a) liver (b) kidney (c) lung (d) all of these
53. Retinol is transported with the
    (a) α_1-globulin (b) α_2-lipoproteins (c) β-globulin (d) γ-lipoproteins
54. Carotenes are transported with the
    (a) proteins (b) lipids (c) lipoproteins (d) minerals
55. The dietary vitamin A which is in the form of vitamin A esters is hydrolysed in the lumen of the intestine by the enzyme
    (a) amylase (b) lipase (c) peptidase (d) nuclease
56. In the blood the vitamin esters are attached to
    (a) α_1-lipoproteins (b) α_2-lipoproteins (c) β-lipoproteins (d) γ-lipoproteins
57. The percentage of vitamin A in the form of esters and stored in the liver is
    (a) 80 (b) 85 (c) 90 (d) 95
58. Preformed vitamin A is supplied by foods such as
    (a) butter (b) eggs (c) fish liver oil (d) all of these
59. One IU is equal to ______ micrograms of β-carotene.
    (a) 0.3 (b) 0.4 (c) 0.6 (d) 0.8
60. The non-protein part of rhodopsin is
    (a) retinal (b) retinol (c) carotene (d) repsin

61. Lumirhodopsin is stable only at a temperature below

(a) –35°C (b) –40°C (c) –45°C (d) –50°C

62. Metarhodopsin is hydrolysed to retinal and opsin at a temperature below

(a) –8°C (b) –10°C (c) –15°C (d) –20°C

63. The normal concentration of vitamin A in blood in IU/dl is

(a) 20–55 (b) 24–60 (c) 30–65 (d) 35–70

64. Continued intake of excessive amounts of vitamin A especially in children produces

(a) irritability (b) anorexia (c) headache (d) all of these

65. Vitamin D_2 is also called

(a) activated ergosterol (b) ergocalciferol
(c) viosterol (d) all of these

66. The poorest source of vitamin D is

(a) eggs (b) butter (c) milk (d) liver

67. The normal concentration of vitamin D in blood in IU/L is

(a) 600–2500 (b) 700–3100 (c) 800–4100 (d) 850–4700

68. The activity of tocopherols is destroyed by

(a) oxidation (b) reduction (c) conjugation (d) all of these

69. Some tocopherols are

(a) terpenoid in structure (b) dionol in structure
(c) isoprenoid in structure (d) farnesyl in structure

70. The methyl groups in the aromatic nucleus of α -tocopherol are

(a) 2 (b) 3 (c) 4 (d) 5

71. Vitamin E is stored in

(a) mitochondria (b) microsomes (c) both (a) and (b) (d) none of these

72. Vitamin E protects the polyunsaturated fatty acids from oxidation by molecular oxygen by the formation of

(a) superoxide (b) peroxide (c) trioxide (d) all of these

73. Tocopherols prevent the oxidation of

(a) vitamin A (b) vitamin D (c) vitamin K (d) vitamin C

74. Vitamin E protects enzymes from destruction in

(a) muscles (b) nerves (c) gonads (d) all of these

75. Vitamin K_2 was originally isolated from
    (a) soyabean (b) oysters (c) putrid fish meal (d) alfalfa
76. The poor sources of vitamin K_1 are
    (a) milk (b) meat (c) fish (d) all of these
77. Vitamin K regulates the synthesis of blood clotting factor
    (a) VII (b) IX (c) X (d) all of these
78. Ascorbic acid can reduce
    (a) 2, 6-dichlorophenolindophenol (b) 2, 4-dinitrobenzene
    (c) 2, 6-dioxypyridine (d) 2, 6-dibromobenzene
79. Sterilized milk is devoid of
    (a) vitamin A (b) vitamin D (c) vitamin C (d) vitamin B_1
80. Vitamin C is required for the metabolism of
    (a) phenlylalanine (b) tryptophan (c) both (a) and (b) (d) none of these
81. The symptoms of scurvy are
    (a) poor healing of wounds (b) loosening of teeth
    (c) anaemia (d) all of these
82. Thiamine is also said to be
    (a) antiberi beri substance (b) antineuritic vitamin
    (c) aneurine (d) all of these
83. Lipoic acid is also termed as
    (a) thioctic acid (b) protogen
    (c) acetate replacement factor (d) all of these
84. Folic acid is also called
    (a) SLR factor (b) pteroylglutamic acid
    (c) liver *Lactobacillus casei* factor (d) all of these
85. Thiamine is oxidized to thiochrome in alkaline solution by
    (a) potassium permanganate (b) potassium ferricyanide
    (c) potassium dichromate (d) potassium chlorate
86. The requirement of vitamin B_1 is increased when metabolism is elevated as in
    (a) fever (b) hyperthyroidism
    (c) increased muscular activity (d) all of these

87. Riboflavin in alkaline solution when exposed to ultraviolet light is converted into lumiflavin which in ultraviolet light has a
    (a) greenish yellow fluorescence (b) bluish yellow fluorescence
    (c) reddish yellow fluorescence (d) light yellow fluorescence

88. The normal concentration of riboflavin in plasma in µg/100 ml is
    (a) 1.5 to 3.5 (b) 2.5 to 4.0 (c) 3.5 to 5.5 (d) 4.5 to 7.5

89. Riboflavin is involved in the regulatory function of some hormones connected with metabolism of
    (a) carbohydrate (b) fat (c) protein (d) minerals

90. FMN is a constituent of the
    (a) Warburg yellow enzyme (b) cytochrome *c* reductase
    (c) *L*-amino acid dehydrogenase (d) all of these

91. The majority of the excess nicotinic acid is excreted in the urine in the form of
    (a) *N*-methyl nicotinamide (b) 6-pyridone of *N*-methyl nicotinamide
    (c) *N*-methyl nicotinic acid (d) all of these

92. Niacin is present in maize in the form of
    (a) niatin (b) niacytin (c) nicotine (d) niacin

93. The normal concentration of niacin in blood in mg/100 ml is
    (a) 0.3 to 0.5 (b) 0.4 to 0.6 (c) 0.5 to 0.8 (d) 0.6 to 0.99

94. Nicotinic acid is essential for the normal functioning of
    (a) skin (b) intestinal tract (c) nervous system (d) all of these

95. Pyridoxidine is a mixture of
    (a) pyridoxine (b) pyridoxal (c) pyridoxamine (d) all of these

96. Pyridoxine produces a coloured compound with
    (a) 2, 6-dichloroquinone chlorimide (b) 2, 6-dichloroquinone
    (c) 2, 4-nitroquinone (d) all of these

97. Pyridoxal phosphate is involved in the desulphuration of
    (a) cysteine (b) homocysteine (c) both (a) and (b) (d) none of these

98. In some subjects pyridoxine deficiency causes
    (a) lymphopenia (b) peripheral neuropathy
    (c) both (a) and (b) (d) all of these

99. In the deficiency states of vitamin B_6, there are inborn errors of metabolism including
    (a) cystathioninuria (b) familial xanthurenic aciduria
    (c) some pyridoxal-responsive anaemias (d) all of these

100. Pantothenic acid exists in the tissues as

(a) β-mercaptoethylamine (b) coenzyme A
(c) pantoic acid (d) β-alanine

101. Pantothenic acid deficiency causes

(a) nausea (b) irritability (c) anaemia (d) all of these

102. Folacin is made up of

(a) pteridine nucleus (b) *p*-aminobenzoic acid
(c) glutamic acid (d) all of these

103. Folic acid coenzymes take part in the synthesis of

(a) purines (b) thymine (c) both (a) and (b) (d) none of these

ANSWERS

1. (d)	2. (a)	3. (d)	4. (a)	5. (c)	6. (c)	7. (c)
8. (a)	9. (c)	10. (a)	11. (a)	12. (a)	13. (c)	14. (c)
15. (d)	16. (a)	17. (a)	18. (d)	19. (a)	20. (d)	21. (a)
22. (c)	23. (d)	24. (d)	25. (c)	26. (c)	27. (a)	28. (a)
29. (c)	30. (a)	31. (b)	32. (a)	33. (d)	34. (d)	35. (c)
36. (b)	37. (a)	38. (a)	39. (c)	40. (c)	41. (b)	42. (b)
43. (d)	44. (d)	45. (d)	46. (c)	47. (a)	48. (a)	49. (b)
50. (b)	51. (c)	52. (d)	53. (a)	54. (c)	55. (b)	56. (c)
57. (d)	58. (d)	59. (c)	60. (a)	61. (d)	62. (c)	63. (b)
64. (d)	65. (d)	66. (c)	67. (b)	68. (a)	69. (a)	70. (b)
71. (c)	72. (b)	73. (a)	74. (d)	75. (c)	76. (d)	77. (d)
78. (a)	79. (c)	80. (c)	81. (d)	82. (d)	83. (d)	84. (d)
85. (b)	86. (d)	87. (a)	88. (b)	89. (a)	90. (d)	91. (d)
92. (b)	93. (c)	94. (d)	95. (d)	96. (a)	97. (c)	98. (c)
99. (d)	100. (b)	101. (d)	102. (d)	103. (c)		

9

BIOENERGETICS

Choose the Correct Answer

1. Respiration is
   (a) breaking down of complex organic substances into simpler substances
   (b) transformation of potential energy into kinetic energy
   (c) liberation of energy
   (d) all of these
2. Respiration is regarded as a/an
   (a) anabolic process (b) catabolic process
   (c) reduction process (d) synthetic process
3. Krebs cycle takes place in
   (a) mitochondria (b) chloroplast
   (c) ribosome (d) endoplasmic reticulum
4. Respiration takes place in
   (a) green parts of the plant only
   (b) all the living cells of the plants
   (c) living and dead cells of the plants
   (d) those parts of the plant which are above the soil
5. Respiration and photosynthesis both require
   (a) glucose (b) cytochromes (c) chlorophyll (d) sunlight
6. Different steps in respiration are controlled by
   (a) enzymes (b) sugars (c) auxins (d) kinins

7. The universal hydrogen acceptor is
   (a) CoA (b) ATP
   (c) FMN (d) NAD (coenzyme)
8. During complete metabolism of glucose, the number of ATP molecules formed from the corresponding number of ADP molecules, or the net gain of energy from one gram of glucose during aerobic respiration, is
   (a) 2 ATP (b) 36 ATP (c) 38 ATP (d) 16 ATP
9. A characteristic feature of ripening of some fruits (banana) is a sudden increase in respiration which is known as
   (a) anthesis (b) photorespiration
   (c) aerobic respiration (d) anaerobic respiration
10. Synthesis of ATP in mitochondria requires
    (a) NADP (b) O_2 (c) FMN (d) pyruvic acid
11. The largest amount of phosphate bond energy is produced in the process of respiration during
    (a) anaerobic respiration (b) tricarboxylic acid cycle
    (c) glycolysis (d) none of these
12. The cycle in which pyruvic acid is broken down in the presence of oxygen is known as
    (a) anaerobic respiration
    (b) glycolysis
    (c) Krebs cycle or tricarboxylic acid cycle
    (d) none of these
13. In hexose monophosphate shunt, the number of CO_2 molecules evolved is
    (a) much lesser than that in glycolysis (b) same as that in glycolysis
    (c) less than that in glycolysis (d) more than that in glycolysis
14. Evolution of CO_2 is more than intake of oxygen
    (a) when fats are respired (b) when glucose is respired
    (c) when sucrose is respired (d) during glycolate cycle
15. An alternate name for Krebs cycle is
    (a) EMP (b) glyoxalate cycle
    (c) citric acid cycle (d) glycolate cycle
16. Pentose-phosphate pathway which is an alternative pathway of respiration was elucidated by
    (a) Blackman (b) Horckar (c) Dickens (d) Warburg
17. In an organism utilizing carbohydrate as its source of energy anaerobically, the RQ is likely to be
    (a) 1.0 (b) 0.9 (c) 0.7 (d) infinity

18. To a living organism which of the following has the greatest amount of available energy per molecule?
    (a) CO_2 (b) H_2O (c) ATP (d) ADP
19. Alcohol is produced during the process of
    (a) photosynthesis (b) aerobic respiration
    (c) anaerobic respiration (d) none of these
20. Energy currency or energy reservoir of the cells is
    (a) AMP (b) RNA (c) DNA (d) ATP
21. ATP synthesis occurs on the
    (a) outer membrane of mitochondrion (b) inner membrane of mitochondrion
    (c) matrix (d) none of these
22. The reactions of Krebs cycle take place
    (a) in the ER (b) in the cytoplasm
    (c) on the surface of mitochondrion (d) in the matrix of mitochondrion
23. Mitochondria are the sites of
    (a) phosphorylation (b) photolysis
    (c) oxidative phosphorylation (d) starch synthesis
24. The site of Krebs cycle is or Krebs cycle takes place in
    (a) lysosomes (b) vesicles of ER
    (c) mitochondrial matrix (d) dictyosomes
25. In instantaneous source of energy is
    (a) protein (b) fat (c) amino acid (d) glucose
26. When a molecule of pyruvic acid is subjected to anaerobic oxidation forming lactic acid, there is
    (a) gain of 2 ATP molecules (b) gain of 4 ATP molecules
    (c) loss of 3 ATP molecules (d) loss of 6 ATP molecules
27. In Krebs cycle the formation of a six-carbon compound occurs by reaction between
    (a) fumaric acid and pyruvic acid (b) oxaloacetic acid and acetyl-CoA
    (c) malic acid and acetyl-CoA (d) succinic acid and pyruvic acid
28. The number of molecules of pyruvic acid formed from one molecule of glucose at the end of glycolysis is
    (a) 2 (b) 3 (c) 4 (d) 1
29. In germinating castor seeds, the RQ is
    (a) zero (b) less than one (c) one (d) more than one

30. The other name for glycolysis is
    (a) TCA pathway (b) HMS pathway (c) carbon pathway (d) EMP pathway
31. Glucose is converted into pyruvic acid in
    (a) HMS pathway (b) Krebs cycle (c) glycolysis (d) TCA pathway
32. The enzyme which converts glucose to glucose 6-phosphate is
    (a) hexokinase (b) glucose synthetase
    (c) glucose 6-phosphatase (d) phosphorylase
33. Which of the following is formed along with ATP in glycolysis?
    (a) $NADPH_2$ (b) $NADH_2$ (c) FAD (d) $FADH_2$
34. RQ for glucose is
    (a) 1 (b) 2 (c) 0.05 (d) 0.5
35. Pyruvate dehydrogenase is used in converting
    (a) glucose to pyruvic acid (pyruvate) (b) pyruvate to glucose
    (c) pyruvate to acetyl-CoA (d) pyruvic acid to lactic acid
36. During hydrolysis from one ATP molecule, the energy released is or the amount of energy given by one molecule of ATP is
    (a) 721 kcal (b) 7600 kcal (c) 1000 kcal (d) 7.3 kcal
37. ______ plays an indispensable role in energy metabolism.
    (a) phosphorus (b) calcium (c) potassium (d) sodium
38. The respiration of young maturing seeds is quite high but as water content decreases during further maturation, respiration
    (a) decreases steadily (b) increases steadily
    (c) stops completely (d) remains high
39. Respiration is
    (a) an exothermic process (b) the complete oxidation of carbohydrate
    (c) anabolic process (d) none of these
40. Fermentation is
    (a) incomplete oxidation of carbohydrate (b) complete oxidation of carbohydrate
    (c) anaerobic respiration (d) none of these
41. Conversion of pyruvic acid into ethyl alcohol is facilitated by the enzyme
    (a) phosphatase (b) carboxylase
    (c) dehydrogenase (d) carboxylase and dehydrogenase

42. In fermentation,

(a) energy is liberated
(b) energy is consumed
(c) energy is stored
(d) energy is neither consumed nor liberated

43. The end product of fermentation is

(a) pyruvic acid and H_2O
(b) H_2O and CO_2
(c) C_2H_5OH and CO_2
(d) CO_2

44. The number of ATP molecules formed during aerobic respiration due to breakdown of glucose molecules is

(a) 18
(b) 28
(c) 38
(d) 4

45. The respiratory quotient (RQ) in Opuntia at night will be

(a) zero
(b) one
(c) less than one
(d) more than one

46. During respiration, pyruvic acid is formed by

(a) TCA cycle
(b) Krebs cycle
(c) glycolysis
(d) none of these

47. Which of the following processes is used in the conversion of pyruvate to acetyl-CoA

(a) oxidative phosphorylation
(b) oxidative dehydration
(c) oxidative decarboxylation
(d) oxidative dehydrogenation

48. Pyruvic acid before entering the Krebs cycle is changed into

(a) citric acid
(b) succinic acid
(c) acetyl-CoA
(d) malic acid

49. If the respirable material in a respiroscope is some organic acid, what will happen to the level of sodium chloride?

(a) It will rise.
(b) It will fall.
(c) It will remain as such.
(d) It will show variation.

50. If CO_2 is given off in respiration, why does the amount of CO_2 in the atmosphere remain relatively constant?

(a) CO_2 is converted in photosynthesis to carbohydrates
(b) CO_2 is split up during photosynthesis
(c) CO_2 is a buffer
(d) CO_2 forms carbonate rocks

51. Hexose monophosphate shunt is

(a) a process by which new products are synthesized
(b) pentose-phosphate pathway
(c) conversion of glucose into pyruvic acid
(d) the sum of all chemical transformation

52. The evolution of carbon dioxide is more than the intake of oxygen when
    (a) fats are respired (b) glucose is respired
    (c) organic acids are respired (d) sucrose is respired

53. High-energy compounds
    (a) are produced when ATP loses two of its phosphate groups
    (b) refer to sugar, the oxidation of which releases large amounts of energy during respiration
    (c) are produced in respiration only
    (d) are compounds which link with exergonic and endergonic processes

54. The principal molecules that store energy are
    (a) DPM (b) DNA
    (c) ATP (d) protein molecules

55. When ATP is converted into ADP, it releases
    (a) enzymes (b) energy (c) hormones (d) secretions

56. Krebs cycle involves the formation of
    (a) ATP from ADP (b) energy from pyruvic acid
    (c) pyruvic acid from glucose (d) lactic acid from glucose

57. ATP is produced during the reaction
    (a) Pyruvic acid $\longrightarrow$ Acetyl-CoA
    (b) Glucose $\longrightarrow$ Glucose 6-phosphate
    (c) Fructose 6-phosphate $\longrightarrow$ Fructose 1,6-diphosphate
    (d) Glucose 6-phosphate $\longrightarrow$ Fructose 6-phosphate

58. RQ is less than 1, when
    (a) sucrose is oxidized (b) fats are oxidized
    (c) organic acids are oxidized (d) glucose is oxidized

59. Oxidative phosphorylation occurs in
    (a) chloroplast (b) peroxisome
    (c) mitochondria (d) centrioles

60. The formation of acetyl-CoA from pyruvic acid is the result of its
    (a) dehydration (b) dephosphorylation
    (c) reduction (d) oxidative decarboxylation

61. Besides the net gain of 2 ATP molecules in glycolysis, which other molecules are simultaneously formed?
    (a) $NADH_2$ (b) $FADH_2$ (c) $NADP_2H_2$ (d) $FAMH_2$

62. Complete oxidation of 1 gram mole of glucose gives rise to
    (a) 6860 cal (b) 68600 cal (c) 686000 cal (d) 6860000 cal
63. In the presence of cyanides, azides and carbon monoxide, the rate of respiration
    (a) increases (b) decreases
    (c) remains the same (d) none of these
64. Which of the following is the product of phosphorylation?
    (a) DPGA (b) PGA
    (c) pyruvic acid (d) fructose 1,6-diphosphate
65. In how many steps is CO_2 released in aerobic respiration?
    (a) one (b) three (c) six (d) twenty
66. In respiration pyruvic acid is
    (a) a result of protein breakdown
    (b) formed only when oxygen is available
    (c) one of the products of Krebs cycle
    (d) broken down into a two-carbon fragment and carbon dioxide
67. Which among the following is the most appropriate reason for storing green coloured apples at low temperature?
    (a) The rate of respiration is reduced.
    (b) The rate of photosynthesis and respiration are reduced.
    (c) Rate of photosynthesis is reduced.
    (d) Respiration and photosynthesis are completely inhibited.
68. Organelles which are regarded as powerhouses of the cell and in which oxidative reactions of the respiratory process take place are
    (a) mitochondria (b) endoplasmic reticulum
    (c) ribosome (d) chloroplast
69. In succulents, the respiratory quotient is always less than one because of
    (a) incomplete reduction (b) incomplete oxidation
    (c) complete oxidation (d) complete reduction
70. Anaerobic catabolism of carbohydrates and glycolysis or EMP pathway involve
    (a) conversion of starch into glucose (b) conversion of glucose into pyruvic acid
    (c) conversion of protein into glucose (d) conversion of glucose into fructose

71. In the process of respiration in plants, 180 g of sugar plus 192 g of oxygen produces
    (a) 528 g of CO_2, 216 g of water and 1348 cal of energy
    (b) 264 g of CO_2 198 g of water and 674 cal of energy
    (c) 132 g of CO_2, 54 g of water and 337 cal of energy
    (d) large amount of CO_2, no water and no energy
72. In aerobic respiration,
    (a) glucose cannot be used directly
    (b) glucose is a preferred substrate
    (c) glucose is an alternative
    (d) any material which burns in air may be used as a substrate
73. The end products of fermentation, when sugars are used as raw materials, are
    (a) CO_2 and H_2O
    (b) C_2H_5OH and CO_2
    (c) H_2O and acetic acid
    (d) alcohol
74. The end products of aerobic respiration in plants are
    (a) H_2O and energy
    (b) CO_2, H_2O and energy
    (c) starch and O_2
    (d) sugar and O_2
75. The incomplete breakdown of sugars in anaerobic respiration results in the formation of
    (a) fructose and water
    (b) glucose and carbon dioxide
    (c) alcohol and carbon dioxide
    (d) water and carbon dioxide
76. Which statement is wrong?
    (a) All angiosperms are with seed development but all seed plants are not angiosperms.
    (b) Stomatal opening is influenced by many factors and by potassium.
    (c) The factors which effect respiration, influence photosynthesis, but the reverse is not possible
    (d) All enzymes are proteins but all proteins are not enzymes.
77. After respiration the conversion of energy is mainly in the form by
    (a) ADP (b) AMP (c) glucose (d) ATP
78. Which of the following forms is the connecting link between glycolysis of citric acid and β oxidation of fatty acids?
    (a) oxaloacetic acid (b) succinic acid (c) acetyl-CoA (d) citric acid
79. The common immediate source of energy in cellular activity is
    (a) ATP (b) NAD (c) RNA (d) DNA

80. ATP is or stands for a/an
    (a) hormone
    (b) protein
    (c) molecule which contains high-energy phosphate bonds
    (d) enzyme which brings about oxidation

81. Oxidation of succinate into fumarate in the Krebs cycle is due to
    (a) addition of oxygen to it
    (b) removal of hydrogen from it
    (c) loss of electron from it
    (d) none of these

82. RQ (respiratory quotient) is represented by
    (a) $\frac{N}{C}$
    (b) $\frac{C}{N}$
    (c) $\frac{O_2}{CO_2}$
    (d) $\frac{CO_2}{O_2}$

83. Most of the energy in the cell is liberated by oxidation of carbohydrates when
    (a) pyruvic acid is converted into acetyl-CoA
    (b) pyruvic acid is converted into CO_2 and H_2O
    (c) sugar is converted into pyruvic acid
    (d) glucose is converted into pyruvic acid

84. Total ATP production during EMP pathway is
    (a) 24 ATP molecules (b) 38 ATP molecules (c) 6 ATP molecules (d) 8 ATP molecules

85. Food is converted into energy in
    (a) nucleolus
    (b) chloroplast
    (c) mitochondria
    (d) nucleus

86. Enzymes taking part in glycolysis are present in or glycolysis takes place in
    (a) cytoplasm
    (b) mitochondria
    (c) vacuole
    (d) none of these

87. Oxidative phosphorylation and photophosphorylation both require ______ as electron carriers.
    (a) oxygen
    (b) carbon dioxide
    (c) water
    (d) cytochromes

88. Krebs cycle is important because
    (a) it produces vitamins
    (b) it produces of amino acids
    (c) it encourages glycolysis
    (d) it produces ATP molecules through oxidative phosphorylation

89. Respiratory quotient refers to
    (a) the number of glucose molecules that break up and produce energy in one hour
    (b) the ratio between the amount of CO_2 produced to the amount of O_2 absorbed
    (c) amount of ATP produced to the amount lost during respiration
    (d) none of these

90. The correct sequence in Krebs cycle is
   (a) Isocitric acid $\longrightarrow$ α-ketoglutaric acid $\longrightarrow$ oxalosuccinic acid
   (b) Isocitric acid $\longrightarrow$ oxalosuccinic acid $\longrightarrow$ α-ketoglutaric acid
   (c) Oxalosuccinic acid $\longrightarrow$ citric acid $\longrightarrow$ α-ketoglutaric acid
   (d) α-ketoglutaric acid $\longrightarrow$ isocitric acid $\longrightarrow$ oxalosuccinic acid

91. The equation $\Delta G = \Delta H - T\Delta S$ depends on
   (a) constant temperature and pressure   (b) constant temperature and volume
   (c) constant pressure and volume   (d) all of these

92. Some autotropic bacteria utilize which of the following reactions?
   (a) $Fe^{+++} \longrightarrow Fe^{++}$ (b) $Fe^{++} \longrightarrow Fe^{+++}$
   (c) $Cu \longrightarrow Cu^{++}$ (d) $Mg^{++} \longrightarrow Mg$

93. Lipmann introduced the symbol Ⓟ indicating
   (a) phosphate bond   (b) low-energy phosphate bond
   (c) high-energy phosphate bond   (d) energy phosphate bond

94. The number of high-energy phosphate contained in ATP is
   (a) 1   (b) 1
   (c) 3   (d) none of the above

95. AMP contains
   (a) high-energy phosphate bond   (b) phosphate bond
   (c) energy phosphate bond   (d) low-energy phosphate bond

96. Free energy is derived from respiratory chain oxidation within
   (a) cytosol   (b) microsomes   (c) mitochondria   (d) lysosomes

97. How many ATP molecules are directly generated at the succinyl thiokinase step of the TCA cycle?
   (a) ~ one   (b) ~ two   (c) ~ three   (d) none of these

98. Phosphagens act as storage forms of
   (a) low-energy phosphate   (b) high-energy phosphate
   (c) energy phosphate   (d) no energy phosphate

99. Phosphagens maintain ATP concentration in
   (a) muscle   (b) liver   (c) kidney   (d) heart

100. When ATP becomes abundant, its concentration can cause
   (a) direct reaction   (b) irreversible reaction
   (c) reverse reaction   (d) all of these

101. The enzyme adenylate kinase is also called

(a) glucokinase (b) pyruvate kinase
(c) phosphoglycerokinase (d) myokinase

102. The activation of long chain fatty acids produces

(a) inorganic phosphate (b) organic phosphate
(c) adenosine diphosphate (d) no phosphate

103. How many P groups are lost as a result of the activation via the pyrophosphate pathway?

(a) ~ one (b) ~ two (c) ~ three (d) none of these

104. Which of the following takes part in phosphorylation in the cell?

(a) UTP (b) GTP (c) CTP (d) All of these

105. Adenylate kinase is a specialized 1

(a) monophosphate kinase (b) diphosphate kinase
(c) triphosphate kinase (d) all of these

106. Oxidases are conjugated proteins having the prosthetic group

(a) magnesium (b) manganese (c) copper (d) iron

107. The cytochrome oxidase complex that is prominent in the ETC is

(a) a_3 (b) aa_3 (c) a (d) none of these

108. Cytochrome oxidase is poisoned by

(a) cyanide (b) sulphide (c) sulphite (d) sulphate

109. Phenolase is an enzyme containing

(a) cobalt (b) iron (c) zinc (d) copper

110. Monoamine oxidase oxidizes

(a) epinephrine (b) norepinephrine (c) glucagon (d) glutathione

111. Aerobic dehydrogenases have the prosthetic group

(a) ATP (b) NAD^+ (c) FAD^+ (d) $NADP^+$

112. Uricase catalyses the oxidation of uric acid to

(a) carbon dioxide (b) ammonia (c) glyoxal (d) allantoin

113. Xanthine dehydrogenase converts purine bases to

(a) uric acid (b) hypoxanthine (c) xanthine (d) urea

114. Alcohol dehydrogenase from liver contains

(a) copper (b) zinc (c) sodium (d) potassium

115. NADP-linked dehydrogenases in the extramitochondrial area are found to synthesize
(a) urea (b) steroid (c) ascorbic acid (d) niacin

116. The endoplasmic recticulum contains cytochrome
(a) *b* (b) *c* (c) *a* (d) P-450

117. Mono-oxygenases are found in the
(a) mitochondria (b) microsomes (c) nuclei (d) cytosol

118. The mitochondrial superoxide dismutase contains
(a) Mg^{++} (b) Zn^{++} (c) Mn^{++} (d) CO^{++}

119. When substrates are oxidized through an NAD-linked dehydrogenase, the P : O ratio is
(a) 1 (b) 2 (c) 3 (d) 4

120. The antibiotic piericidin A inhibits the site of respiratory chain
(a) I (b) II (c) III (d) all of the above

121. The uncoupling agent of oxidative phosphorylation is
(a) antimycin (b) dicoumarol (c) barbiturates (d) oligomycin

122. The oxidation and phosphorylation in intact mitochondria is completely blocked by
(a) streptomycin (b) gentamicin (c) puromycin (d) oligomycin

123. The high-energy compound is
(a) UDPG (b) ATP
(c) ADP (d) arginine phosphate

124. The RQ of mixed diets is about
(a) 0.72 (b) 0.75 (c) 0.80 (d) 0.85

125. The BMR is influenced by
(a) sex (b) climate (c) habit (d) all of these

126. The BMR is subnormal in
(a) Addison's disease (b) hyperthyroidism
(c) hypothyroidism (d) all of these

127. The BMR is below normal in
(a) starvation (b) lipoid nephrosis
(c) both of (a) and (b) (d) none of these

128. The BMR is above normal in
(a) diabetes insipidus (b) leukaemia
(c) polycythemia (d) all of the above

129. When lard, glucose and protein are combined, the percentage of SDA is less than the sum of their individual SDA by
    (a) 20 (b) 22 (c) 24 (d) 25
130. When protein and lard are combined, the percentage of SDA is less than the sum of their individual SDA by
    (a) 45 (b) 50 (c) 54 (d) 58
131. The SDA is decreased more by
    (a) carbohydrate (b) fat (c) protein (d) all of these
132. When different amino acids are fed to animals, high SDA is found to be produced by
    (a) glycine (b) alanine (c) phenylalanine (d) all of these
133. The internal secretion of glands has no direct influence upon the SDA of
    (a) protein (b) fat (c) carbohydrate (d) all of these
134. Thyroidectomy in animals reduces the SDA of
    (a) carbohydrate (b) fat (c) both (a) and (b) (d) none of these
135. A diet rich in ______ is unfavourable for heavy muscular work.
    (a) carbohydrate (b) protein (c) fat (d) all of these
136. More heat is generated in the performance of work with
    (a) carbohydrate (b) fat (c) both (a) and (b) (d) none of these
137. When new tissue is formed, the usual SDA is not exerted by
    (a) carbohydrates (b) fat (c) vitamins (d) protein
138. A man of 70 kg weight and 180 cm height with a surface area of 1.8 square metre, aged 30 and having a BMR of 40 will have a basal metabolism of
    (a) 1750 calories (b) 1800 calories (c) 1875 calories (d) 1925 calories
139. ATP formation leads to the expenditure of
    (a) 8000 calories (b) 9000 calories (c) 10000 calories (d) 12000 calories
140. In long bicycle races, in 24 hours, there is the expenditure of
    (a) 8000 calories (b) 9000 calories (c) 10000 calories (d) 12000 calories
141. Sunday's consumption of food is ______ % above the average for the six days.
    (a) 24 (b) 28 (c) 32 (d) 36
142. Women have a basal metabolic rate lower than that of men by
    (a) 7–10% (b) 8–12% (c) 10–15% (d) 12–20%

143. The average supplement for muscular work in calories is
(a) 400 (b) 600 (c) 800 (d) 1000

144. The daily average caloric requirement of the adult male is
(a) 2000 (b) 2200 (c) 2500 (d) 3000

145. If the NPU (Net Protein Utilization) is low, the requirements for proteins are
(a) low (b) high (c) moderate (d) none of these

146. The relatively essential amino acid is
(a) glycine (b) serine (c) alanine (d) histidine

147. Vegetable proteins are almost associated with a large amount of
(a) minerals (b) vitamins (c) carbohydrate (d) fat

148. Animal proteins are associated with fat but very little carbohydrate except
(a) milk (b) peas (c) beans (d) potatoes

149. The foodstuff having the "protein-sparing" effect is
(a) carbohydrate (b) fat
(c) both (a) and (b) (d) none of these

150. Protein has a more catalytic function in the form of
(a) hormones (b) enzymes (c) antigens (d) antinogens

151. Proteins differ in "biological value" depending on their contents of
(a) non-essential amino acids (b) semi-essential amino acids
(c) essential amino acids (d) all of these

152. Good quality proteins include
(a) soyabeans (b) peanuts (c) potatoes (d) all of these

153. High quality proteins include
(a) eggs (b) dairy products (c) kidney (d) all of these

154. Proteins are also important sources of
(a) nitrogen (b) sulphur (c) phosphorus (d) all of these

155. The protective substances in a diet are
(a) first class protein (b) mineral elements (c) vitamins (d) all of these

156. The bulk of the food is reduced by
(a) carbohydrate (b) fat (c) protein (d) minerals

157. The daily consumption of fat in grams is usually over
(a) 50 (b) 60 (c) 80 (d) 90

158. Animal fats that have a better biological fatty acid solubility are
(a) A (b) D (c) both (a) and (b) (d) none of these

159. The major symptoms of essential fatty acid deficiency are
(a) a scaly dermatitis (b) hair loss (c) poor wound healing (d) all of these

160. In societies where fat is the principal food for energy intake, the population tends to develop
(a) coronary heart disease (b) obesity (c) cancer of the bowel (d) all of these

161. Carbohydrate is the cheapest source of food, and for producing energy, it can be readily
(a) digested (b) absorbed (c) utilized (d) all of these

ANSWERS

1. (d)	2. (b)	3. (a)	4. (b)	5. (b)	6. (a)	7. (d)
8. (b)	9. (d)	10. (b)	11. (b)	12. (c)	13. (d)	14. (d)
15. (c)	16. (d)	17. (d)	18. (c)	19. (c)	20. (c)	21. (b)
22. (d)	23. (c)	24. (c)	25. (d)	26. (d)	27. (b)	28. (a)
29. (b)	30. (d)	31. (c)	32. (a)	33. (b)	34. (a)	35. (c)
36. (d)	37. (a)	38. (a)	39. (a)	40. (a)	41. (d)	42. (a)
43. (c)	44. (c)	45. (a)	46. (c)	47. (c)	48. (c)	49. (b)
50. (a)	51. (b)	52. (c)	53. (d)	54. (c)	55. (b)	56. (b)
57. (b)	58. (b)	59. (c)	60. (d)	61. (a)	62. (c)	63. (b)
64. (d)	65. (b)	66. (d)	67. (a)	68. (a)	69. (b)	70. (b)
71. (b)	72. (b)	73. (b)	74. (b)	75. (c)	76. (c)	77. (d)
78. (c)	79. (a)	80. (c)	81. (d)	82. (b)	83. (d)	84. (c)
85. (a)	86. (d)	87. (d)	88. (d)	89. (b)	90. (b)	91. (a)
92. (b)	93. (c)	94. (b)	95. (d)	96. (c)	97. (a)	98. (b)
99. (a)	100. (c)	101. (d)	102. (a)	103. (b)	104. (d)	105. (a)
106. (c)	107. (b)	108. (a)	109. (d)	110. (a)	111. (c)	112. (d)
113. (a)	114. (b)	115. (b)	116. (d)	117. (b)	118. (c)	119. (c)
120. (a)	121. (b)	122. (d)	123. (a)	124. (d)	125. (d)	126. (a)
127. (c)	128. (d)	129. (b)	130. (c)	131. (b)	132. (d)	133. (a)
134. (c)	135. (b)	136. (c)	137. (d)	138. (b)	139. (a)	140. (c)
141. (d)	142. (a)	143. (b)	144. (d)	145. (b)	146. (d)	147. (c)
148. (c)	149. (a)	150. (b)	151. (c)	152. (d)	153. (d)	154. (d)
155. (d)	156. (b)	157. (c)	158. (c)	159. (d)	160. (d)	161. (d)

10

CARBOHYDRATE METABOLISM

CHOOSE THE CORRECT ANSWER

1. The energy-rich compound that is produced during biological oxidation of glucose is
   (a) adenosine triphosphate (b) pyruvic acid
   (c) acetoacetate (d) adenosine monophosphate
2. In submerged leaves, respiration takes place through
   (a) stomata present on ventral surface of leaf
   (b) general surface of the leaf
   (c) stomata present on dorsal surface of the leaf
   (d) stomata present on both surfaces of the leaf
3. The respiratory substrates used during respiration are carbohydrates, proteins, organic acids and fats which are used in the order,
   (a) fats, carbohydrates, protein, organic acid
   (b) carbohydrates, fats, proteins, organic acid
   (c) carbohydrates, proteins, organic acid
   (d) protein, organic acid, carbohydrates
4. What would you suggest an athlete to take before going to participate in an event?
   (a) butter (b) sugar (c) bread (d) cheese
5. One-day-old cane sugar juice was found by the customer to have an alcoholic smell. It may be due to
   (a) aerobic respiration (b) fermentation
   (c) putrefaction (d) fungal growth

6. Which of the following is important in carbohydrate and fat metabolism?
   (a) pyruvic acid (b) citric acid
   (c) acetyl coenzyme A (d) carbon dioxide
7. Organelles which are regarded as "powerhouses" of the cell and in which the oxidative reactions of the respiratory process take place are
   (a) mitochondria (b) endoplasmic reticulum
   (c) ribosomes (d) chloroplasts
8. Conversion of 2-phosphoglyceric acid into 2-phosphoenol acid requires
   (a) Na^+ (b) K^+ (c) Ca^{++} (d) Mg^{++}
9. During conversion of one pyruvic acid molecule into one molecule of acetyl-CoA, how many molecules of CO_2 are produced?
   (a) four (b) three (c) two (d) one
10. Conversion of oxaloacetic acid to citric acid requires one molecule of
    (a) CO_3 (b) CO_2 (c) CO (d) H_2O
11. Which of the following compounds mediates the flow of electrons from reduced flavoproteins to respiratory cytochromes?
    (a) cytochrome a_6 (b) coenzyme (c) cytochrome b_6 (d) cytochrome *a*
12. A fresh red berry kept in a cupboard can perform
    (a) photosynthesis (b) transpiration (c) respiration (d) both (a) and (c)
13. Pyruvic acid has ______ carbon atoms.
    (a) 5 (b) 4 (c) 3 (d) 2
14. Life without air is
    (a) anaerobic (b) impossible
    (c) reduction (d) free from oxidative damage
15. Fishes in an aquarium thrive better if green plants are growing there because they
    (a) can feed on them (b) like green surroundings
    (c) inhale oxygen released by green plants (d) inhale CO_2 released by green plants
16. If soaked pea seeds are kept in four respiratory flasks, they would germinate best if the flask contains
    (a) hydrogen (b) oxygen (c) nitrogen (d) carbon dioxide
17. The synthesis of ATP in both photosynthesis and respiration is essentially an oxidation process involving the removal of energy from
    (a) cytochromes (b) phytochromes (c) electrons (d) oxygen

18. In hexose monophosphate shunt, the net formation of ATP molecules is
    (a) 38 (b) 36 (c) 35 (d) 34
19. Energy released per gram would be
    (a) highest when rice starch is the respiratory substrate
    (b) highest when wheat starch is the respiratory substrate
    (c) highest when potato starch is the respiratory substrate
    (d) same in all the cases
20. Respiration is accompanied by
    (a) decrease in size (b) decrease in weight
    (c) increase in weight (d) no change in weight
21. Zymase is
    (a) enzyme complex (b) acetaldehyde dehydrogenase
    (c) pyruvate dehydrogenase (d) pyruvate decarboxylase
22. Phosphorylation of glucose with the help of ATP and hexokinase produces
    (a) glucose 1,6-diphosphate (b) glucose 1-phosphate
    (c) glucose 6-phosphate (d) fructose 1,6-diphosphate
23. The molecules formed from complete oxidation of fructose 1,6-diphosphate is
    (a) ATP (b) NADP (c) CO_2 and H_2O (d) NADH
24. Glycolysis
    (a) produces ATP (b) uses ATP
    (c) produces NADH (d) all of these
25. ATP molecules produced respectively by $NADH_2$ and $FADH_2$ during electron transport are
    (a) 1 and 1 (b) 2 and 3 (c) 3 and 2 (d) 3 and 3
26. Tricarboxylic acids of Krebs cycle are
    (a) citric acid, isocitric and malic acid
    (b) citric acid, isocitric acid and oxalosuccinic acid
    (c) succinic acid, fumaric acid and citric acid
    (d) oxalosuccinic acid, citric acid and α-ketoglutaric acid
27. How many ATP molecules are synthesized by the citric acid cycle, in aerobic respiration of a glucose molecule?
    (a) 38 (b) 36 (c) 24 (d) 12
28. In muscles, anaerobic conditions change pyruvic acid to
    (a) $C_3H_6O_3$ (b) C_2H_4O (c) $C_3H_4O_3$ (d) C_2H_5OH

29. RQ of germinating castor seed is
   (a) 1 (b) >1 (c) <1 (d) 0
30. Woody plants have structures which allow air to diffuse into the respiring cells. The structures are called
   (a) stomata (b) pores (c) lenticels (d) leaf scars
31. If a starved plant is provided with hexose, the rate of respiration will
   (a) increase (b) decrease
   (c) decrease or increase (d) no change
32. Mineral activator needed for the enzyme carboxylase of TCA cycle is
   (a) Mo (b) Mg (c) Mn (d) Fe
33. Which of the scientific aspects would you assign to plant physiology?
   (a) cell and cell division (b) evergreen forests of India
   (c) embryo culture (d) respiratory activities in plants
34. The link between carbohydrate and fat metabolism or between glycolysis, Krebs cycle and β-oxidation is
   (a) pyruvic acid (b) oxaloacetate (c) acetyl-CoA (d) citric acid
35. Which 5-carbon organic acid of the Krebs cycle is a key compound in the nitrogen metabolism of a cell?
   (a) fumaric acid (b) citric acid
   (c) α-ketoglutaric acid (d) oxalosuccinic acid
36. Fermentation
   (a) begins with glucose (b) gives a net gain of only 2 ATP
   (c) occurs in cytosol (d) all of these
37. The greatest contributor of electrons to the electron transport system is
   (a) glycolysis (b) oxygen
   (c) Krebs cycle (d) transition reaction
38. Substrate-level phosphorylation takes place in the
   (a) Krebs cycle and transition reaction
   (b) glycolysis and Krebs cycle
   (c) glycolysis and the electron transport system
   (d) electron transport system and the transition reaction

39. Fatty acids are broken down to
    (a) pyruvate molecules, which take electron to the transport system
    (b) acetyl groups, which enter the Krebs cycle
    (c) amino acids which excrete ammonia
    (d) all of these

40. Of the 36 ATP molecules that are produced during the complete breakdown of glucose, most are formed by
    (a) substrate level phosphorylation (b) chemiosmotic phosphorylation
    (c) the electron transport system (d) both (a) and (c)

41. In eukaryotes, the net gain of ATP in the complete oxidation of one molecule of glucose is
    (a) 20 ATP (b) 56 ATP (c) 36 ATP (d) 46 ATP

42. Glyoxysomes are concerned with the metabolism of
    (a) proteins (b) fats (c) carbohydrates (d) all of these

43. The value of RQ decreases in germinating seeds
    (a) when there is a shift from fat to carbohydrate respiration
    (b) when there is a shift from protein to carbohydrate respiration
    (c) when there is a shift from aerobic respiration to anaerobic, the respiration becoming totally anaerobic
    (d) when there is a shift from carbohydrate to fat respiration

44. Energy released during aerobic respiration is more than the energy released during anaerobic respiration. How many times more energy is released in the former than the latter?
    (a) 19 times (b) 18 times (c) 36 times (d) 8 times

45. Glycolysis occurs in the
    (a) cytosol (b) mitochondria (c) Golgi complex (d) ribosomes

46. Which is the final electron acceptor in respiration?
    (a) hydrogen (b) oxygen (c) cytochrome (d) dehydrogenase

47. Glycolysis and Krebs cycle are linked by
    (a) acetyl-CoA (b) citric acid (c) pyruvic acid (d) oxaloacetic acid

48. RQ is less than one in the case of
    (a) glucose (b) fats (c) organic acid (d) sucrose

49. The site of glycolysis in a cell is
    (a) chromosome (b) nucleus (c) cytoplasm (d) chloroplast

50. Respiration can be defined as a/an
   (a) anabolic process by which animal cells utilize oxygen and carbon to form ATP
   (b) anabolic process by which animal cells utilize oxygen, produce carbon dioxide and convert the released energy to ATP
   (c) catabolic process by which animal cells utilize carbon dioxide and convert the released energy to ATP
   (d) catabolic process by which animal cells utilize oxygen, produce carbon dioxide and convert the released energy to ATP

51. The formula for the process of glycolysis is
   (a) $2H_2O + 6CO_2 \longrightarrow C_6H_{12}O_6 + 6O_2$
   (b) $C_6H_{12}O_6 \longrightarrow 2C_3H_4O_3 + 4H$
   (c) $C_6H_{12}O_6 + 6CO_2 \longrightarrow 6CO_2 + 6H_2O$
   (d) none of these

52. By the oxidation of the one molecule of pyruvate by mitochondrial respiration, the molecules of ATP generated are
   (a) 38 (b) 30 (c) 8 (d) 15

53. Plants, but not animals, can convert fatty acids to sugars by a series of reactions known as the
   (a) glyoxylate cycle (b) photosynthesis (c) Krebs cycle (d) glycolysis

54. When yeast ferments glucose, the products are
   (a) water and CO_2 (b) methanol and water
   (c) methanol and CO_2 (d) ethanol and CO_2

55. The respiratory substrate yielding maximum number of ATP molecules among the following is
   (a) amylose (b) betagenic amino acid
   (c) glucose (d) glycogen

56. The major role of phosphorus in plant metabolism is
   (a) to generate metabolic energy
   (b) to evolve oxygen during photosynthesis
   (c) to evolve carbon dioxide during respiration
   (d) to create anaerobic conditions

57. Biological oxidation in Krebs cycle involves
   (a) N_2 (b) CO_2 (c) O_2 (d) SO_2

58. Oxidative phosphorylation occurs during the process of
   (a) protein synthesis (b) N_2-fixation (c) respiration (d) transpiration

59. Mitochondria supply the necessary biological energy by
 (a) breaking down of sugar (b) oxidizing substrates of TCA cycle
 (c) reducing NADP (d) breaking down of protein
60. EMP pathway and pentose-phosphate pathway
 (a) may occur simultaneously (b) occur together
 (c) do not occur simultaneously (d) are mutually exclusive
61. For each molecule of glucose respired, TCA cycle must rotate
 (a) five times (b) four times (c) thrice (d) twice
62. Which component of ATPase is embedded in mitochondrial membrane?
 (a) F_0 (b) F_1 (c) both (a) and (b) (d) none of these
63. RQ of malic acid is
 (a) 0.7 (b) 1 (c) 1.33 (d) 4
64. The enzymes of oxidative phosphorylation are found in the
 (a) endoplasmic reticulum (b) chloroplast
 (c) mitochondria (d) Golgi bodies
65. Vinegar is synthesized from alcohol by
 (a) *Lactobacillus* bacteria (b) *Bacillus aceti* bacteria
 (c) *Azotobacter* (d) *Rhizobium*
66. What is common between photosynthesis and respiration?
 (a) cytochromes (b) light (c) H_2O (d) temperature
67. In which of the following processes, is pyruvic acid the end product?
 (a) Krebs cycle (b) glycolysis
 (c) fermentation (d) oxidative phosphorylation
68. The process of respiration and photosynthesis have one thing in common, viz.,
 (a) energy (b) cytochrome (c) chlorophyll (d) enzyme
69. One of the products of aerobic respiration is
 (a) malic acid (b) lactic acid (c) pyruvic acid (d) ethyl alcohol
70. The end product of anaerobic respiration in plants is
 (a) pyruvic acid (b) malic acid (c) ethyl alcohol (d) methyl alcohol
71. Which of the following products is formed during glycolysis of glucose?
 (a) pyruvic acid (b) carbon dioxide (c) citric acid (d) ethanol

72. Which of the following groups of scientists discovered the EMP pathway of glycolysis?
    (a) Embden, Meyerhof and Parnas
    (b) Emerson, Hoffman and Peterson
    (c) Embden, Morrison and Picher
    (d) Avery, Macleod and McCarthy

73. Hexose monophosphate pathway (HMS) is shunt of glycolysis. It is common in the animal cells of the liver and adrenal cortex though found in prokaryotes also. This pathway occurs in
    (a) cytoplasm and produces 36 ATP only
    (b) both cytoplasm and mitochondria and produces 38 ATP
    (c) cytoplasm and produces 35 ATP directly
    (d) mitochondria and produces 35 ATP directly

74. Pyruvic acid in formed in which phase of respiration?
    (a) EMS pathway
    (b) Krebs cycle
    (c) anaerobic respiration
    (d) ETS

75. Respiration of starved leaves (consuming proteins) is called
    (a) aerobic respiration
    (b) oxidative phosphorylation
    (c) anaerobic respiration
    (d) protoplasmic respiration

76. If germinating seeds are kept in a bottle which is connected with a tube containing calcium water, the latter will turn
    (a) red
    (b) yellowish
    (c) unchanged
    (d) milky

77. Maximum amount of energy in a cell is released by oxidation of carbohydrates when
    (a) CO_2 and H_2O are formed from pyruvic acid
    (b) pyruvic acid is formed
    (c) C_2H_3OH and CO_2 are formed
    (d) acetyl-CoA is formed from pyruvic acid

78. How many ATP molecules are formed as electrons move across the ETC from $FADH_2$ to O_2?
    (a) 0
    (b) 2
    (c) 3
    (d) 1

79. Enzymes involved in oxidative decarboxylation of pyruvic acid are found in
    (a) oxysomes
    (b) F_0-F_1 particles
    (c) cytoplasm
    (d) matrix of mitochondria

80. Oxidative phosphorylation (synthesis of ATP during respiration in mitochondrion) requires O_2 and occurs in
    (a) stroma
    (b) inner membrane of mitochondria
    (c) outer membrane of mitochondria
    (d) matrix of mitochondria

81. Which cycle operates in photorespiration?
    (a) Calvin cycle
    (b) HSK cycle
    (c) glycolate cycle
    (d) Krebs cycle

82. In which of the following is the value of RQ greater than the respiratory substrate
    (a) castor seeds (b) fat/fatty acid
    (c) malic acid/organic acid (d) gram seeds

83. Seeds can be preserved if they are stored
    (a) after drying in steam (b) absolutely fresh
    (c) after boiling in water (d) with minimum moisture

84. Action of ATPase needs the presence of
    (a) Mg^{++} and K^{++} (b) Cu^{++} and Fe^{++} (c) Na^{+} and K^{+} (d) Ca^{++} and Mg^{++}

85. Flow of electrons in ETCs is in the order $NADH_2 \longrightarrow FMN \longrightarrow FeS$ protein $\longrightarrow CoQ \longrightarrow$ cyt $b \longrightarrow$ cyt $c_1 \longrightarrow$ cyt $c \longrightarrow$ cyt $a \longrightarrow$ cyt $a_3 \longrightarrow O_2$. At three steps ATP is formed (oxidative phosphorylation). Where does II ATP is formed?
    (a) between cyt c_1 and cyt c (b) between cyt a and cyt a_3
    (c) between cyt b and cyt c_1 (d) between $NADH_2$ and FMN

86. Krebs cycle was discovered by Krebs in pigeon muscles in 1940. Which step is called Gateway step or link reaction in respiration?
    (a) ETS terminal oxidation (b) formation of acetyl-CoA
    (c) glycolysis (d) citric acid formation

87. According to Mitchell's chemiosmotic coupling theory of ATP formation, passage of $2e^-$ from $NADH_2$ over ETC releases energy to push out 3 pairs (proton pump). These protons then pass inwardly through F_1 head (ATPase) to form ATP. How many ATP are formed when one pair of protons return to inner chamber through F_1 of F_0–F_1 particle?
    (a) 2 ATP (b) 36 ATP (c) 1 ATP (d) 3 ATP

88. In Krebs cycle,
    (a) OAA undergoes 4 oxidations and 2 decarboxylations
    (b) acetyl coenzyme A undergoes 4 oxidations and 2 decarboxylations
    (c) pyruvic acid undergoes 4 oxidations and 2 decarboxylations
    (d) TCA undergoes 4 oxidations and 4 decarboxylations

89. Which of the following is formed through phosphorylation in glycolysis?
    (a) 1,3-biphosphoglyceric acid (b) fructose 1,6-biphosphate
    (c) both a and b (d) GAP

90. In Krebs cycle the first produced is citric acid which is a 6-carbon compound. It is formed by a condensing irreversible reaction between
    (a) OAA and acetyl-CoA (b) OAA and pyruvic acid
    (c) pyruvic acid and acetyl-CoA (d) OAA and citrate synthetase

91. Amphibolic cycle that occurs only in aerobic condition is
    (a) Krebs cycle (b) E-D pathway (c) glycolysis (d) EMP pathway

92. Which step of glycolysis requires pyruvic kinase, Mg^{++}, K^{+} and ADP?
    (a) cleavage of fructose 1,6-biphosphate
    (b) conversion of PEP into pyruvic acid
    (c) conversion of 3 PGA into 2 PGA
    (d) none of these

93. One molecule of glucose requires 2 ATP to get phosphorylated to form fructose 1-6 biphosphate in glycolysis. How many ATP are used in the same process if the substrate is fructose?
    (a) 0 (b) 1 (c) 2 (d) 4

94. Which of the following steps in glycolysis is irreversible?
    (a) PEP $\longrightarrow$ Pyruvic acid (b) Fructose 6-PO_4 $\longrightarrow$ Fructose 1,6-biPO_4
    (c) Glucose $\longrightarrow$ Glucose 6-PO_4 (d) all of these

95. Which of these statements is wrong for glycolysis?
    (a) CO_2 is produced. (b) ATP is formed.
    (c) ATP is used. (d) Substrate level phosphorylation occurs.

96. The atom within each cytochrome molecule that actually accepts and releases electron is
    (a) Mg (b) Zn (c) C (d) Fe

97. Oxaloacetic acid (OAA) is ______ carbon compound.
    (a) 5 (b) 6 (c) 4 (d) 2

98. In the formation of acetyl-CoA from pyruvic acid in mitochondria, pyruvic acid gets
    (a) decarboxylated (b) oxidized
    (c) both (a) and (b) (d) reduced and isomerized

99. The importance of glycolysis is that
    (a) it provides building material for the synthesis of cellular components
    (b) it degrades glucose to generate ATP
    (c) both (a) and (b)
    (d) none of these

100. Chemosynthetic bacteria resemble the photosynthetic bacteria in
    (a) synthesizing assimilatory power
    (b) synthesizing carbohydrate through Calvin cycle
    (c) oxidizing inorganic compounds for synthesizing ATP
    (d) none of these

101. Ratio of ATP in aerobic and anaerobic respiration is
(a) 4 : 1 (b) 18 : 1 (c) 18 : 2 (d) 36 : 1

102. More energy is stored in per unit weight of
(a) protein (b) glucose (c) fat (d) carbohydrate

103. Importance of Krebs cycle is
(a) to accelerate glycolysis (b) production of amino acid
(c) production of ATP (d) production of vitamins and amino acids

104. Ratio of CO_2 produced in aerobic and anaerobic respiration is
(a) 4 : 1 (b) 1 : 1 (c) 2 : 1 (d) 3 : 1

105. In ETC, electrons move from
(a) higher ψ to lower ψ (b) higher (–) to higher (+) potential
(c) higher (+) to lower (–) redox potential (d) none of these

106. Conversion of a ketoglutaric acid into succinic acid is a step of
(a) ED pathway (b) TCA cycle (c) HMP pathway (d) EMP pathway

107. KCN (potassium cyanide) stops cell activity by
(a) coagulating protein carriers
(b) blocking trachea
(c) preventing transfer of e^- from cytochrome a_3 to O_2
(d) preventing diffusion of O_2

108. In the first step of glycolysis, glucose is phosphorylated. This phosphate group comes from
(a) ATP (b) nucleotide (c) ADP (d) inorganic PO_4

109. How many stages does aerobic breakdown of glucose require?
(a) 4 (b) 3 (c) 2 (d) 1

110. Glyceraldehyde 3-phosphate (PG Ald or GAP) is oxidized during glycolysis. What happens to the hydrogen and electron that are removed during its oxidation?
(a) They oxidize NAD^+. (b) They reduce NAD^+.
(c) They are removed. (d) They are transferred to pyruvic acid.

111. The cut surface of an apple turns brown when exposed. If the cut surface is dipped in ascorbic acid (vitamin C), browning does not occur because
(a) ascorbic acid prevents release of polyphenols
(b) ascorbic acid prevents oxidation of cut surface
(c) ascorbic acid inhibits oxidase enzyme
(d) ascorbic acid overcomes injury

112. Before entering into Krebs cycle, pyruvic acid undergoes oxidative decarboxylation to form
(a) acetyl-CoA (b) CoA (c) citric acid (d) OAA

113. Where do glycolysis and TCA (Kreb's cycle) occur in a cell?
(a) glycolysis in cytoplasm and TCA cycle in oxysomes of mitochondrion
(b) glycolysis in cytosol and TCA cycle in stroma of chloroplast
(c) glycolysis in cytosol and TCA cycle in matrix of mitochondria
(d) glycolysis in matrix of mitochondrion and TCA cycle on inner membrane

114. The production of ATP by oxidative phosphorylation is driven by energy from
(a) cytochromes (b) $NADH_2$
(c) CoA (d) diffusion of protons in mitochondrion

115. At the end of Krebs cycle, most of the energy removed from glucose molecule is transferred to
(a) citric acid (b) ATP
(c) OAA (d) $NADH_2/FADH_2$

116. Mitochondria are powerhouses (miniature batteries) of the cell. They are the seats of
(a) cellular respiration
(b) internal respiration
(c) biological respiration
(d) all of the above but never with external respiration

117. The main and intermediate source of bioenergy which is obtained by a cell from catabolic reaction and is an important means of transfer of energy is
(a) glycogen (b) pyruvic acid (c) ATP (d) glucose

118. In ETC, ATP is not formed in which of the following steps?
(a) cyt a-a_3 (b) cyt b-cyt c
(c) cyt c-cyt a (d) $NADH_2 \longrightarrow FAD$

119. The first 5-C dicarboxylic acid in Krebs cycle is
(a) acetyl-CoA (b) citric acid
(c) α-ketoglutaric acid (d) OAA

120. Climacteric rise is a phenomenon encountered in respiration of
(a) aquatics (b) fleshy fruits (c) all fruits (d) all plants

121. In Krebs cycle, the H^+ removed at succinate level is accepted by
(a) FAD (b) FMN (c) NAD (d) ADP

122. Phosphorylation of hexose sugar takes place with the help of
(a) TPN (b) NADP (c) NAD^{++} (d) ATP and Mg^{++}

123. The number of ATP molecules synthesized in photorespiration is

(a) 0 (b) 2 (c) 16 (d) 34

124. Pasteur effect is

(a) reduction in oxidation rate even in the presence of plenty of O_2

(b) shifting of aerobic conditions to anaerobic or vice versa

(c) conversion of ADP to ATP

(d) sterilization at 63°C for 30 minutes

125. Glyoxylate cycle (conversion of fats into sugars) is a modified

(a) HMP pathway (b) EMP pathway (c) TCA cycle (d) none of these

126. Proteins enter into Krebs cycle through

(a) OAA (b) α-ketoglutarate (c) both (a) and (b) (d) none of these

127. If fructose 1,6-biphosphate is oxidized in aerobic respiration, the number of ATP molecules produced will be

(a) 38 (b) 36 (c) 32 (d) 20

128. The correct sequence of events in Krebs cycle is

(a) Acetyl-CoA ⟶ citric acid ⟶ α-ketoglutaric acid ⟶ succinic acid ⟶ fumaric acid ⟶ malic acid ⟶ OAA

(b) Acetyl-CoA ⟶ citric acid ⟶ malic acid ⟶ α-ketoglutaric acid ⟶ succinic acid ⟶ OAA

(c) Acetyl-CoA ⟶ citrate ⟶ pyruvate ⟶ α-ketoglutarate ⟶ succinate ⟶ malate ⟶ fumarate ⟶ OAA

(d) none of these

129. RQ is measured with Ganong's respiroscope. If the Hg level remains constant during the experiment, the RQ value comes to unity. When it shows rise in Hg level, then

(a) RQ value will be infinity (b) value of RQ will be 0

(c) value of RQ will be < 1 (d) RQ value will be unity

130. When production of CO_2 is more than intake of O_2, the respiratory substrate is

(a) fat/protein (b) glucose (c) sucrose (d) organic acid

131. ATP synthesis occurs in mitochondria during cellular respiration. In the energy transfer process from ADP to ATP, donor of phosphate is

(a) glucose 6-phosphate (b) creatine phosphate

(c) inorganic phosphate (d) GTP

132. O_2 during terminal oxidation in ETS gets reduced to H_2O. Here cytochrome oxidase (cyt a_3) present which has both Fe^{++} and Cu^{++}. The oxygen here

(a) acts as terminal acceptor of both electron and proton
(b) acts as terminal acceptor of proton (H^+)
(c) acts as terminal acceptor of e^-
(d) is molecular and accepts only two hydrogen atoms

133. Match List I with list II and select the correct answer using the codes given below the list.

List I	**List II**
A. Conversion of glucose to pyruvate	(1) Glycolysis
B. Conversion of pyruvate to acetyl-CoA	(2) Oxidative decarboxylation
C. Breakdown of acetyl-CoA to yield reduced coenzymes	(3) Krebs cycle
D. Synthesis of ATP	(4) Electron transport system

Codes

	A	B	C	D
(a)	3	4	1	2
(b)	3	4	2	1
(c)	4	3	2	1
(d)	3	1	4	2

134. The very first step in the formation of acetyl-CoA is the

(a) formation of a complex between thiamine pyrophosphate and pyruvate
(b) release of the acetyl group from lipoic acid CoA
(c) decarboxylation of pyruvate
(d) transfer of electron from reduced lipoic acid to NAD

135. The glyoxylate cycle is a modification of

(a) TCA cycle or Krebs cycle (b) Calvin cycle
(c) glycophosphate pathway (d) oxidative pentose-phosphate pathway

136. The proton pumps provide energy to cell through

(a) pH gradient and electropotential gradient
(b) ATP and ADP
(c) auxins and hormones
(d) carbohydrates and lipids

137. During the glyoxylate cycle, four-carbon oxaloacetate is generated from two-carbon acetate. The correct sequence in which the intermediate compounds appear from citrate is
    (a) citrate, succinate, fumarate, glyoxylate
    (b) citrate, fumarate, glyoxylate, malate
    (c) citrate, isocitrate, glyoxylate, malate
    (d) citrate, isocitrate, malate, glyoxylate

138. Krebs cycle, also called TCA or citric acid cycle, is also called metabolic sink as it is
    (a) a common pathway for carbohydrates and organic acids only
    (b) a common pathway for carbohydrates, fats and proteins (amino acids)
    (c) a common pathway for carbohydrates and fats only
    (d) none of these

139. Breakdown of pyruvic acid in Krebs cycle occurs in
    (a) anaerobic respiration (b) fermentation
    (c) aerobic respiration (d) mitochondria

140. Krebs cycle occurs in
    (a) chloroplast (b) ribosomes (c) cytoplasm (d) mitochondria

141. Mitochondria are the sites for
    (a) carboxylation (b) transpiration
    (c) oxidative phosphorylation (d) photophoshorylation

142. How many ATP will be produced during the production of one molecule of acetyl-CoA from one molecule of pyruvic acid?
    (a) 3 ATP (b) 5 ATP (c) 8 ATP (d) 38 ATP

143. At the end of glycolysis, which one of the following changes takes place?
    (a) starch into glucose (b) proteins into glucose
    (c) glucose into fructose (d) glucose into pyruvic acid

144. Cytochrome functions in
    (a) oxidation of glucose (b) production of energy
    (c) transfer of electrons (d) growth

145. The importance of Krebs cycle is in the production of
    (a) Acetyl-CoA (b) water (c) ATP (d) ADP

146. Vant Hoff's law states that
    (a) respiration rate is constant irrespective of change in temperature
    (b) respiration stops with increase of temperature beyond 10°C

(c) respiration rate increases two or three times for every rise of 10°C
(d) respiration rate decreases two or three times for every rise of 10°C

147. EMP (Embden, Meyerhof and Parnas) is the biochemical pathway in
(a) glycolysis (b) Krebs cycle (c) photosynthesis (d) none of these

148. Anaerobic process after glycolysis is called
(a) Krebs cycle (b) TCA cycle (c) Calvin cycle (d) none of these

149. The high boiling point of water is advantageous to living organisms because
(a) the environment seldom reaches the boiling point of water
(b) organisms can easily boil off enough water to keep themselves cool
(c) it allows organisms to spread heat evenly throughout their bodies
(d) organisms can absorb a great deal of heat before they reach the boiling point

150. The mechanism of ATP formation both in chloroplast and mitochondria is explained by
(a) relay pump theory of Goldewski (b) Munch's pressure/mass flow model
(c) chemiosmotic theory of Mitchell (d) Cholodny Went's model

151. Respiratory enzymes are located in
(a) mitochondrial matrix (b) permitochondrial space
(c) cristae (d) outer membrane

152. Fermentation is represented by the equation
(a) $C_6H_{12}O_6 \longrightarrow 6H_2O + 6O_2 + 673$ kcal
(b) $6CO_2 + 12H_2O \xrightarrow{\text{Light}} 6H_2O + O_2 + C_6H_{12}O_6$
(c) $6CO_2 + 6H_2O \xrightarrow{\text{Chlorophyll}} C_6H_{12}O_6 + 6O_2$
(d) $C_6H_2O_6 \longrightarrow 2C_2H_5OH + 2CO_2 + 18$ kcal

153. Heat energy of plants is measured in
(a) grams (b) decibels (c) calories (d) pounds

154. The highenergy phosphate produced in Krebs cycle when succinyl-CoA is converted into succinate is
(a) ATP (b) GTP (c) CTP (d) UTP

155. Anaerobic respiration takes place in
(a) vacuole (b) cytoplasm (c) nucleus (d) ribosome

156. Enzymes α-ketoglutrate dehydrogenase bring about
(a) reduction only (b) oxidation only
(c) decarboxylation only (d) oxidation and decarboxylation

157. Maximum calories are produced by

(a) vitamins (b) proteins (c) carbohydrates (d) fats

158. The process of release of energy by chemical breakdown of glucose in the absence of oxygen is termed as

(a) anaerobiosis (b) aerobiosis (c) ornithine cycle (d) deamination

159. RQ value of 4 may be expected for the complete oxidation of

(a) glucose (b) oxalic acid (c) malic acid (d) tartaric acid

160. Mitochondrial matrix contains the enzymes for

(a) respiration (b) TCA cycle
(c) glycolysis and TCA cycle (d) TCA cycle and electron transport

161. What would happen if $NADH_2$ is oxidized to form water in a single step?

(a) 3 ATP are produced
(b) most of the energy would be wasted in heat form
(c) 5 ATP are produced
(d) cell would burn

162. In eukaryotes, the electron transport chain and enzymes of Krebs or TCA cycle are found in

(a) nucleolus (b) endoplasmic reticulum
(c) mitochondria (d) ribosomes

163. Lactic acid fermentation is characterized by lack of production of

(a) ATP (b) $NADH_2$ (c) CO_2 and $NADH_2$ (d) CO_2

164. The number of glucose molecules required to produce 38 ATP molecules under anaerobic conditions by a yeast cell is

(a) 38 (b) 19 (c) 1 (d) 2

165. The food products are kept in cold storage or under refrigeration because

(a) it becomes very tasty
(b) can be used in off-season
(c) food products maintain their freshness, longevity, taste, etc. due to least respiration
(d) they remain very cool

166. The number of oxygen atoms utilized during aerobic oxidation of one molecule of pyruvate is

(a) 12 (b) 6 (c) 5 (d) 3

167. Fermentation is anaerobic production of

(a) protein and acetic acid (b) alcohol, lactic acid or similar compounds
(c) ethers and acetones (d) respiration and lipoproteins

168. Protein is mostly the respiratory substrate in
    (a) floating respiration
    (b) protoplasmic respiration
    (c) anaerobic respiration
    (d) respiration of oily seeds

169. Oxidative phosphorylation involves simultaneously oxidation and phosphorylation to finally form
    (a) pyruvate
    (b) ATP
    (c) DPN
    (d) NADP

170. Which of the following is the key intermediate compound linking glycolysis to the Krebs cycle?
    (a) ATP
    (b) acetyl-CoA
    (c) NADH
    (d) malic acid

171. In Krebs cycle, FAD participates as electron acceptor during conversion of
    (a) succinyl-CoA to succinic acid
    (b) α-ketoglutarate to succinyl-CoA
    (c) succinic acid to fumaric acid
    (d) fumaric acid to malic acid

172. The fact that the major link in transfer of chemical energy is the ATP–ADP system was first described by Lipman in
    (a) 1940
    (b) 1950
    (c) 1960
    (d) 1970

173. The end products of fermentation are
    (a) O_2 and acetaldehyde
    (b) O_2 and ethyl alcohol
    (c) CO_2 and ethyl alcohol
    (d) CO_2 and acetaldehyde

174. The inner wall of mitochondria contains
    (a) enzymes of Krebs cycle
    (b) enzymes of respiratory chain
    (c) DNA
    (d) ribosomes

175. The energy released by one molecule of glucose during aerobic respiration is
    (a) 521 cal
    (b) 573 cal
    (c) 621 cal
    (d) 673 cal

176. Oxidative phosphorylation occurs during
    (a) photosynthesis
    (b) respiration
    (c) transpiration
    (d) protein synthesis

177. The chemical linking glycolysis and the Krebs cycle is
    (a) pyruvic acid
    (b) acetyl-CoA
    (c) lactic acid
    (d) succinic acid

178. Plants die during winter due to frost because
    (a) photosynthesis does not occur at low temperature
    (b) respiration stops
    (c) there is dehydration and mechanical injury in the tissue
    (d) rate of respiration increases

179. Cytochromes present in cells are
    (a) carbon acceptors
    (b) hydrogen acceptors
    (c) nitrogen acceptors
    (d) oxygen acceptors

180. The enzymes which take part in glycolysis are found in
   (a) mitochondria and cytoplasm (b) cytoplasm
   (c) endoplasmic reticulum (d) vacuole

ANSWERS

1. (a)	2. (b)	3. (b)	4. (b)	5. (b)	6. (c)	7. (a)
8. (d)	9. (d)	10. (d)	11. (b)	12. (d)	13. (c)	14. (a)
15. (c)	16. (b)	17. (c)	18. (c)	19. (d)	20. (b)	21. (a)
22. (c)	23. (c)	24. (d)	25. (c)	26. (b)	27. (c)	28. (a)
29. (c)	30. (c)	31. (a)	32. (c)	33. (d)	34. (c)	35. (c)
36. (d)	37. (c)	38. (b)	39. (b)	40. (d)	41. (c)	42. (b)
43. (d)	44. (b)	45. (a)	46. (b)	47. (a)	48. (b)	49. (c)
50. (d)	51. (b)	52. (d)	53. (a)	54. (d)	55. (d)	56. (a)
57. (c)	58. (c)	59. (b)	60. (a)	61. (d)	62. (b)	63. (c)
64. (c)	65. (b)	66. (a)	67. (b)	68. (b)	69. (c)	70. (c)
71. (a)	72. (a)	73. (a)	74. (a)	75. (d)	76. (d)	77. (a)
78. (b)	79. (d)	80. (b)	81. (c)	82. (b)	83. (d)	84. (c)
85. (c)	86. (b)	87. (c)	88. (b)	89. (c)	90. (a)	91. (a)
92. (c)	93. (c)	94. (d)	95. (a)	96. (d)	97. (c)	98. (c)
99. (c)	100. (b)	101. (b)	102. (a)	103. (c)	104. (d)	105. (b)
106. (b)	107. (c)	108. (a)	109. (a)	110. (b)	111. (c)	112. (a)
113. (c)	114. (d)	115. (d)	116. (d)	117. (c)	118. (c)	119. (c)
120. (b)	121. (a)	122. (d)	123. (a)	124. (b)	125. (c)	126. (c)
127. (a)	128. (a)	129. (c)	130. (d)	131. (c)	132. (a)	133. (a)
134. (c)	135. (a)	136. (b)	137. (c)	138. (b)	139. (c)	140. (c)
141. (c)	142. (a)	143. (b)	144. (c)	145. (c)	146. (c)	147. (a)
148. (d)	149. (d)	150. (c)	151. (a)	152. (d)	153. (c)	154. (c)
155. (b)	156. (d)	157. (d)	158. (a)	159. (b)	160. (b)	161. (a)
162. (c)	163. (c)	164. (b)	165. (c)	166. (c)	167. (b)	168. (b)
169. (b)	170. (b)	171. (a)	172. (a)	173. (c)	174. (b)	175. (d)
176. (b)	177. (b)	178. (c)	179. (b)	180. (b)		

11

LIPID METABOLISM

CHOOSE THE CORRECT ANSWER

1. What is the normal level of ketone bodies in the blood?
   (a) up to 3 mg% (b) up to 4 mg% (c) up to 5 mg% (d) none of these
2. A fatty acid is degraded step by step by the sequential removal of how many carbon units?
   (a) 2 (b) 3 (c) 4 (d) 8
3. Glucogenic fatty acid is
   (a) acetic acid (b) propionic acid
   (c) both (a) and (b) (d) none of these
4. The chief fat depots of the body are
   (a) subcutaneous fatty tissue (b) retroperitoneal fatty tissue
   (c) intermuscular fatty tissue (d) all of these
5. How many moles of ATP are formed by one turn of β-oxidation?
   (a) 2 (b) 3 (c) 4 (d) 5
6. Disease of lipid metabolism is
   (a) Gaucher's disease (b) Niemann–Pick disease
   (c) Tay–Sach's disease (d) Fabry's disease
   (e) all of these
7. Which type of lipoprotein contains maximum cholesterol?
   (a) LDL (b) VLDL (c) HDL1 (d) HDL9

8. Unsaturated fatty acid with three double bonds in its structure is
   (a) linoleic acid (b) palmitoleic acid (c) oleic acid (d) none of these
9. Unsaturated fatty acid with four double bonds in its structure is
   (a) arachidonic acid (b) palmitoleic acid (c) oleic acid (d) none of these
10. Lecithins contain
    (a) glycerol (b) fatty acids and phosphoric acid
    (c) choline (d) all of these
11. Cephalins contain
    (a) glycerol (b) fatty acids
    (c) phosphoric acid and ethanolamine (d) all of these
12. Which vitamin prevents rancidity?
    (a) vitamin E (b) vitamin C (c) vitamin D (d) none of these
13. Unsaturated fatty acid with one double bond in its structure is
    (a) palmitoleic acid (b) oleic acid (c) both (a) and (b) (d) none of these
14. Unsaturated fatty acid with two double bonds in its structure is
    (a) linoleic acid (b) palmitoleic acid (c) oleic acid (d) none of these
15. Cerebrosides are lipids which contain
    (a) carbohydrate (galactose or glucose) (b) one fatty acid
    (c) sphingosine (d) all of these

ANSWERS

1. (a)	2. (a)	3. (c)	4. (d)	5. (d)	6. (e)	7. (a)
8. (a)	9. (a)	10. (d)	11. (d)	12. (a)	13. (c)	14. (a)
15. (d)						

12

BIOMEMBRANE

Choose the Correct Answer

1. Which of these are membrane proteins?
   (a) permeases (b) cellulases (c) lyases (d) all of these
2. Large protein molecules which project beyond the lipid bilayer on both sides and help water-soluble materials to pass through the membrane are
   (a) extrinsic proteins (b) peripheral proteins
   (c) integral proteins (d) both (a) and (b)
3. Cell membrane is visible under
   (a) electron microscope (b) optical microscope
   (c) both (a) and (b) (d) oil immersion lens
4. Fluid mosaic is related to structure of
   (a) plasma membrane (b) cell wall
   (c) cytoplasm (d) plasmodesmata
5. Exchange of the materials between, in and out of cells and other cells is facilitated by
   (a) middle lamella (b) cell wall (c) plasmalemma (d) plasmodesmata
6. Average thickness of plasmalemma is
   (a) 0.25 nm (b) 2.5 nm (c) 0.75 nm (d) 7.5 nm
7. Cell membrane is
   (a) unilaminar (b) bilaminar (c) trilaminar (d) quadrilaminar

8. Liquid food drinking is
   (a) pinocytosis (b) phagocytosis (c) imbibition (d) none of these
9. ATP molecules combine with carrier molecules and allow passage of substances
   (a) along concentration gradient (b) against concentration gradient
   (c) both (a) and (b) (d) along with solvent
10. Cell membranes are
    (a) selectively permeable (b) undergo dynamic changes
    (c) take part in phagocytosis (d) all of these
11. The latest and most acceptable model of cell membranes is
    (a) lamellar model (b) fluid mosaic model
    (c) micellar model (d) unit membrane concept
12. Cell membrane is composed of
    (a) phospholipid (b) nucleoprotein (c) polysaccharides (d) lipoprotein
13. The percentage of lipids and proteins in plasma membrane is
    (a) 25% lipid and 75% protein (b) 75% lipid and 25% protein
    (c) 42% lipid and 58% protein (d) none of the above
14. According to lamellar model, the correct sequence in plasma membrane is
    (a) lipid-protein-protein-lipid (b) lipid-protein-lipid-protein
    (c) protein-lipid-protein lipid (d) protein-lipid-lipid-protein
15. The three layers of cell membrane are
    (a) protein–phospholipid–protein
    (b) phospholipid–protein–protein
    (c) phospholipid–protein–phospholipid
    (d) protein–phospholipid–carbohydrate
16. A biomembrane is made of proteins, lipids and
    (a) carbohydrates (b) RNA (c) DNA (d) hormones
17. Cell ghosts are
    (a) lymphocytes (b) thrombocytes (c) choanocytes (d) erythrocytes
18. A cell membrane has
    (a) middle electron-dense layer (b) middle electron-transparent layer
    (c) outer electron-transparent layer (d) inner electron-transparent layer

19. Which of the following forms the continuous part of the cell membrane?
    (a) proteins (b) carbohydrates (c) lipids (d) all the above
20. Which of the following constitutes the maximum percentage of cell membrane?
    (a) proteins (b) lipids (c) carbohydrates (d) none of these
21. The most abundant lipid of the cell membrane is
    (a) cholesterol (b) phospholipid (c) glycolipid (d) cerebroside
22. Membrane structure is mainly maintained by
    (a) hydrophobic attractions (b) hydrophilic attractions
    (c) van der Waal's forces (d) ionic bonds
23. In fluid mosaic model
    (a) phospholipid monolayer is sandwiched between two protein layers
    (b) phospholipid monolayer is present on the top of a protein layer
    (c) phospholipid bilayer is present on the top of a protein layer
    (d) proteins are embedded at places in phospholipid bilayer
24. Carbohydrates occurring in the cell membrane are
    (a) glycogen and other polysaccharides (b) oligosaccharides
    (c) mucopolysaccharides (d) monosaccharides
25. Carbohydrates occur attached to membrane
    (a) inner proteins and both layers of phospholipids
    (b) extrinsic proteins of both surfaces
    (c) outer proteins and outer lipids
    (d) inner proteins and inner lipids
26. Carbohydrates present in cell membrane function in
    (a) attachment to microfilaments (b) attachment to microtubules
    (c) cell to cell recognition (d) transport of substances
27. Integral proteins of plasmalemma occur
    (a) on the outer surface (b) on the inner surface
    (c) on both the surface (d) in the phospholipid matrix
28. In membrane phospholipids, there are
    (a) one polar head and two nonpolar tails (b) two polar heads and one nonpolar tail
    (c) one nonpolar head and two polar tails (d) two nonpolar heads and one polar tail

29. Amphipathy means
    (a) hydrophilic molecule
    (b) hydrophobic molecule
    (c) both hydrophilic and hydrophobic groups in same molecule
    (d) all of these

30. Hydrophobic end of phospholipid has
    (a) affinity for water
    (b) repulsion for water
    (c) affinity for metabolites
    (d) repulsion for metabolites

31. The fluid-mosaic model of cell membrane
    (a) is made of mosaic of two types of fluids
    (b) is quasifluid and can flow from one cell to another
    (c) has phospholipid bilayer with proteins occurring at places
    (d) has protein bilayer with phospholipids occurring at places

32. Extrinsic proteins of cell membrane are
    (a) present superficially and are easily separable
    (b) present superficially but are not separable
    (c) attached to intrinsic proteins but are easily separable
    (d) attached to intrinsic proteins and are not easily separable

33. Main function of plasma membrane is to
    (a) control cell movements
    (b) control cell activities
    (c) maintain cell shape and size
    (d) regulate exchange of materials

34. Beet root kept in benzene for some time and then dipped in water will
    (a) swell up
    (b) dissolve partially
    (c) become extra red
    (d) lose colour

35. Heating will make the cell membrane
    (a) permeable
    (b) impermeable
    (c) partially permeable
    (d) semipermeable

36. Membrane fluidity was proved by
    (a) Frye and Edidin
    (b) Singer and Nicolson
    (c) Helleir and Hoffmann
    (d) Gorter and Grendel

37. Fusion of differently coloured mouse and human cells was undertaken by Frye and Edidin by means of
    (a) heating
    (b) polyethylene glycol
    (c) fluorescein
    (d) rhodamine

38. Membrane fluidity
    (a) decreases with rise in temperature
    (b) increases with lowering of temperature
    (c) decreases with lowering of temperature
    (d) is not affected by temperature changes

39. Adsorptive dyes used by Frye and Edidin to prove membrane fluidity were
    (a) fluorescein and rhodamine (b) fast green and safranin
    (c) janus green and phloroglucinol (d) methyl green and pyronin

40. Microvilli are meant for
    (a) slowing down of cell movements (b) helping in cell movements
    (c) rapid osmosis (d) increasing absorptive area

41. Microvilli are
    (a) outfolds of cell membrane (b) infolds of cell membrane
    (c) both (a) and (b) (d) submembrane vacuoles

42. Microvilli are supported by
    (a) collagen fibres (b) microfilaments
    (c) microtubules (d) connective tissue

43. Surface of a microvillus bears
    (a) pores for absorption (b) adsorptive sites for phagocytosis
    (c) glycocalyx (d) both (a) and (b)

44. Brush border is due to
    (a) microvilli (b) cilia (c) stereocilia (d) glycocalyx

45. A eukaryotic flagellum/cilium is covered by
    (a) protein sheath (b) plasmalemma sheath
    (c) nine doublet fibres (d) nine singlet fibres

46. Stereocilia are
    (a) membrane invaginations (b) cilia capable of performing rotatory motion
    (c) membrane evaginations (d) both (b) and (c)

47. Inpushing of plasmalemma forms in bacteria
    (a) lomasome (b) lamellasome (c) nexus (d) mesosome

48. Transfer cells are characteristic of
    (a) Monera (b) Protista (c) Metazoa (d) Metaphyta

49. Gap junction is characterized by
    (a) fusion of plasmalemmae of two adjacent cells
    (b) presence of connections
    (c) occurrence of thickenings
    (d) tonofibrils

50. In gap junction, two cells are interconnected by
    (a) hydrophilic channels (b) membranous outgrowths
    (c) undulations (d) phloem

51. Intercellular bridges of animal cells take part in
    (a) transfer of materials (b) adherence
    (c) transfer of stimuli (d) increasing surface area

52. Interdigitations are
    (a) membrane outgrowths of adjacent cells fitting into one another
    (b) meant for increased adherence
    (c) useful for increasing exchange
    (d) all of these

53. Desmosomes occur in
    (a) epithelia subjected to disruption (b) cells subjected to pressure
    (c) absorptive tissues (d) excretory tissues

54. In the area of desmosome, the adjacent cells have
    (a) intercellular thickening material (b) transmembrane linkers
    (c) intracellular thickenings and tonofibrils (d) all of these

55. Instead of intercellular thickening, collagen fibrils are present in
    (a) terminal bars (b) hemidesmosomes
    (c) septate desmosomes (d) tight junctions

56. Septate desmosomes occur in
    (a) amphibians (b) reptiles (c) invertebrates (d) birds

57. Cell membranes are fused to form occluding areas in
    (a) tight junctions (b) gap junctions (c) terminal bars (d) interdigitations

58. Desmosomes without ion fibrils and discoid thickenings are
    (a) belt desmosomes (b) terminal bars
    (c) zonulae adherentes (d) all of these

59. Plasma membranes of adjacent cells get thickened at certain regions which are called
    (a) desmosomes (b) peroxisomes (c) terminal bars (d) secondary wall

60. Plasmalemma is
    (a) permeable (b) selectively permeable
    (c) nonpermeable (d) semipermeable

61. In cell membranes
    (a) small molecules permeate more readily than larger molecules
    (b) membranes are selectively permeable
    (c) permeation is related to size and charge of solute
    (d) all of these

62. For passage of water, plasma membrane is
    (a) permeable (b) impermeable
    (c) semipermeable (d) selectively permeable

63. Transport molecules of cell membrane are
    (a) lipids (b) phospholipids (c) glycolipids (d) proteins

64. Water enters the cells by
    (a) pinocytosis (b) phagocytosis (c) endosmosis (d) absorption

65. Movement of substances across cell membrane is controlled by
    (a) size of permeating particles (b) permeability of membrane
    (c) membrane proteins (d) all of these

66. Red blood corpuscles placed in sodium chloride solution will show
    (a) bursting (b) crenation (c) plasmolysis (d) turgidity

67. What is the requirement for operation of osmosis?
    (a) semipermeable membrane (b) permeable membrane
    (c) impermeable membrane (d) selectively permeable membrane

68. A plant cell placed in hypotonic solution will
    (a) shrink (b) undergo plasmolysis
    (c) swell up and become turgid (d) swell up and burst

69. An animal cell placed in hypertonic solution undergoes
    (a) plasmolysis (b) crenation
    (c) bursting (d) swelling to become turgid

70. In the thistle funnel experiment, osmosis stops automatically after some time due to
    (a) equalization of osmotic concentration
    (b) equalization of solvent
    (c) equalization of solute
    (d) development of hydrostatic pressure
71. Salting of playing grounds kills weeds due to
    (a) plasmolysis
    (b) nonavailability of water
    (c) exclusion of air
    (d) both (b) and (c)
72. Diffusion occurs from
    (a) lower partial pressure to higher partial pressure
    (b) higher partial pressure to lower partial pressure
    (c) lower diffusion pressure to higher diffusion pressure
    (d) lower electrochemical potential to higher electrochemical potential
73. Facilitated diffusion resembles passive diffusion in
    (a) requirement of energy
    (b) requirement of permeases
    (c) nonrequirement of energy
    (d) nonrequirement of permeases
74. Facilitated diffusion is similar to active transport in requiring
    (a) carriers
    (b) energy
    (c) both (a) and (b)
    (d) no carriers but energy
75. Intact raisins kept in water swell up due to
    (a) endosmosis
    (b) exosmosis
    (c) plasmolysis
    (d) exocytosis
76. A living animal cell placed in water bursts after some time while amoeba and other protists do not do so because of
    (a) absence of endosmosis in amoeba
    (b) ability of amoeba to throw out extra water
    (c) nondifference of osmotic concentration of amoeba and external water
    (d) all of these
77. Which one of the following requires energy?
    (a) pinocytosis
    (b) phagocytosis
    (c) Na^+–K^+ pump
    (d) all the above
78. Facilitated diffusion is similar to active transport in showing
    (a) movement against concentration gradient
    (b) selective permeability
    (c) saturation effect
    (d) both (b) and (c)

79. Endocytosis means
    (a) phagocytosis (b) pinocytosis (c) both (a) and (b) (d) endosmosis
80. The process of taking in liquid material by infolding of membrane is known as
    (a) phagocytosis (b) osmosis (c) active transport (d) pinocytosis
81. The movement of substances across cell membrane from lower concentration to higher concentration is called
    (a) osmosis (b) passive transport
    (c) active transport (d) facilitated transport
82. Taking in of food particles or foreign bodies through cell membrane is
    (a) phagocytosis (b) pinocytosis (c) osmosis (d) active transport
83. A number of different types of molecules move together through the membrane, the phenomenon being called
    (a) bulk flow (b) facilitated diffusion
    (c) diffusion (d) ion pump
84. Excretory products pass out of the cell through
    (a) pinocytosis (b) phagocytosis (c) exocytosis (d) endocytosis
85. Movement of ions or molecules against prevailing direction of electrochemical gradient is
    (a) diffusion (b) passive transport (c) active transport (d) pinocytosis
86. Cell movements through membrane undulations occur in
    (a) amoeba (b) leucocytes (c) euglena (d) fibroblasts
87. The plasma membrane is made up of
    (a) carbohydrates and fats (b) fats and proteins
    (c) proteins and carbohydrates (d) derived carbohydrates
88. Plasmalemma is another name for
    (a) cell wall (b) middle lamella
    (c) microfibrils (d) plasma membrane
89. Every living cell has a
    (a) chloroplast (b) cell membrane (c) cell wall (d) food vacuole
90. Liquid food drinking is
    (a) pinocytosis (b) phagocytosis
    (c) imbibition (d) none of the above

91. Carbohydrates are present in the plasmalemma in the form of
    (a) glycolipids and glycoproteins (b) cellulose
    (c) hemicellulose (d) starch

92. Average thickness of unit membrane is
    (a) 75 Å (b) 250 Å (c) 25 Å (d) 5 Å

93. The latest model for plasma membrane is
    (a) lamellar model (b) unit membrane model
    (c) fluid mosaic model (d) molecular lipid model

94. According to fluid mosaic model, plasma membrane is composed of phospholipids and
    (a) oligosaccharides (b) hemicellulose
    (c) integral proteins (d) extrinsic proteins and intrinsic proteins

95. Singer and Nicolson's fluid mosaic model biomembrane differs from Robertson's membrane in
    (a) arrangement of lipids (b) number of lipid layers
    (c) arrangement of proteins (d) absence of proteins

96. Plasma membrane
    (a) controls cell activities
    (b) controls cell shape and size
    (c) regulates inward and outward flow of materials
    (d) stores hormones

97. Plasma membrane is made of
    (a) proteins and carbohydrates (b) proteins and lipids
    (c) proteins, lipids and carbohydrates (d) proteins, some nucleic acid and lipids

98. Extrinsic proteins in Singer and Nicolson model of cell membrane are
    (a) tightly attached to intrinsic protein and cannot be easily separated
    (b) tightly attached to intrinsic protein and can be easily separated
    (c) superficially present but cannot be separated easily
    (d) superficially present and can be easily separated

99. Fluid mosaic model of cell membrane proposes that a lipid bilayer
    (a) has embedded protein only
    (b) has proteins on the outer surface only
    (c) is coated with proteins on both the surface
    (d) has embedded or superficial proteins

100. The difference between permeability and bulk transport across the cell membrane is related to
    (a) molecular weight of substance
    (b) solvent system
    (c) structure of molecules allowed passage
    (d) solvent system as well as structure of molecules

101. Cell recognition is due to
    (a) lipid portion of cell membrane
    (b) carbohydrate portion of glycoproteins
    (c) protein portion of glycoprotein
    (d) both carbohydrate and protein portion of glycoprotein

102. Proteins, lipids and carbohydrates are present in the cell membrane. Which of the following is true?
    (a) Carbohydrates are minimum. (b) Carbohydrates are maximum.
    (c) Lipids are minimum. (d) All the three are in equal proportion.

103. Carrier molecules facilitating transport across cell membrane are
    (a) proteinaceous (b) fatty (c) starch (d) alkaloids

104. Thickness of biomembrane is
    (a) 70–90 Å (b) 7–9 nm (c) 7–9 mμ (d) all the above

105. Pinocytosis is
    (a) ingestion of solid particles by plasma membrane
    (b) ingestion of liquid particles
    (c) change of permeability of plasmalemma towards ions
    (d) both (a) and (b)

106. Size of the molecules that can pass through plasma membrane is
    (a) 1–15 Å (b) 8–10 Å (c) 10–13 Å (d) 15–75 Å

107. Plasmalemma of animals cells is elastic due to the presence of
    (a) proteins (b) carbohydrates (c) lipids (d) microfilaments

108. Basic components of cell membrane are
    (a) proteins (b) proteins and carbohydrates
    (c) proteins and phospholipids (d) phospholipids and carbohydrates

109. Cell membrane is
    (a) semipermeable (b) permeable (c) impermeable (d) inelastic

110. Correct sequence of protein (P) and lipid (L) in cell membrane is (as per lamellar model)
   (a) L–P–L–P (b) L–P–P–L (c) P–L–L–P (d) P–P–L–L

111. Cell recognition and adhesion occur due to biochemicals of cell membranes named
   (a) proteins (b) lipids
   (c) proteins and lipids (d) glycoproteins and glycolipids

112. Robertson's model of cell membrane is similar to that of Danielli and Davson's in
   (a) type of proteins (b) lamellar structure
   (c) permeases (d) carrier particles

113. Cell membrane consists of (as per lamellar model)
   (a) carbohydrate–protein–lipid (b) lipid – protein–lipid
   (c) protein–lipid–protein (d) phospholipid–glycoprotein–protein

114. Which of the following statements regarding differentiation between active and passive transport is true?
   (a) Active transport does not need energy while passive transport requires energy.
   (b) Active transport is connected with cations and passive transport with anions.
   (c) Active transport is nonselective while passive transport is selective.
   (d) Active transport requires energy while passive transport requires concentration gradient.

115. Facilitated diffusion requires
   (a) enzymes and energy (b) carriers but no energy
   (c) carriers and energy (d) receptors and energy

116. Phagocytosis was first seen by
   (a) Huxley (b) Strasburger (c) Haeckel (d) Metchnikoff

117. Desmosomes are concerned with
   (a) cell adherence (b) cytolysis (c) cell division (d) cellular excretion

118. Cellular membranes occur in
   (a) eukaryotes (b) prokaryotes (c) akaryotes (d) all of these

119. Action potential of the outer surface of plasmalemma is generally
   (a) neutral (b) positive (c) negative (d) variable

120. Two animal cells are interconnected by
   (a) plasmodesmata (b) cell wall
   (c) desmosomes (d) plasma membrane

121. Plasmalemma is made of
   (a) single protein layer
   (b) single lipid layer
   (c) single lipid layer and two protein layers
   (d) single protein and single lipid layer

122. Active transport across biomembrane involves
   (a) production of ATP
   (b) requirement of energy
   (c) production of toxin
   (d) release of energy

123. Carrier proteins take part in
   (a) passive transport of ions
   (b) active transport of ions
   (c) water transport
   (d) water evaporation

124. Cell membrane spends energy in
   (a) diffusion of gases
   (b) movement of solute from lower to higher concentration
   (c) movement of solute from higher to lower concentration
   (d) movement of solvent

125. Most substances are transported across the membrane by the process of
   (a) diffusion (b) osmosis (c) active transport (d) both (a) and (b)

126. Phospholipids are
   (a) amphibolic (b) amphipathic (c) hydrophobic (d) hydrophilic

127. Thread-like protoplasmic projections found on the free surface of absorptive cells are
   (a) cilia (b) microfilaments (c) microvilli (d) plasmodesmata

128. Ion carriers occur in
   (a) cell wall (b) cell membrane (c) cytoplasm (d) nucleus

129. Cell eating is
   (a) phagocytosis (b) pinocytosis (c) diffusion (d) osmosis

130. Active transport occurs
   (a) against concentration gradient and requires ATP
   (b) against concentration gradient but does not require ATP
   (c) along concentration gradient but requires ATP
   (d) along concentration gradient but does not require ATP

131. Which side of the membrane is hydrophilic?
   (a) outer (b) inner (c) interface (d) all of the above

ANSWERS

1. (a)	2. (c)	3. (a)	4. (a)	5. (c)	6. (d)	7. (c)
8. (a)	9. (b)	10. (c)	11. (b)	12. (d)	13. (c)	14. (d)
15. (a)	16. (a)	17. (d)	18. (b)	19. (c)	20. (a)	21. (b)
22. (a)	23. (d)	24. (b)	25. (c)	26. (c)	27. (d)	28. (a)
29. (c)	30. (b)	31. (c)	32. (c)	33. (d)	34. (d)	35. (a)
36. (a)	37. (b)	38. (c)	39. (a)	40. (d)	41. (a)	42. (b)
43. (c)	44. (a)	45. (b)	46. (c)	47. (d)	48. (d)	49. (b)
50. (a)	51. (c)	52. (d)	53. (a)	54. (d)	55. (b)	56. (c)
57. (a)	58. (d)	59. (a)	60. (b)	61. (b)	62. (a)	63. (d)
64. (c)	65. (d)	66. (a)	67. (a)	68. (c)	69. (b)	70. (d)
71. (a)	72. (b)	73. (c)	74. (a)	75. (a)	76. (b)	77. (d)
78. (d)	79. (c)	80. (d)	81. (c)	82. (a)	83. (a)	84. (c)
85. (c)	86. (d)	87. (b)	88. (d)	89. (b)	90. (a)	91. (d)
92. (a)	93. (a)	94. (c)	95. (d)	96. (c)	97. (c)	98. (c)
99. (d)	100. (d)	101. (c)	102. (b)	103. (a)	104. (a)	105. (d)
106. (b)	107. (a)	108. (c)	109. (c)	110. (a)	111. (c)	112. (d)
113. (b)	114. (c)	115. (d)	116. (b)	117. (d)	118. (a)	119. (d)
120. (c)	121. (c)	122. (c)	123. (b)	124. (b)	125. (b)	126. (c)
127. (b)	128. (c)	129. (b)	130. (a)	131. (a)		

13

DNA

CHOOSE THE CORRECT ANSWER

1. In the nucleotide sequence of the ______ molecule, code words exist for each amino acid. This is referred to as the genetic code.
   (a) mRNA (b) tRNA (c) rRNA (d) none of these
2. The adapter molecules translating the code words into the amino acid sequence are the ______ molecules.
   (a) tRNA (b) mRNA (c) rRNA (d) none of these
3. The process by which the cell translates information by the machinery of the nucleotide sequence of an mRNA into the sequence of amino acids of the corresponding specific protein is said to be
   (a) translation (b) transfer (c) derepression (d) repression
4. The rough endoplasmic reticulum is the factory of
   (a) polyribosomes (b) ribosomes (c) microbodies (d) microsomes
5. Code words consisting of 2 nucleotides each can provide only ______ specific code words.
   (a) 16 (b) 12 (c) 10 (d) 8
6. Each code word consists of a sequence of ______ nucleotides.
   (a) 3 (b) 4 (c) 5 (d) 6
7. 64 triplet codes in mRNA with 4 nucleotides are to provide for ______ amino acids and three codons do not code for specific amino acids and these have been termed nonsense codons.
   (a) 20 (b) 25 (c) 30 (d) 40

8. Transcription is decreased due to the decreased amount of the primary transcript and of mature ______ PEPCK which in turn is directly related to the decreased rate of PEPCK synthesis.
   (a) mRNA (b) tRNA (c) rRNA (d) none of these
9. More than ______ specific mRNAs are effected by insulin.
   (a) 100 (b) 80 (c) 50 (d) 20
10. Amino acids are activated by the enzyme aminoacyl-tRNA synthetase in the presence of
   (a) ATP (b) GTP (c) EF-2 (d) EF-1
11. The binding of the mRNA to the 40S ribosomal subunit requires the presence of
   (a) IF-3 (b) IF-2 (c) EF-1 (d) all of these
12. Many ribosomes on the same ______ molecule form a polysome.
   (a) mRNA (b) tRNA (c) rRNA (d) none of these
13. The complex formed by anticodon of tRNA and IF-I attaches the 60S ribosomal subunits with the release of
   (a) IF-1 (b) IF-2 (c) EF-3 (d) all of these
14. The complete ribosome contains the mRNA at which site?
   (a) P site (b) A site (c) Both (a) and (b) (d) None of these
15. During the process of initiation, the complete ______ ribosome is formed.
   (a) 80S (b) 40S (c) 60S (d) 10S
16. EF-I forms a complex with ______ and the entering aminoacyl-tRNA.
   (a) GTP (b) ATP (c) EF-1 (d) EF-2
17. The alpha amino group of the new aminoacyl-tRNA in the A site combines with the carboxyl group of the peptidyl-tRNA occupying the P site in the presence of peptidyl transferase of the ribosomal subunits.
   (a) 60S (b) 40S (c) 20S (d) none of these
18. The newly formed peptidyl-tRNA at the A site is translocated into the vacated P site by
   (a) EF-2 (b) GTP (c) Both (a) and (b) (d) none of these
19. After multiple cycles of elongation, the nonsense codon of mRNA appears in the
   (a) A site (b) P site (c) Both (a) and (b) (d) none of these
20. The releasing factor hydrolyses the bond between peptide and the ______ occupying the P site
   (a) tRNA (b) mRNA (c) rRNA (d) none of these
21. On hydrolysis and release, the SOS ribosome dissociates into its 40S and ______ subunits which are then recycled.
   (a) 60S (b) 40S (c) 20S (d) none of these

22. The releasing factors are

    (a) proteins (b) ribosomes (c) mRNA (d) none of these

23. The antibiotic puromycin which has structural similarity with that of tyrosinyl-tRNA is incorporated via the _______ on the ribosome.

    (a) A site (b) P site (c) both (a) and (b) (d) none of these

24. Which of the following is inactivated by diphtheria toxin and thereby inhibits mammalian protein synthesis?

    (a) EF-1 (b) EF-2 (c) IF-1 (d) IF-2

25. A single ribosome is capable of translating about _______ codons in _______ seconds into a protein with a molecular weight of 40,000.

    (a) 400; 10 (b) 400; 20 (c) 300; 10 (d) 300; 20

26. Gene regulations are

    (a) positive regulation (b) negative regulation
    (c) both (a) and (b) (d) none of these

27. A double negative has the effect of acting as a

    (a) positive (b) negative (c) neutral (d) none of these

28. The gene is the smallest segment of the DNA molecule containing about _______ nucleotides.

    (a) 600 (b) 500 (c) 300 (d) 400

29. When a pair of alleles carries genes with the same characteristic, say tallness, then the individual is said to be

    (a) homozygous (b) heterozygous (c) ribosomes (d) none of these

30. In a pair of alleles if one carries the gene for tallness and the other for shortness, then the offspring is

    (a) heterozygous (b) homozygous (c) both (a) and (b) (d) none of these

31. Lac operon is nothing but a/an

    (a) structural gene (b) operator gene (c) both (a) and (b) (d) none of these

32. The regulator gene induces the synthesis of protein macromolecules and the process is called

    (a) repression (b) derepression (c) translation (d) transfer

33. The phenomenon in which the operon becomes active because the suppressor system is itself inactivated is called

    (a) derepression (b) repressor (c) translation (d) transfer

34. The operator locus is a region of double-stranded DNA of _______ base pairs long with a 2-fold rotational symmetry in a region that is _______ base pairs long.

    (a) 27; 21 (b) 21; 27 (c) 21; 21 (d) 27; 27

35. The operator locus lies between the promoter site at which the DNA-dependent RNA polymerase attaches to commence transcription and the beginning of the
    (a) Z-gene (b) structural gene (c) operator gene (d) none of these
36. Normally ______ repressor molecules are present and one or two operator loci per cell are also present.
    (a) 20–40 (b) 20–30 (c) 20–50 (d) 10–20
37. The binding of the inductor to the repressor molecule involves the amino acid residues in positions
    (a) 74; 75 (b) 74; 78 (c) 74; 76 (d) none of these
38. The larger hnRNA contains exons that will form the mature
    (a) mRNA (b) tRNA (c) rRNA (d) none of these
39. Exons are spliced to form
    (a) mRNA (b) tRNA (c) rRNA (d) none of these
40. 61 codons code for ______ amino acids. So there must be degeneracy in the genetic code.
    (a) 20 (b) 30 (c) 40 (d) 80
41. Every amino acid except ______ is represented by several codons.
    (a) methionine (b) pyrimidine (c) purine (d) none of these
42. Some of the proteins synthesized by the attached ______ are packaged by the Golgi apparatus.
    (a) polysomes (b) ribosomes (c) amino acids (d) all of these
43. For any specific codon, only a single ______ is indicated. So the genetic code is unambiguous.
    (a) amino acid (b) pyrimidine (c) purine (d) none of these
44. Mitochondria require only ______ tRNA molecules to read their genetic code, whereas cytoplasmic translation system passes a full complement of ______ tRNA species
    (a) 22; 31 (b) 21; 27 (c) 74; 75 (d) none of these
45. When the first base in anticodon is ______ or ______, the pairing with third base in codon is ______ or ______
    (a) C or A; G or A (b) G or A ; C or A (c) G or A; G or A (d) C or A; C or A
46. When the first base in the anticodon is 1, the third base may be
    (a) A (b) C (c) U (d) none of these
47. The codons for glycine which can form a base pair for one anticodon is/are
    (a) GGU (b) GGC (c) GGA (d) CCI
48. The chain-terminating codons is/are
    (a) UAA (b) UAG (c) GAG (d) all of these

49. An alteration in the sequence of bases in a gene resulting in an altered gene product is called
(a) mutation (b) translation (c) transfer (d) repressor

50. In translocation, a given pyrimidine is changed to another ______ or a given purine is changed to the other ______
(a) pyrimidine; purine (b) purine; pyrimidine
(c) amino acid; purine (d) none of these

51. In transitions, a purine is changed to a
(a) purine (b) pyrimidine (c) amino acid (d) none of these

52. A nonsense codon may appear that would result in the premature termination of ______ incorporation into a peptide chain.
(a) amino acid (b) purine (c) pyrimidine (d) none of these

53. Which of the following eukaryotic initiation factors does the process of initiation need?
(a) tRNA (b) rRNA (c) mRNA (d) all of these

54. The binding of two eukaryotic initiation factors eIF-3 and iIF-1A to the 40S subunit favours dissociation of the 80S ribosome into its ______ subunit.
(a) 40S (b) 60S (c) both (a) and (b) (d) none of these

55. Reassociation is also prevented by the binding of eIF-3 to the subunits
(a) 60S (b) 40S (c) 20S (d) none of these

56. The eIF-2 consists of subunits
(a) α (b) β (c) both (a) and (b) (d) none of these

57. Which of the following is phosphorylated by a cAMP-independent protein kinase, which is activated when a cell is under stress?
(a) eIF-2α (b) eIF-3α (c) eIF-3 (d) eIF-1A

58. The methyl guanosyl triphosphate cap helps the binding of mRNA to the ______ preinitiation complex.
(a) 40S (b) 60S (c) 20S (d) none of these

59. ATP hydrolysis is required in the association of mRNA with the ______ preinitiation complex.
(a) 40S (b) 60S (c) 20S (d) none of these

60. The eIF-3 is a key protein since it binds mRNA to ______ ribosomal subunits.
(a) 60S (b) 40S (c) 20S (d) none of these

61. eIF-4A is an RNA-dependent ATPase that helps unwind the
(a) RNA (b) p 220 (c) both (a) and (b) (d) none of these

62. Which of the following is important since cap binding is a rate-limiting step in translation.
    (a) 4E (b) 4E-BP2 (c) 4E-3P1 (d) none of these
63. The phosphorylation of 4E and activation of the protein are caused by
    (a) insulin (b) mitogenic growth factors
    (c) both (a) and (b) (d) none of these
64. A family of proteins 4E-BP1 and 4E-BP2 binds to and inactivates
    (a) 4E (b) 4E-BP1 (c) 4E-BP2 (d) none of these
65. The hydrolysis of the ATP bound to eIF-2 by eIF-5 is caused by the binding of the 60S ribosomal subunits to the initiation complex and ______.
    (a) 40S (b) 60S (c) 20S (d) none of these
66. The association of which of the following subunits forms the 80S ribosome?
    (a) 60S (b) 40S (c) both (a) and (b) (d) none of these
67. Elongation is a cyclic process consisting of several steps catalysed by
    (a) eEF (b) 4E (c) 4E-BP1 (d) 4E-BP2
68. The aminoacyl-tRNA binds to the A site of the complete ______ ribosome requiring proper codon recognition.
    (a) 80S (b) 60S (c) 40S (d) 20S
69. The amino acyl-tRNA enters the A site with the release of
    (a) eEF-1 α -GDP (b) phosphate (c) both (a) and (b) (d) none of these
70. The alpha amino group of the new aminoacyl-tRNA in the A site has a nucleophilic attack on the esterified carboxyl group of the ______ occupying the P site.
    (a) peptidyl-tRNA (b) peptidyl-mRNA (c) peptidyl-rRNA (d) none of these
71. Which of the following has a direct role in protein synthesis?
    (a) RNA (b) tRNA (c) mRNA (d) rRNA
72. The amino acid on the amino ______ is already activated for which further energy is required.
    (a) acyl-mRNA (b) acyl-tRNA (c) acyl-rRNA (d) none of these
73. Which of the following is/are responsible for the translocation of the newly formed peptidyl-tRNA at the A site into the empty P site?
    (a) eEF-2 (b) GTP (c) eIF-3 (d) eIF-4A
74. Which of the following is hydrolysed to AMP which is equivalent to the hydrolysis of two ATPs to two ADPs and phosphates in the charging of the tRNA molecule with the amino acyl moiety?
    (a) ATP (b) GTP (c) both (a) and (b) (d) none of these

75. The hydrolysis of GTP is caused by the translocation of the newly formed peptidyl-tRNA in the A site into the P site by
   (a) eEF-2 (b) eEF-3 (c) eIF-2 (d) eIF-5

76. The formation of one peptide bond causes the hydrolysis of
   (a) two ATP molecules (b) two GTP molecules
   (c) both (a) and (b) (d) none of these

77. In the ______ ribosome, 18 amino acids are incorporated per second.
   (a) prokaryotic (b) eukaryotic (c) both (a) and (b) (d) none of these

78. The termination signal in the A site is recognized by the eRR signal sequence which are ______ that hydrolyse the peptidyl-tRNA bond when a nonsense codon occupies the A site.
   (a) proteins (b) amino acids (c) ribosomes (d) none of these

79. The 80S ribosome dissociates into its 40S and 60S subunits upon
   (a) hydrolysis (b) release (c) both (a) and (b) (d) none of these

80. Ferritin synthesis is stimulated by elemental iron by the release of a cytoplasmic______that binds to 5′ nontranslated region of protein ferrite.
   (a) mRNA (b) tRNA (c) rRNA (d) none of these

81. Some viruses inhibit host cell protein synthesis by preventing the association of mRNA with the ______ ribosome.
   (a) 40S (b) 80S (c) 60S (d) 20S
   (e) rRNA (f) none of these

82. A single mammalian ribosome can synthesize about ______ peptide bonds per minute.
   (a) 200 (b) 150 (c) 100 (d) 50

83. Polysomes actively synthesizing ______ can exist as free particles in the cellular cytoplasm.
   (a) proteins (b) ribosomes (c) amino acids (d) all of these

84. Protein synthesis on all ribosomes is inhibited by puromycin which is structurally similar to
   (a) tyrosinyl-tRNA (b) tyrosinyl-mRNA
   (c) tyrosinyl-rRNA (d) none of these

85. Cycloheximide inhibits peptidyltransferase in the 60S ribosomal subunit in eukaryotes by binding to the component
   (a) rRNA (b) mRNA (c) tRNA (d) none of these

86. Diphtheria toxin catalyses the ADP-ribosylation found in mammalian cells.
   (a) eFF-2 (b) eFF-1 (c) eRF-1 (d) eEF-2

ANSWERS

1. (a)	2. (a)	3. (a)	4. (a)	5. (d)	6. (a)	7. (a)
8. (a)	9. (a)	10. (a)	11. (a)	12. (a)	13. (d)	14. (c)
15. (a)	16. (a)	17. (a)	18. (c)	19. (a)	20. (a)	21. (a)
22. (a)	23. (a)	24. (b)	25. (a)	26. (c)	27. (a)	28. (b)
29. (a)	30. (a)	31. (c)	32. (a)	33. (a)	34. (a)	35. (a)
36. (a)	37. (a)	38. (a)	39. (a)	40. (a)	41. (a)	42. (a)
43. (a)	44. (a)	45. (a)	46. (d)	47. (c)	48. (d)	49. (a)
50. (a)	51. (b)	52. (a)	53. (a)	54. (c)	55. (a)	56. (c)
57. (a)	58. (a)	59. (a)	60. (b)	61. (c)	62. (a)	63. (c)
64. (a)	65. (a)	66. (c)	67. (a)	68. (a)	69. (c)	70. (a)
71. (a)	72. (b)	73. (c)	74. (a)	75. (a)	76. (b)	77. (a)
78. (a)	79. (a)	80. (a)	81. (a)	82. (c)	83. (a)	84. (a)
85. (a)	86. (a)					

14

MOLECULAR GENETICS

CHOOSE THE CORRECT ANSWER

1. The integrated use of biochemistry, microbiology, and engineering sciences in order to achieve technological application of the capabilities of microorganisms, cultured tissue cells and parts thereof is known as

   (a) molecular genetics (b) tissue culture
   (c) biotechnology (d) ELISA

2. A technique called genetic engineering involves breakage of a DNA molecule at two desired places to isolate a specific DNA segment and its insertion in another DNA molecule at a desired position, and the product thus obtained is called

   (a) recombinant DNA (b) foreign DNA
   (c) recon (d) none of these

3. Which of the following is defined as autonomous elements, whose genomes exist in the cell as extrachromosomal units?

   (a) clone (b) plasmid (c) recombinant DNA (d) genes

4. Who among the following developed a reproducible technique for tissue culture and brought about a true beginning of animal cell and tissue culture?

   (a) Ross Harrison (1907) (b) Alex Carrel (1912)
   (c) Freshney (1987) (d) Cockinig (1936)

5. In animals if the tissue explant maintains its structure and function in culture, it is described as

   (a) cell culture (b) tissue culture
   (c) organotypic culture (d) organ culture

6. If cells in culture reassociate to create a three-dimensional structure, irrespective of the tissue from which it was derived, it is described as a
   (a) histotypic culture (b) histological culture
   (c) histochemical culture (d) hybrid

7. The beginning of plant tissue culture was made by ______ in 1998 when he successfully cultured fully differentiated individual plant cells isolated from different tissues in several plant species.
   (a) Gautheret (b) White (c) Nobecourt (d) Haberlandt

8. During 1934–1939, due to discovery of the importance of auxins and B-vitamins, the foundation of plant tissue culture was laid down by three scientists, namely,
   (a) Gautheret, White and Nobecourt (b) Gautheret, Haberlandt and Nobecourt
   (c) White, Haberlandt and Nobecourt (d) Gautheret, Wilkins and Nobecourt

9. Plant cell culture is also being utilized for producing valuable products including secondary metabolites through ______, which also utilizes microbes.
   (a) bio-transformation (b) bio-information
   (c) bio-engineering (d) bio-screening

10. The examples of the use of cell cultures in biotransformation include
    (a) *Digitalis lanata, Datura* (b) mouse
    (c) microbes (d) eucalyptus, orchid

11. The phenomenon of the reversion of mature cells to the meristematic state leading to the formation of callus is called
    (a) redifferentiation (b) dedifferentiation
    (c) cellular totipotency (d) regeneration

12. The component cells of the callus have the ability to form a whole plant, a phenomenon described as
    (a) redifferentiation (b) regeneration (c) totipotency (d) dedifferentiation

13. A property that is found only in plant cells and not in animal cells is named as
    (a) transformation (b) cellular totipotency (c) redifferentiation (d) regeneration

14. The process where any cell of the embryo sac or any part of the sporophyte may give rise to an embryo is called
    (a) embryo differentiation (b) embryogenesis
    (c) somatic embryogenesis (d) clonal propagation

15. Somatic embryogenesis in culture was demonstrated for the first time in
    (a) citrus (b) *Mangifera indica*
    (c) *Brassica campestris* (d) *Daucus carota*

16. A variety of plant species can be conveniently propagated through the technique of cell, tissue or organ culture. This is popularly described as
    (a) clonal propagation (b) micropropagation
    (c) both (a) and (b) (d) none of these

17. Regeneration from callus, leaf explants or plant protoplasts leads to the generation of considerable variation, described as
    (a) somaclonal variation (b) biological variation
    (c) trans-variation (d) none of the above

18. The term used for embryo culture, ovule culture and ovary culture, since in each case the objective is to rescue the embryo, is
    (a) *in vitro* rescue (b) embryo rescue (c) ovule rescue (d) ovary rescue

19. Triploid plants are useful for production of
    (a) seedless fruits (b) nutritious fruits (c) both (a) and (b) (d) none of these

20. Which of the following can also be used for micropropagation, particularly in cases where embryos develop from nucellar cells?
    (a) callus (b) nucleus (c) endosperm (d) embryo

21. The most popular and effective method for long-term storage of cell cultures involves storage at a very low temperature using liquid nitrogen and this is described as
    (a) preservation (b) cold storage (c) conservation (d) cryopreservation

22. One of the very popular methods for production of haploids is through culturing. Which of the following develops on artificial culture medium?
    (a) anthers or micropores (b) nucellus
    (c) callus (d) none of these

23. Who among the following demonstrated in 1960 that naked cells called protoplasts can be obtained through enzymatic degradation of cell walls?
    (a) M.S. Swaminathan (b) Heslop–Harrrison
    (c) C.E. Cocking (d) none of the above

24. Which of the following is particularly ideal for isolation of protoplasts in large quantities?
    (a) mature cell suspensions (b) young cell suspensions
    (c) callus (d) meristematic tissue

25. The isolated protoplasts can be cultured by
    (a) bulbosum technique (b) chromosome doubling
    (c) hanging drop technique (d) Bergmann's cell plating technique

26. Which of the following is particularly significant for hybridization between species of genera, which cannot be made to cross by conventional methods of sexual hybridization?
    (a) somatic embryogenesis (b) somaclonal variation
    (c) distant hybridization (d) somatic hybridization

27. In somatic hybridization, the induced fusion of protoplasts requires a suitable agent known as
    (a) fusogen (b) mutagen (c) homokaryon (d) none of the above

28. In inactivated animals, which of the following is needed to induce fusion?
    (a) protoplast (b) polio virus (c) bacteria (d) sendai virus

29. Hybrids in which nucleus is derived from one parent and cytoplasm is derived from both are called
    (a) sexual hybrids (b) cybrids (c) somatic hybrids (d) haploids

30. In higher plants, the cultured protoplasts can be used for gene transfer followed by regeneration leading to the production of
    (a) hybrids (b) transgenic plants (c) haploids (d) somatic hybrids

31. Plant proteases like ______ from papaya are used as deworming medicine for humans and domestic animals.
    (a) ficin (b) papain (c) trypsin (d) thrombin

32. Which of the following, isolated from bacteria, fungi, pancreas and malt, is used in textile industry as softening agent for starched clothes?
    (a) oxidases (b) peroxidases (c) amylases (d) none of the above

33. Which of the following are often added in canned fruit juices and in wine, since they hydrolyse the pectin in making the juice or wine clear?
    (a) pectinases (b) peroxidases (c) alcalases (d) amylases

34. Which of the following is used in the manufacture of chocolate-covered berries and other such candies?
    (a) invertase (b) pectinase (c) trypsin (d) lactase

35. Which of the following is used for the prevention of lactose crystals in ice-cream?
    (a) glucose isomerase (b) lactase (c) invertase (d) xylanase

36. Which of the following plant toxins is used as immunotoxin?
    (a) recin (b) ficin (c) trypsin (d) papain

37. Which of the following drugs is designed for treatment of ulcers?
    (a) rifamcin (b) trimethoprim (c) cimetidine (d) propanolol

38. Which of the following is used in the improvement of cellular activities by manipulation of enzymatic, transport and regulatory functions of the cell with the use of recombinant DNA technology?

(a) metabolic engineering (b) genetic engineering
(c) biotechnology (d) none of the above

39. Which of the following is generally produced by fermentation of some sugar such as rice products, with the help of the yeast, *Saccharomyces cerevisiae*?

(a) xylol (b) carbohydrate (c) ethanol (d) methanol

40. Acetone, butanol, butyric acid and diisopropanol are obtained through fermentation of starch, molasses, sucrose, wood hydrolysates and pentoses using bacteria belonging to the genus

(a) *Streptomyces* (b) *Clostridium* (c) *Bacillus* (d) *Saccharomyces*

41. Which of the following is the most important group of industrial enzymes, which hydrolyse starch?

(a) diastases (b) alpha-amylases (c) proteases (d) lipases

42. In the energy sector, which of the following can be obtained by microbial production to meet food and fodder problems?

(a) single-cell proteins (SCP) (b) simple-cell proteins (SCP)
(c) somatic-cell proteins (SCP) (d) all the above

43. Which of the following term refers not only to the isolated cell protein but also to any microbial biomass from both uni- and multicellular bacteria, yeasts, filamentous fungi or algae, which can be used as food or feed additives?

(a) single cell protein (b) biomass
(c) substrate (d) none of the above

44. Which of the following organisms are inoculated in the sewage for sewage treatment process?

(a) single cell cultures (b) microbial cultures
(c) starter cultures (d) sewage cultures

45. Bacterial activity is implicated in the weathering, leaching and deposition of mineral ores. This has led to a relationship between biotechnology and metallurgy, creating a new discipline known as

(a) biohydrometallurgy (b) bio-metallurgy
(c) biomineralization (d) all the above

46. The area of biotechnology which utilizes the large-scale and long-term activity of bacteria causing metals to be deposited as insoluble oxides and sulphides is called

(a) biohydrometallurgy (b) biomineralization
(c) biotransformation (d) bioinformatics

47. Copper is leached from its ores due to the activity of a bacterium
    (a) *Thiobacillus ferroxidans* (b) *Thiobacillus thuringiensis*
    (c) *Agrobacterium* (d) none of the above
48. Which of the following results from interaction of extruded metabolic products with extraneous metal ions in the surrounding environment?
    (a) viral mineralization (b) bacterial mineralization
    (c) algal mineralization (d) none of these
49. Which of the following is defined as biologically active products or microbial inoculants of bacteria, algae and fungi (separately or in combination), which may help in biological nitrogen fixation for the benefit of plants?
    (a) biomasses (b) biochemicals (c) bio-fixatives (d) biofertilizers
50. Which of the following, when rendered in a viable form due to the interaction of microorganisms with plants, is a good biofertilizer?
    (a) organic fertilizers (b) biochemicals
    (c) biologically active products (d) none of the above
51. Which of the following is able to enter into symbiotic relationship with legumes and fix atmospheric nitrogen thereby increasing the crop productivity?
    (a) *E. coli* (b) rhizobia (c) *Azotobacter* (d) *Azospirillum*
52. Which of the following is an asymbiotic free-nitrogen fixer?
    (a) *Bacillus polymyxa* (b) Rhizobium
    (c) *Azolla* (d) none of the above
53. Together with blue-green algae, the main source of algal biofertilizer is
    (a) legume (b) *Ficus* (c) biota (d) *Azolla*
54. Which of the following is a symbiotic association of fungi with roots of plants, so that nutrients absorbed from soil by the fungus are released to the host cells and in turn, the fungus takes its food requirements from the host?
    (a) mycorrhiza (b) mycobiont (c) phycobiont (d) none of the above
55. Which of the following is found on the roots of forest trees and absorb nitrogen, phosphorus, potassium and calcium?
    (a) endomycorrhizae (b) ectomycorrhizae
    (c) mycobiont (d) none of the above
56. Which of the following is found in the roots of most fruits and other horticultural crops and particularly helps in phosphorus nutrition?
    (a) blue-green algae (b) ectomycorrhizae
    (c) endomycorrhizae (d) none of the above

57. Which of the following is generally composed of bacterial spores, crystal proteins and inert filters?
    (a) bioinsecticides (b) biofertilizers
    (c) biogas (d) mycorrhiza
58. Bioinsecticides based on the mutant strains of which of the following will be ecologically safe?
    (a) *Pseudomonas putida* (b) *Bacillus thuringiensis*
    (c) *Thiobacillus* (d) none of the above
59. Which of the following can be very useful as an insecticide?
    (a) nuclear polyhedrosis virus (b) *Bacillus papillae*
    (c) *Nomuraea rileyi* (d) all of these
60. The nitrogen-fixing efficiency of Rhizobia can be increased by transfer of useful ______ genes and other related genes involved in nodulation, nitrogenase activity and other related functions.
    (a) *nif* (b) *hup* (c) *ice* (d) *tra*
61. Nieman–Pick disease is due to the deficiency of the enzyme
    (a) sphigomyelinase (b) hexosaminidase
    (c) ceraminodase (d) none of these
62. Deficiency of enzyme β-glucosidase causes
    (a) Gaucher's disease (b) Fabry's disease
    (c) ceraminodase (d) hexosaminidase
63. The smallest enzyme of the cell is
    (a) peroxidase (b) catalase (c) zymase (d) amylase
64. Pure crystals of pepsin and trypsin were obtained by
    (a) Northrop (b) Kuhne (c) Buchner (d) all of the above
65. Restriction endonuclease was discovered by
    (a) Arber (b) Nathan (c) Smith (d) all of the above
66. All enzymes are proteins except RNase-P. This statement was given by
    (a) Lehninger (b) Hansen and Cech (c) CarlNurberg (d) Kornberg
67. Galactosemia is due to deficiency of
    (a) galactose 1-phosphate uridyltranferase (b) cytidylase
    (c) both (a) and (b) (d) none of these
68. In which cell organelle is maximum enzyme present?
    (a) mitochondria (b) chloroplast (c) ribosome (d) none of these

69. Value of K_m lies between
    (a) 10^{-1} to 10^{-6} molal
    (b) 10^{-6} to 10^{-9} molal
    (c) 10^{-9} to 10^{-11} molal
    (d) none of these
70. K_m is defined as
    (a) substrate concentration at which the chemical reaction attains half its maximum velocity
    (b) substrate concentration at which the chemical reaction attains twice its maximum velocity
    (c) both (a) and (b)
    (d) none of these
71. Which of the following groups does an active site contain?
    (a) NH_2
    (b) COOH
    (c) SH
    (d) OH
72. The fastest acting enzyme carbonic anhydrase has its turn over number as
    (a) 30
    (b) 10,000
    (c) 5 million
    (d) 36 million
73. Methaemoglobinemia is due to defect in the enzyme
    (a) methaemoglobin
    (b) methaemoglobin reductase
    (c) hydrolases
    (d) none of these
74. Enzymes generally have
    (a) same pH and temperature optima
    (b) same pH but different temperature optima
    (c) different pH but same temperature optima
    (d) different pH and different temperature optima
75. Nitrate reductase enzyme forms
    (a) N_2
    (b) amino acids
    (c) NO_2
    (d) NO_3
76. Which of the following lack enzymes?
    (a) algae
    (b) fungi
    (c) bacteria
    (d) virus
77. Which of the following is an iron–porphyrin complex?
    (a) cytochrome
    (b) FAD
    (c) CoA
    (d) NAD
78. FAD or FMN is a co-enzyme. Which vitamin is incorporated into its structure?
    (a) vitamin C
    (b) vitamin B_1
    (c) vitamin B_6
    (d) vitamin B_2 (riboflavin)
79. The ratio of the enzyme molecules to substrate molecules can be as high as
    (a) 1 : 1000
    (b) 1 : 100,000
    (c) 1 : 1000,000
    (d) 1 : 50,000
80. During enzyme action,
    (a) value of K_m is low
    (b) value of K_i is low
    (c) value of K_m is high
    (d) value of K_i is high

81. Which of the following inactivates an enzyme by changing the enzyme shape?
(a) non-competitive inhibitor (b) competitive inhibitor
(c) coenzyme (d) all the above

82. Which segment of Figure 'X' represents the usable energy yield?

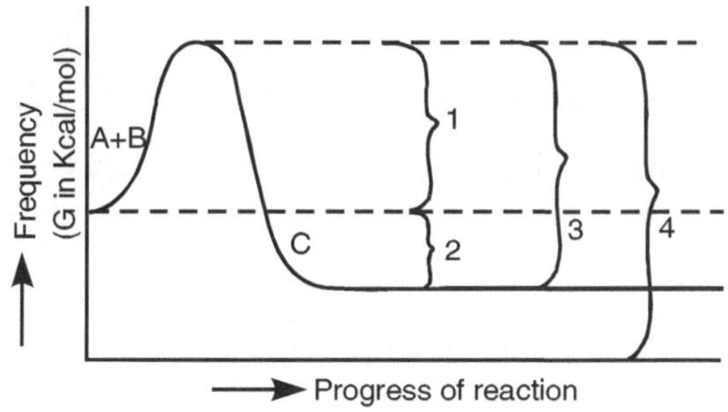

(a) 1 (b) 2 (c) 3 (d) 4

83. K_i indicates
(a) competitive inhibition (b) denaturation of enzyme
(c) reaction velocity (d) all of the above

84. Presence of enzyme in the urine is called
(a) enzymuria (b) enzymopenria (c) enzymopathy (d) none of the above

85. A protein having both structural and enzymic character is
(a) myosin (b) myoglobin (c) actin (d) trypsin

86. The enzyme which joins with the no-proteinic part to form a functional enzyme is
(a) holoenzyme (b) prosthetic group (c) apoenzyme (d) coenzyme

87. Enzymes which help in electron transfer are
(a) cytochrome (b) isomerase (c) proteinase (d) all of the above

88. Induced-fit theory of enzyme action was given by
(a) Koshland (b) Sumner (c) Fischer (d) Hershey and Chase

89. Which of the following amino acids are found over the active site of aldolase enzyme?
(a) glycine (b) histidine (c) alanine (d) all of the above

90. Activation energy is required for
(a) exothermal reactions (b) endothermal reactions
(c) enzyme-mediated reactions (d) initiating biochemical reaction

91. Ribozyme is
(a) RNA with enzyme activity (b) RNA with sugar
(c) RNA with phosphate (d) RNA with extra phosphate

92. Enzymes are polymers of
    (a) fatty acids (b) amino acids (c) sugars (d) inorganic phosphates
93. Acetyl coenzyme belongs to
    (a) synthetase (b) isomerase (c) hydrolase (d) transferase
94. cAMP-mediated 'Cascade Model' of enzyme regulation was proposed by
    (a) Fischer (b) Sutherland (c) Sumner (d) Koshland
95. Histidine decarboxylase catalyses breaking covalent bonds between (C—C) atoms of histidine to give rise to CO_2 and histamine. This shows that the enzyme histidine decarboxylase belongs to
    (a) hydrolases (b) lyases (c) isomerases (d) ligases
96. Michaelis constant (K_m) may be defined as the substrate conc. at which the velocity is
    (a) V_{max} (b) 1/2 V_{max} (c) V_{min} (d) $V = 0$
97. Sulpha drug is a structural analogue to
    (a) *p*-aminobenzoic acid (b) vitamin riboflavin
    (c) folic acid (d) cytochrome oxidase
98. Cyanide kills an animal by inhibiting
    (a) *p*-aminobenzoic acid (b) vitamin riboflavin
    (c) folic acid (d) cytochrome oxidase
99. Induced-fit theory was given by
    (a) Emil Fischer (b) Sumner (c) Buchner (d) Koshland
100. Michaelis constant was given by
    (a) Victor Henri (b) Leoner Michaelis (c) Mand Menten (d) both (b) and (c)
101. Esterases belong to the category of
    (a) lyases (b) ligases (c) transferases (d) hydrolases
102. Nitrate reductase enzyme forms
    (a) N_2 (b) amino acids (c) NO_2 (d) NO_3
103. Lactic dehydrogenase (LDH) that takes part in catalysis of pyruvate is an example of
    (a) isoenzyme (b) zymogen (c) coenzyme (d) apoenzyme
104. A protein having both structural and enzymatic traits is
    (a) myosin (b) collagen (c) trypsin (d) actin
105. ELISA test is used to
    (a) separate viral RNA (b) purify proteins
    (c) isolate DNA sequence (d) identity specific proteins

106. Which one of the following is without coenzyme activity?
(a) vitamin E (b) thiamine (c) biotin (d) riboflavin

107. Which of the following has coenzyme activity?
(a) nicotinamides (b) purine (c) pyrimidine (d) both (b) and (c)

108. The activity of succinate dehydrogenase is inhibited by
(a) malonate (b) pyruvate (c) glycolate (d) both (b) and (c)

109. A coenzyme is
(a) an organic non-proteinaceous group that is essential for enzyme activity
(b) an organic or inorganic group that is essential for enzyme activity
(c) the same enzyme found in different organs or tissues
(d) one that shares functions of another enzyme

110. Which of the following is required for enzyme action?
(a) low K_m (b) high K_m (c) low K_i (d) high K_i

111. The enzyme which converts glucose to glucose 6-phosphate is
(a) phosphorylase (b) glucose 6-phosphatase
(c) hexokinase (d) glucose synthetase

112. The respiratory enzymes are located in
(a) mitochondrial matrix (b) perimitochondrial space
(c) cristae (d) outer membrane

113. Enzymes taking part in glycolysis are present in
(a) mitochondria (b) perimitochondrial space
(c) cristae (d) outer membrane

114. ETC and TCA enzymes occur in
(a) ribosomes (b) endoplasmic reticulum
(c) mitochondria (d) cytoplasm and nucleus

115. Which of the following enzymes is absent in mitochondria?
(a) aconitase (b) malic dehydrogenase
(c) fumarase (d) hexokinase

116. Which of the following transfers electron to ETS?
(a) phytochrome (b) FeS (c) cytochromes (d) both (b) and (c)

117. Excess of ATP inhibits respiration by inhibiting which of the following enzymes?
(a) phosphoglyceric acid (b) hexokinase
(c) pyruvic decarboxylase (d) aldolase

118. In glycolysis, enzyme enolase produces
   (a) phosphoglyceric acid (b) phosphoenol pyruvate
   (c) phosphoglyceraldehyde (d) pyruvate
119. Steps in respiration are controlled by
   (a) enzymes (b) substrates (c) hormones (d) body juices
120. The enzyme connected with the opening and closing of stomata is
   (a) α-amylase (b) pyruvic kinase
   (c) PEP oxygenase (d) RuDP carboxylase
121. In C_3 plants, the first carbon dioxide acceptor enzyme is
   (a) PEPCO (b) RuBP carboxylase
   (c) RuBP oxygenase (d) oxidase
122. PEP, the first CO_2 acceptor in C_4 cycle, is a
   (a) 3-C compound (b) 4-C compound (c) 5-C compound (d) 6-C compound
123. The enzyme required in early CO_2 fixation in C_4 cycle is
   (a) RuBP carboxylase (b) RuBP oxygenase
   (c) PEP carboxylase (d) PGA dehyrogenase
124. In sugarcane, $^{14}CO_2$ is fixed in malic acid. The enzyme is
   (a) RuBP carboxylase (b) PEP carboxylase
   (c) ribulose phosphate kinase (d) fructose phosphatase
125. C_4 plants belong to
   (a) Gramineae (b) monocots (c) dicots (d) both monocots and dicots
126. Which of these is required for photosynthetic evolution of oxygen?
   (a) F-enzyme (b) Z-enzyme (c) Co-enzyme (d) M-enzyme
127. Which group of three of the following five statements (1–5) contain all three statements correct regarding beri beri?
   (A) A crippling disease prevalent among the native population of sub-Saharan Africa
   (B) A deficiency disease caused by lack of thiamine (vitamin B_1)
   (C) A nutritional disorder in infants and young children when the diet is persistently deficient in essential protein
   (D) Occurs in those countries where the staple diet is polished rice
   (E) The symptoms are pain from neuritis, paralysis, muscle wasting, progressive oedema, mental deterioration and finally heart failure.
   (a) B, D and E (b) A, B and D (c) A, C and E (d) B, C and E

128. The catalytic efficiency of two different enzymes can be compared by

(a) the formation of the product (b) the pH of optimum value
(c) the *Km* value (d) the molecular size of the enzyme

129. Enzymes, vitamins and hormones can be classified into a single category of biological chemicals, because all of these

(a) help in regulating metabolism
(b) are exclusively synthesized in the body of a living organism as at present
(c) are conjugated proteins
(d) enhance oxidative metabolism

130. An organic substance bound to an enzyme and essential for its activity is called

(a) isoenzyme (b) coenzyme (c) holoenzyme (d) apoenzyme

131. The figure given below shows three velocity–substrate concentration curves for an enzyme reaction. What do the curves i, ii and iii depict respectively?

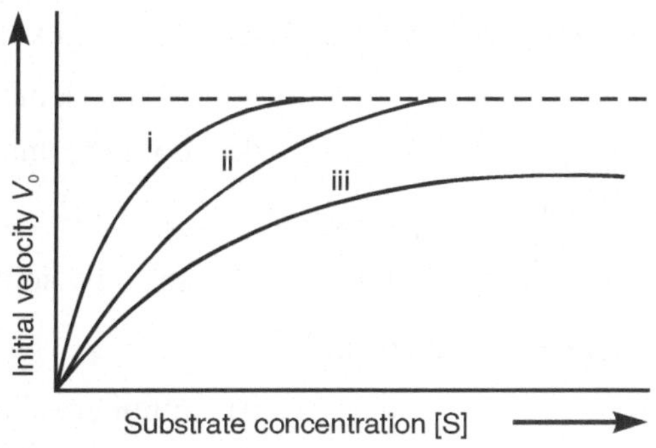

(a) i. Normal enzyme reaction
ii. Competitive inhibition
iii. Non-competitive inhibition

(b) i. Enzyme with an allosteric modulator added
ii. Normal enzyme activity
iii. Competitive inhibition

(c) i. Enzyme with an allosteric stimulator
ii. Competitive inhibitor added
iii Normal enzyme reaction

(d) i. Normal enzyme reaction
ii. Non-competitive inhibitor added
iii. Allosteric inhibitor added

132. Feedback inhibition of enzyme is affected by
   (a) end product (b) enzyme
   (c) substrate (d) temperature/other external factors

133. UbQ (Ubiquinone) is
   (a) a protein coenzyme (b) non-protein coenzyme
   (c) an activator (d) a protein cofactor

134. In the reaction $\text{Glucose} + \text{ATP} \xrightarrow{\text{hexokinase}} \text{Glucose 6-phosphate} + \text{ADP}$, hexokinase is
   (a) a lyase (b) a ligase (c) an isomerase (d) a transferase

135. The temperature range at which enzymes are maximum functional is
   (a) 20–30°C (b) 30–45°C (c) 15–25°C (d) 40–65°C

136. Enzyme required to cut the double strand of DNA in genetic engineering is
   (a) DNase (b) reverse transcriptase
   (c) restriction endonuclease (d) ligase

137. Accumulation of glucose 6-PO_4 inhibits the activity of glucohexokinase. This is an example of
   (a) competitive inhibition (b) feedback inhibition
   (c) allosteric inhibition (d) both (b) and (c)

138. Enzyme ptyalin (salivary amylase) works best in ______ medium.
   (a) slightly acidic (b) slightly alkaline (c) strongly acidic (d) slightly neutral

139. Turn-over number of an enzyme depends on its
   (a) molecular weight (b) number of active sites
   (c) size of enzyme (d) concentration of enzyme

140. Polynucleotide chain/nucleic acid can be hydrolysed/fragmented by enzymes called
   (a) nucleases (b) polymerases (c) proteases (d) ligases

141. Who conceived the idea that cell comes only from pre-existing cells?
   (a) Rudolf Virchow (b) Robert Hook (c) Robert Brown (d) none of these

142. The doubling and halving, the cycle of cell growth and division is called
   (a) cell division (b) cell cycle (c) amitosis (d) none of these

143. The phases of cell cycle are
   (a) G1 and G2 (b) S and M (c) G1, S, G2 and M (d) none of these

144. Phases G1, S and G2 comprise one phase which is called
   (a) prophase (b) interphase (c) first phase (d) none of these

145. Amitosis, mitosis and meiosis respectively are also called as
   (a) karyokinesis, meiotic and reduction divisions
   (b) amitotic, mitotic and meiotic divisions
   (c) reduction, mitotic and direct cell divisions
   (d) none of these

146. Usually protozoan cell divides
   (a) mitotically (b) meiotically (c) amitotically (d) none of these

147. Somatogenesis is
   (a) mitosis (b) amitosis (c) meiosis (d) direct cell division

148. The different phases of mitosis are
   (a) prophase, metaphase, anaphase and telophase
   (b) telophase, metaphase and prophase
   (c) prophase, anaphase and telophase
   (d) none of these

149. In mitosis the number of chromosomes present in a cell remains
   (a) half in the daughter cells (b) double in the daughter cells
   (c) same in the daughter cells (d) none of these

150. Meiosis occurs in the
   (a) somatic cells (b) gonads
   (c) both (a) and (b) (d) none of these

151. In which stage of prophase-I does chiasma formation occur?
   (a) diplotene (b) leptotene (c) pachytene (d) zygotene

152. Replication occurs if DNA is limited to
   (a) S phase (b) M phase (c) G1 phase (d) G2 phase

153. Principally, chromosomes are composed of
   (a) proteins (b) DNA (c) RNA (d) DNA and proteins

154. Principally, division of chromosomal materials takes place in
   (a) amitosis (b) mitosis (c) meiosis (d) all of these

155. Meiosis results in the production of
   (a) four daughter cells having only half the number of chromosomes
   (b) two cells that are similar qualitatively and quantitatively
   (c) both (a) and (b)
   (d) none of the above

156. For the study of chromosome morphology, which of the following mitotic stages is the best stage?
   (a) pachytene (b) anaphase (c) metaphase (d) interphase
157. At G1 stage, the uncoiling of chromosomes allow
   (a) protein synthesis (b) DNA synthesis (c) RNA synthesis (d) none of these
158. The number of linkage groups in man is
   (a) 46 (b) 44 (c) 23 (d) 22
159. B. Lee is famous for the discovery of
   (a) chromosome (b) polyploidy (c) mutation (d) none of these
160. DNA polymerase was isolated by
   (a) Atman (b) Kornberg (c) Lutz (d) none of these
161. The genes are arranged in
   (a) chromosomes (b) ribosomes (c) lysosomes (d) none of these
162. A fungus or bacterium which can grow on a minimal medium is called
   (a) autotroph (b) prototroph (c) heterotroph (d) none of these
163. Sutton and Boveri are famous for
   (a) mutation (b) chromosome theory of heredity
   (c) linkage (d) none of these
164. Which event is common to mitosis and meiosis?
   (a) separation of paired chromosomes (b) DNA synthesis
   (c) formation of diploid cells (d) none of these
165. Each chromosome becomes longitudinally double at
   (a) interphase (b) late metaphase (c) anaphase (d) none of these
166. Prokaryotic cell is divided
   (a) mitotically (b) amitotically (c) meiotically (d) none of these
167. Which process of cell division is of longer duration?
   (a) mitosis (b) amitosis (c) meiosis (d) none of these
168. What materials are synthesized in the G1 phase of cell division?
   (a) RNA (b) DNA (c) chromosomes (d) none of these
169. Centromere or kinetochore is
   (a) genetic material (b) pachytene
   (c) threads of chromosome (d) none of these

170. During which of the following stages does synapsis occur?

(a) zygotene (b) pachytene (c) leptotene (d) none of these

171. Cell lineage theory was proposed by

(a) Hofmister (b) Schultze (c) Altman (d) R. Virchow

172. Mitosis was discovered by

(a) Walter Fleming (b) Strasburger
(c) Alexander Fleming (d) Rudolf Virchow

173. The term "Mitosis" was coined by

(a) Strasburger (b) Farmer and Moore
(c) Walter (d) Rudolf Virchow

174. Mitosis occurs commonly in

(a) pollen grains (b) spore mother cells
(c) vegetative cells (d) xylem vessels

175. The best part of a plant to observe mitosis is

(a) anther (b) root apex (c) stem apex (d) cambium

176. DNA replication occurs in

(a) G1 phase (b) G2 phase (c) M-phase (d) S-phase

177. In meristematic cells, interphase occurs

(a) initially (b) terminally
(c) between successive divisions (d) permanently

178. With regard to biochemical activities, which of these phases is very active?

(a) anaphase (b) interphase (c) cytokinesis (d) metaphase

179. Anastral mitosis is characteristic of

(a) higher animals (b) lower animals (c) higher plants (d) all organisms

180. Spindle fibres are made up of

(a) cellulose (b) proteins (c) lipids (d) pectin

181. Equatorial plate refers to

(a) a fluid plate (b) the cell plate
(c) chromosomes arranged as plate (d) the phragmoplast

182. Centromeric division occurs during

(a) prophase (b) metaphase (c) anaphase (d) interphase

183. The shortest phase is
   (a) interphase (b) metaphase (c) anaphase (d) telophase
184. Daughter nuclei are reorganized in
   (a) interphase (b) prophase (c) metaphase (d) telophase
185. Phragmoplast is related to
   (a) cell elongation
   (b) cytokinesis
   (c) assemblage of chromosomes at metaphase plate
   (d) division of nucleus
186. Material of cell plate is synthesized by
   (a) phragmoplast (b) protoplast (c) Golgi complex (d) nucleus
187. Growth of cell plate is
   (a) centripetal (b) centrifugal (c) acentric (d) concentric
188. Which one is not possible due to mitosis?
   (a) wound healing (b) regeneration
   (c) grafting (d) reduction of chromosome number
189. The phase which is genetically most inactive is
   (a) anaphase (b) interphase (c) prophase (d) telophase
190. Despiralization takes place during
   (a) prophase (b) metaphase (c) anaphase (d) telophase
191. Reduction of chromosome number was first identified by
   (a) Strasburger (b) August Weismann
   (c) J.B. Farmer and J.E Moore (d) Walter Flemming
192. The term "meiosis" was coined by
   (a) Walter Flemming (b) Strasburger
   (c) J.B.Sumner and Moore (d) J.B. Farmer and J.E Moore
193. In plants, meiosis was discovered by
   (a) Farmer and Moore (b) Strasburger
   (c) Weismann (d) Virchow
194. Cell that undergoes meiosis is
   (a) meiocyte (b) endocyte (c) mitocyte (d) oocyte

195. Meiocyte in thallophytes (lower plants) is
   (a) a zygote (b) a spore mother cell
   (c) a gamete mother cell (d) an archesporial cell

196. In plants, meiosis occurs in
   (a) roots (b) leaves (c) stems (d) anthers and ovules

197. Best structure in plants to observe meiosis is
   (a) root apex (b) zygote (c) anther (d) ovule

198. The number of chromosomes is halved during
   (a) mitosis (b) meiosis (c) amitosis (d) endomitosis

199. Prolonged initial interphase is seen in
   (a) permanent cell (b) meristematic cell
   (c) meiocyte (d) zygote of higher plants

200. "Bouquet stage" in meiosis is seen at
   (a) leptotene (b) zygotene (c) pachytene (d) diplotene

201. Pairing of homologous chromosomes occurs in
   (a) leptotene (b) zygotene (c) pachytene (d) diplotene

202. Selective pairing of homologous chromosomes results in the formation of
   (a) bivalents (b) synapsis (c) chiasmata (d) heterokaryon

203. The two chromatids of a metaphase chromosome represent
   (a) replicated chromosomes to be separated at anaphase
   (b) homologous chromosomes of a diploid set
   (c) nonhomologous chromosomes joined at the centromere
   (d) maternal and paternal chromosomes joined at the centromere

204. Selective pairing of homologous chromosomes is called
   (a) proterminal synapsis (b) apoterminal synapsis
   (c) random synapsis (d) linkage

205. Exchange of chromatin between non-sister chromatids is called
   (a) linkage (b) crossing over (c) diakinesis (d) disjunction

206. Crossing over occurs during
   (a) leptotene (b) zygotene (c) pachytene (d) diplotene

207. Crossing over occurs at
    (a) two-strand stage
    (b) single strand stage
    (c) four-strand stage
    (d) strand stage
208. Enzymes that help in crossing over are
    (a) endonuclease and ligase
    (b) restriction endonuclease and ligase
    (c) endonuclease and exonuclease
    (d) polymerase and endonuclease
209. Crossing over results in
    (a) recombination of non-linked genes
    (b) recombination of linked genes
    (c) deletion of genes
    (d) addition of genes
210. Chiasmata is clearly visible in ______ stage.
    (a) pachytene (b) diplotene (c) diakinesis (d) anaphase-I
211. Displacement of intermediary chiasmata to poles is called
    (a) terminalization (b) translocation (c) synapsis (d) transcription
212. The best phase to count chromosomes is
    (a) metaphase (b) diakinesis (c) zygotene (d) anaphase-II
213. Bivalents are oriented at equatorial plane during
    (a) metaphase (b) metaphase-I (c) metaphase-II (d) diakinesis
214. Movement of homologous chromosomes of bivalent to opposite poles is called
    (a) disjunction (b) crossing over (c) terminalization (d) synapsis
215. Disjunction of homologous chromosomes occurs in
    (a) metaphase-I (b) diakinesis (c) anaphase-II (d) telophase-I
216. Genomes migrate to opposite poles by the end of
    (a) anaphase-I (b) telophase-I (c) diakinesis (d) cytokinesis
217. Daughter nuclei formed by the end of telophase I are
    (a) diploid (b) haploid (c) polyploid (d) ploidy not known
218. Total number of spindles formed in meiocyte is
    (a) 1 (b) 2 (c) 3 (d) 4
219. Main significance of meiosis is
    (a) maintenance of constant chromosome number
    (b) maintenance of genetic continuity and equality
    (c) recombination of genes
    (d) formation of gametes

220. Number of meiotic divisions necessary to form 50 pollen grains is
(a) 12 (b) 50 (c) 100 (d) 125

221. Number of meiotic divisions required to produce 100 seeds is
(a) 25 (b) 50 (c) 100 (d) 125

222. The sequence of cell cycle is
(a) S, G, G2, M (b) S, M, G, G2 (c) G1, S, G2, M (d) M, G, G2, S

223. Mitosis means
(a) division of nucleus only (b) division of cytoplasm only
(c) both nuclear and cytoplasmic division (d) chromosomes become chromatids

224. During prophase, which of the following is not possible?
(a) Prochromosomes become chromosomes.
(b) Nuclear envelope and nucleolus disappear.
(c) Each chromosome shows two chromatids.
(d) Chromosomes become chromatids.

225. The stage of mitosis in which chromosomes begin to separate and move to two opposite poles of a dividing cell is
(a) prophase (b) metaphase (c) anaphase (d) telophase

226. Nuclear envelope is
(a) reorganized from pieces of previous nuclear envelope
(b) synthesized newly by nucleolus
(c) reorganized from RER
(d) reorganized from Golgi complex

227. A barrel-shaped structure organized at the interzonal region by remaining spindle fibres at the end of telophase is
(a) tonoplast (b) phragmoplast (c) mitoplast (d) tyloses

228. Genetic integrity of the organism is maintained by
(a) mitosis (b) meiosis
(c) endomitosis (d) free nuclear division

229. The number of meiotic cycles occurring in a meiocyte is
(a) 1 (b) 2 (c) 3 (d) Infinite

230. Meiosis in plant occurs, when there is a change from
(a) gametophyte to sporophyte (b) sporophyte to sporophyte
(c) sporophyte to gametophyte (d) gametophyte to gametophyte

231. Chromosome replicates once and nucleus divides twice in

(a) mitosis (b) amitosis (c) meiosis (d) endomitosis

232. During first nuclear division of the meiocyte,

(a) the number of chromosomes remain unaltered
(b) the number of chromosomes is halved
(c) the number of chromosomes is doubled
(d) the number of chromosomes is tripled

233. The most lengthy phase during M phase of meiotic cycle is

(a) interphase (b) prophase-I (c) anaphase-I (d) leptotene

234. Repulsion of homologous chromosomes is initiated during

(a) pachytene (b) diplotene (c) diakinesis (d) anaphase-I

235. Chromosomes with both chromatids migrate to opposite poles during

(a) anaphase-I (b) anaphase-II (c) anaphase (d) telophase

236. The process which is antithecal to fertilization is

(a) mitosis (b) meiosis
(c) vegetative propagation (d) apomixis

237. The fact that the new cells arise from pre-existing cells by division was proposed by

(a) Strasburger (b) Walter Fleming
(c) Rudolf Virchow (d) Schleiden and Schwann

238. Mitosis in plants was discovered by

(a) Strasburger (b) W. Fleming (c) R. Virchow (d) Van Benden

239. A cell is about to divide. Before doing so, the amount of DNA in it

(a) halves (b) remains same (c) doubles (d) degenerates

240. The so-called resting phase is not really a resting phase. What is that phase?

(a) interphase (b) M-phase (c) S-phase (d) telophase

241. Changes occurring in this phase are almost reverse that of prophase.

(a) metaphase (b) interphase (c) M-phase (d) telophase

242. Main significance of mitosis is

(a) formation of two identical cells
(b) vegetative propagation
(c) wound healing
(d) maintenance of same chromosomal number in sexually reproducing organisms

243. After mitosis, the quantity of DNA in the nucleus of each daughter cell will be
   (a) half as much as in parent nucleus
   (b) same as that in parent nucleus
   (c) twice as much as in parent nucleus
   (d) highly variable

244. Cell division
   (a) involves coordination of various cellular elements
   (b) can occur at any time
   (c) is a highly coordinated process
   (d) is not a precise one

245. Daughter chromosomes appear first in
   (a) prophase (b) metaphase (c) anaphase (d) telophase

246. Every sexually reproducing organism starts its life cycle with
   (a) zygote (b) spore (c) embryo (d) gamete

247. Intermediary meiosis occurs in
   (a) algae (b) spore (c) embryo (d) gamete

248. In flowering plants, meiosis occurs at the time of
   (a) germination of seeds
   (b) formation of spores
   (c) fertilization
   (d) formation of gametes

249. Meiosis is characterized by
   (a) two nuclear divisions with chromosomes dividing twice
   (b) two nuclear divisions with chromosomes dividing once
   (c) two nuclear divisions with chromosomes dividing four times
   (d) one nuclear division with chromosomes dividing once

250. Beadlike structures that appear in leptotene chromosomes are called
   (a) nucleosomes (b) chromomeres (c) telomeres (d) knobs

251. The number of chromatin fibres in a leptotene chromosome is
   (a) 1 (b) 2 (c) 4 (d) many

252. Regions of crossing over are indicated morphologically by
   (a) chromomeres (b) chiasmata (c) centromeres (d) telomeres

253. The number of spindle apparati formed during the formation of one hundred microspores from microspore mother cells is
   (a) 25 (b) 75 (c) 50 (d) 100

254. The number of mitotic divisions required to produce 16 cells at the stem tip is
   (a) 4 (b) 8 (c) 15 (d) 16

255. How many generations of mitotic divisions occur to form 128 cells?
(a) 64 (b) 32 (c) 16 (d) 7

256. When a pea plant pollen mother cell undergoes meiotic division, what is the number of chromatids present in anaphase-I nucleus at one pole?
(a) 7 (b) 14 (c) 16 (d) 28

257. The number of meiotic divisions required to produce 100 pollen grains is
(a) 100 (b) 50 (c) 25 (d) 10

258. The number of meiotic divisions required to produce 100 seeds is
(a) 25 (b) 50 (c) 100 (d) 125

259. When a cell with 16 chromosomes undergoes meiosis, how many chromosomes are found at each metaphase plate in meiosis-II?
(a) 16 (b) 8 (c) 4 (d) 32

260. DNA replication occurs in
(a) G1 phase (b) G2 phase (c) S phase (d) M phase

261. Mitosis occurs in
(a) pollen grains (b) vegetative cells (c) embryo (d) ovule

262. The stage of mitosis during which nucleolus disappears is
(a) interphase (b) anaphase (c) metaphase (d) prophase

263. Spindle apparatus is formed during this stage of mitosis
(a) prophase (b) metaphase (c) anaphase (d) telophase

264. In mitosis, the chromosomes migrate towards the poles during
(a) metaphase (b) anaphase (c) telophase (d) interphase

265. In which phase of mitosis do the two chromatids of a chromosome separate from each other?
(a) anaphase (b) telophase (c) metaphase (d) interphase

266. Which of the following is called phragmoplast?
(a) cell plate (b) cell wall
(c) barrel-shaped spindle fibres (d) mother cell wall

267. Cell division is favoured by
(a) auxins (b) cytokinins (c) meiocytes (d) meristems

268. If a bacterial cell divides once in every minute, it takes one hour to fill a cup. How much time will it take to fill half of the cup?
(a) 30 min (b) 59 min (c) 58 min (d) 15 min

269. Meiosis can be studied in

(a) root tip (b) shoot apex (c) anther lobes (d) lateral meristem

270. Meiotic divisions are present in

(a) all diploid cells (b) spore mother cells
(c) somatic cells (d) sex cells

271. Meiosis takes place in

(a) apical meristem (b) intercalary meristem
(c) reproductive cells (d) vegetative cells

272. Meiosis in a plant takes place when there is a change

(a) from gametophyte to sporophyte (b) from sporophyte to gametophyte
(c) from gametophyte to gametophyte (d) from sporophyte to sporophyte

273. Meiosis is characterized by

(a) two successive divisions of cytoplasm and nuclei
(b) two successive divisions of cytoplasm and nuclei accompanied by only one replication of chromosomes
(c) one division of cytoplasm and nuclei accompanied by replication of chromosomes twice
(d) one division of cytoplasm and nuclei accompanied by replication of chromosomes once

274. The largest phase of meiosis is

(a) prophase-I (b) metaphase-I (c) anaphase-I (d) telophase-I

275. Which of the following occurs only during meiosis?

(a) pairing of homologous chromosomes (b) cytokinesis
(c) separation of duplicated strands (d) disappearance of nucleus

276. In meiosis, pairing of homologous chromosomes takes place during

(a) pachytene (b) diplotene (c) diakinesis (d) zygotene

277. Exchange of sections of chromatids during synapsing of chromosomes during meiosis-I is termed as

(a) synapsis (b) crossing over (c) disjunction (d) none of these

278. Homologous chromosomes exchange segments with blocks of genes during

(a) pachytene (b) zygotene (c) diplotene (d) leptotene

279. Crossing over of chromosomes during meiosis leads to

(a) mutation (b) sex determination
(c) new gene combination (d) extinction of species

280. Chiasma represents the site of
   (a) crossing over (b) synapsis (c) linkage (d) none of the above
281. In the first meiotic division, in which of the following stages does each chromosome undergo longitudinal division to give rise to two sister chromatids?
   (a) zygotene (b) diplotene (c) diakinesis (d) pachytene
282. As a result of linkage and crossing over,
   (a) chromatids become chromosomes
   (b) nuclear divisions come to a close
   (c) chromosomes become half in number
   (d) exchange of chromosomal material takes place
283. In meiosis the nuclear membrane breaks down, spindle is organized and centromeres become attached to spindle fibres during
   (a) diplotene (b) pachytene (c) diakinesis (d) zygotene
284. During meiosis the disjunction of chromosomes can be seen in
   (a) anaphase-I (b) anaphase-II (c) metaphase-I (d) metaphase-II
285. Mendelian recombinations are due to
   (a) crossing over (b) synapsis (c) disjunction (d) none of these
286. In meiosis, the number of chromosomes is reduced to half at the end of
   (a) metaphase-I (b) telophase-I (c) telophase-II (d) metaphase-II
287. During the first division of the pollen mother cell,
   (a) the number of chromosomes remains unaltered
   (b) the number of chromosomes is halved
   (c) the number of chromosomes is doubled
   (d) the number of chromosomes is tripled
288. Colchicine arrests any mitotically dividing cell at metaphase by
   (a) preventing the polymerization of tubulin into microtubules of the spindle
   (b) preventing the movement of spindle fibres as a result of which chromosomes fail to go to the opposite poles
   (c) destroying the contractile property of the spindle protein which eventually coagulates
   (d) dislodging chromosomes from the fibrous elements of the spindle and consequently the chromosomes remain confined to the equatorial plane
289. During meiotic cell division, genetic recombinations between two homologous chromosomes are facilitated by
   (a) movement of centromeres (b) extrusion of polar bodies
   (c) movement of centrioles (d) formation of synaptonemal complex

290. Duplication of chromosomes without division of the nucleus resulting in an increased number of chromosomes within the cell is referred to as

(a) endomitosis (b) amitosis (c) polyteny (d) mitosis

291. The process of penetration of cancer cells into vessels of the circulatory or lymphatic systems which allows them to move throughout the body, establishing secondary tumours is called

(a) oncogenesis (b) metastasis (c) lymphoma (d) hybridoma

292. Match List-I with List-II and select the correct answer using the codes given below the lists:

List-I	**List-II**
A. Pairing of homologous chromosomes	1. Chiasmata
B. Actual interchange of segments	2. Synaptonemal complex
C. Protein body formed between paired homologues	3. Synapsis
D. The cross-shaped configuration visible at diplotene between homologues	4. Crossing-over

Codes

	A	B	C	D
(a)	4	3	1	2
(b)	3	4	1	2
(c)	3	4	2	1
(d)	4	3	2	1

293. During cell division, mitochondria

(a) accumulate near the spindle and form rings around the myofibrillar bands in the muscle cell
(b) split into two equal halves in order to migrate to each daughter cell
(c) arrange themselves at the equator and twist to form coils
(d) migrate to terminal poles on long threads and split vertically

294. Before cell division, the entire DNA content of the cell gets doubled during interphase. This doubling takes place

(a) throughout the interphase
(b) at the beginning of interphase
(c) at the end of interphase
(d) somewhere during the middle of interphase

295. When a cell in G1 phase is successfully fused with another cell is S phase

(a) cleavage (just as in zygote cells) would start in the fused cell
(b) DNA replication (onset of S phase) would start in the G1 nucleus
(c) DNA replication would be stopped in the S phase nucleus
(d) each of the two nuclei would continue to remain in its original pre-fusion phase

296. What is the most important morphological peculiarity of polytene chromosomes of salivary cells of drosophila?
   (a) size
   (b) length
   (c) G-and P-Q-RS bands/interbands
   (d) easily observable features

297. Mitotic chromosomes
   (a) have high rate of protein synthesis
   (b) have high rate of DNA synthesis
   (c) show maximum chromosome condensation
   (d) have homologous recombination

298. The mitosis-promoting factor (MPF), which propels an animal cell into division stage, has two essential subunits: one is catalytic and the other is regulatory. Which one of the following is the regulatory subunit to MPF?
   (a) cyclin
   (b) CDK (cyclin-dependent kinase)
   (c) proline
   (d) ubiquitin

299. For the production of monoclonal antibody, which one of the following cell lines is used?
   (a) Chinese hamster ovary (CHO) cells
   (b) myeloma cells
   (c) HeLa cells
   (d) fibroblast cells

300. Colchicine influences
   (a) DNA replication
   (b) organization of spindle
   (c) chromosome condensation
   (d) chromosome division

301. Plane of cell wall formation in dividing cell is determined by
   (a) ER
   (b) microtubules
   (c) microfilaments
   (d) Golgi apparatus

302. Histone protein synthesis occurs during
   (a) G1 phase
   (b) G2 phase
   (c) S phase
   (d) prophase

303. Segregation of Mendelian factors (Aa) occurs during
   (a) diplotene
   (b) anaphase I
   (c) zygotene/pachytene
   (d) anaphase II

304. Disjunction is
   (a) chromosome aberration involving deletion
   (b) modification of gene action by nonallelic gene
   (c) separation of homologous chromosomes at anaphase
   (d) incompatibility of genes

305. Chromosome end is called
   (a) telomere
   (b) centromere
   (c) satellite
   (d) metamere

306. Poleward movement of dyads occurs during
    (a) anaphase (b) anaphase I (c) anaphase II (d) telophase
307. Exchange of paternal and maternal chromosome material during cell division is
    (a) dyad formation (b) bivalent formation
    (c) crossing over (d) synapsis
308. If mitotic division is restricted to G1 phase of cell, the condition is known as
    (a) G2 phase (b) S phase (c) G0 phase (d) M phase
309. Radioactive thymidine is fed to cells about to enter S phase. It will make radioactive
    (a) euchromatin (b) heterochromatin
    (c) neither euchromatin nor heterochromatin (d) both euchromatin and heterochromatin
310. Synthesis of macromolecules for multiplication of cell organelles occurs in
    (a) G1 phase (b) S phase (c) G2 phase (d) all
311. Who demonstrated meiosis as reduction division?
    (a) Weismann (b) Edward van Beneden
    (c) J.B. Farner and Moore (d) Strasburger and Winiwater
312. Stain used for study of cell division is
    (a) colchicine (b) Sudan black (c) acetocarmine (d) Congo red
313. Biologically active DNA was synthesized de novo by
    (a) Watson (b) Sinsheimer (c) Kornberg (d) Itabura
314. The stage between meiosis I and meiosis II is
    (a) interphase (b) interkinesis (c) I phase (d) diakinesis
315. Zygotic meiosis occurs in
    (a) *Pinus* (b) yeast (c) *Chlamydomonas* (d) *Dryopteris*
316. Plant cell lacks
    (a) centrioles (b) asters (c) spindle fibres (d) both (a) and (b)
317. Best stage to observe shape, size and number of chromosomes is
    (a) interphase (b) metaphase (c) prophase (d) telophase
318. Name the stage of mitosis in which chromosomes are arranged at the equator of
    (a) anaphase (b) metaphase (c) prophase (d) telophase
319. In mitosis chromosome duplication occurs during
    (a) interphase (b) prophase (c) late prophase (d) late telophase

320. Characteristic 'bouquet formation' occurs at which substage of meiosis?
(a) leptotene (b) zygotene (c) pachytene (d) diplotene

321. Synaptonemal complex (SC) was first observed by
(a) Flemming (b) Moses (c) Strasburger (d) none of these

322. In *Saccharomyces*, meiosis occurs in
(a) ascopore (b) ascus (c) zygote (d) pseudomycelium

323. PMC (pollen mother cell) is the best structure to show
(a) chromosome elongation (b) chiasmata
(c) cell fusion (d) centrosomes

324. Correct sequence of different substages of prophase I of meiosis is
(a) leptotene, zygotene, pachytene, diplotene, diakinesis
(b) zygotene, leptotene, diplotene, pachytene, diakinesis
(c) leptotene, pachytene, zygotene, diplotene, diakinesis
(d) leptotene, diplotene, pachytene, diakinesis, zygotene

325. In meiosis, the chromatids separate during
(a) anaphase I (b) metaphase II (c) anaphase II (d) metaphase II

326. When DNA and chromosome replicate without the division of nucleus, such a change is known as:
(a) endomitosis (b) allopolyploidy (c) linkage (d) crossing over

327. C-Mitosis is
(a) abnormal mitosis in colchicine-treated cells
(b) mitosis in collenchyma
(c) mitosis having circular spindle
(d) division of circular DNA

328. Chiasmata was discovered by
(a) Morgan in Drosophila (b) Beadle in *Neurospora*
(c) Janssen in amphibians (d) Hammerling in *Acetabularia*

329. The spindle microtubules are polar, the orientation is
(a) (+) ends towards the poles (b) (–) ends towards the poles
(c) (+) and (–) both at the equator (d) (+) and (–) both at the poles

330. How many times will mitotic division take place to produce 512 cells from a single parent cell?
(a) 9 (b) 256 (c) 158 (d) 150

331. The number of chromosome pairs at diakinesis in a meiocyte is 12, the chromosome number expected in each nucleus after first and second cycle of division could respectively be

(a) 12 and 6 (b) 12 and 12 (c) 6 and 12 (d) 24 and 12

332. RNA and proteins are synthesized in which phase of cell cycle?

(a) G1 phase (b) G2 phase (c) S phase (d) all of these

333. At what point do spindle fibres attach to chromosomes during cell division?

(a) kinetochore (b) chromonema (c) centrosome (d) none of these

334. The phase of cell cycle during which DNA polymerase is functionally active is

(a) G2 (b) S (c) M (d) G1

335. Choose the figure which correctly shows the anaphase-a stage?

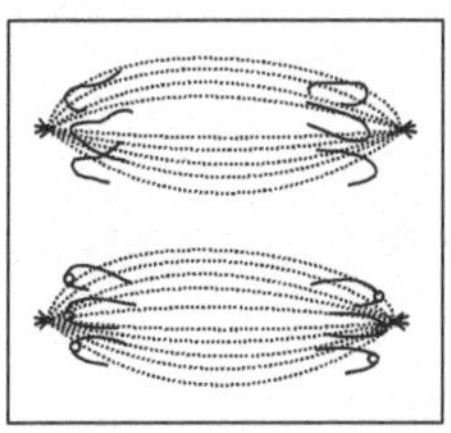
(a)

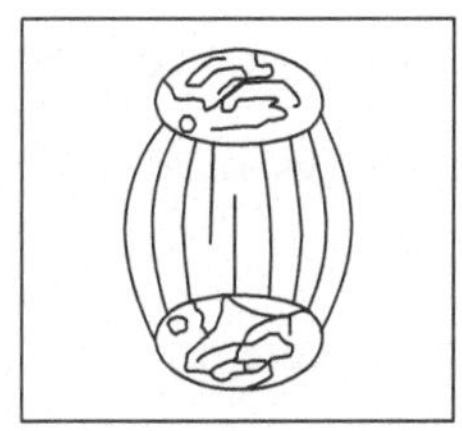
(b)

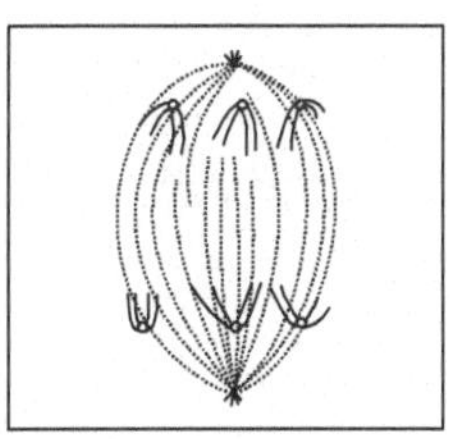
(c)

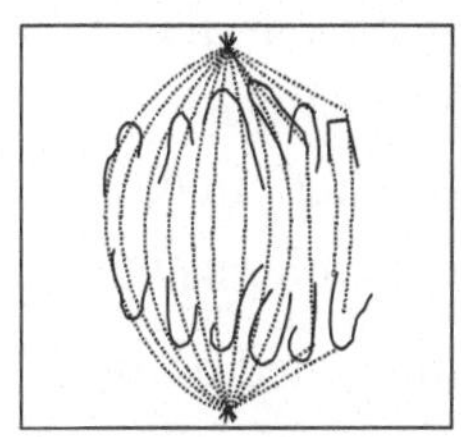
(d)

336. Which is the recent and most accurate technique for preparation of karyotype?

(a) banding technique (b) FISH
(c) McFISH (d) flow cytometry

337. The pairing of homologous chromosomes during zygotene is called

(a) synapse (b) terminalization (c) synapsis (d) crossing over

338. Temporary inactivation of one chromosome of a pair is

(a) facultative heterochromatin (b) constitutive heterochromatin
(c) inactivated chromatin (d) none of the above

339. Idioblast is
   (a) plant cell different from others
   (b) plant cell having cell inclusions
   (c) both (a) and (b)
   (d) animal cell different from others

340. If the centromere is terminal, the chromosome is
   (a) metacentric (b) submetacentric (c) telocentric (d) acrocentric

341. Chromosomes were first seen by
   (a) Hofmeister (b) Waldeyer (c) Walter S. Sutton (d) Crick and Watson

342. During staining, chromosomes get stained with
   (a) eosine (b) borax carmine (c) acetocarmine (d) safranin

343. Balbiani rings are
   (a) uncoiling of chromonemata
   (b) coiling of chromonemata
   (c) enlargements of centromere
   (d) none of these

344. Banding technique used in the case of plant chromosomes are
   (a) G and R (b) C and G (c) C and N (d) Q and G

345. A chromosome carrying centromere at one end is
   (a) acrocentric (b) telocentric (c) metacentric (d) submetacentric

346. Chromosomes in bacterial cell can be 1 to 3 in number and
   (a) are always circular
   (b) are always linear
   (c) can be either circular or linear but never both within the same cell
   (d) can be circular as well as linear within the same cell

347. Bead-like thickened portions of leptotene chromosomes are
   (a) puffs (b) chromosomes (c) centromeres (d) genes

348. Chromosomes can be counted best at the stage of
   (a) telophase (b) late anaphase (c) metaphase (d) late prophase

349. In meiosis I, the centromere undergoes
   (a) division between anaphase and interphase
   (b) division between prophase and metaphase
   (c) division, but the daughter chromosomes do not separate
   (d) no division

350. The number of chromatids at metaphase is
   (a) two each in mitosis and meiosis
   (b) two in mitosis and one in meiosis
   (c) two in mitosis and four in meiosis
   (d) one in mitosis and two in meiosis

351. Mitosis is absent in
   (a) zygote (b) germinal cell (c) bone cell (d) none of these
352. In meiosis, centromere divides during
   (a) prophase I (b) metaphase I (c) anaphase I (d) anaphase II
353. Colchicine interferes with cell division because of
   (a) splitting of chromosomes (b) nonpairing of chromosomes
   (c) double replication of chromosomes (d) no formation of spindle
354. Longest stage in M-phase of cell cycle is
   (a) prophase (b) metaphase (c) anaphase (d) telophase
355. During cell cycle, two molecules of DNA are present in chromosome during
   (a) G1 phase (b) beginning of S-phase
   (c) G2 phase (d) end of M-phase
356. Number of DNA strands present in chromosomes during G2 phase is
   (a) one (b) two (c) four (d) eight
357. In metaphase I, chromosomes are
   (a) in tetrad stage (b) in dyad stage
   (c) diploid (d) attracted to each other

ANSWERS

1. (c)	2. (a)	3. (b)	4. (a)	5. (c)	6. (a)	7. (d)
8. (a)	9. (a)	10. (a)	11. (b)	12. (a)	13. (b)	14. (c)
15. (d)	16. (c)	17. (a)	18. (b)	19. (a)	20. (c)	21. (d)
22. (a)	23. (c)	24. (b)	25. (d)	26. (d)	27. (a)	28. (d)
29. (b)	30. (b)	31. (b)	32. (c)	33. (a)	34. (a)	35. (b)
36. (a)	37. (c)	38. (a)	39. (c)	40. (b)	41. (b)	42. (a)
43. (a)	44. (c)	45. (a)	46. (b)	47. (a)	48. (b)	49. (d)
50. (a)	51. (b)	52. (a)	53. (d)	54. (a)	55. (b)	56. (c)
57. (a)	58. (b)	59. (d)	60. (a)	61. (a)	62. (a)	63. (a)
64. (a)	65. (d)	66. (a)	67. (a)	68. (a)	69. (a)	70. (a)
71. (c)	72. (d)	73. (a)	74. (c)	75. (c)	76. (d)	77. (a)
78. (d)	79. (c)	80. (b)	81. (a)	82. (b)	83. (a)	84. (a)
85. (a)	86. (c)	87. (a)	88. (a)	89. (d)	90. (d)	91. (a)
92. (b)	93. (a)	94. (b)	95. (b)	96. (b)	97. (a)	98. (d)
99. (d)	100. (d)	101. (a)	102. (c)	103. (a)	104. (a)	105. (d)

106. (a)	107. (a)	108. (a)	109. (a)	110. (c)	111. (c)	112. (a)
113. (b)	114. (c)	115. (d)	116. (d)	117. (a)	118. (b)	119. (a)
120. (c)	121. (b)	122. (a)	123. (c)	124. (b)	125. (d)	126. (b)
127. (a)	128. (c)	129. (a)	130. (b)	131. (a)	132. (a)	133. (b)
134. (d)	135. (b)	136. (c)	137. (d)	138. (a)	139. (b)	140. (a)
141. (a)	142. (b)	143. (c)	144. (b)	145. (b)	146. (c)	147. (a)
148. (a)	149. (c)	150. (a)	151. (a)	152. (a)	153. (d)	154. (c)
155. (a)	156. (c)	157. (b)	158. (b)	159. (b)	160. (b)	161. (a)
162. (b)	163. (b)	164. (b)	165. (b)	166. (b)	167. (c)	168. (b)
169. (b)	170. (a)	171. (d)	172. (a)	173. (c)	174. (c)	175. (b)
176. (d)	177. (c)	178. (b)	179. (c)	180. (b)	181. (c)	182. (c)
183. (c)	184. (d)	185. (b)	186. (c)	187. (b)	188. (d)	189. (c)
190. (d)	191. (b)	192. (d)	193. (b)	194. (a)	195. (a)	196. (d)
197. (c)	198. (b)	199. (c)	200. (a)	201. (b)	202. (a)	203. (a)
204. (c)	205. (b)	206. (c)	207. (c)	208. (a)	209. (b)	210. (b)
211. (a)	212. (b)	213. (b)	214. (a)	215. (c)	216. (a)	217. (b)
218. (c)	219. (a)	220. (b)	221. (d)	222. (c)	223. (c)	224. (d)
225. (c)	226. (c)	227. (b)	228. (a)	229. (a)	230. (c)	231. (c)
232. (b)	233. (b)	234. (b)	235. (a)	236. (c)	237. (c)	238. (a)
239. (c)	240. (a)	241. (d)	242. (a)	243. (b)	244. (c)	245. (c)
246. (a)	247. (c)	248. (b)	249. (b)	250. (b)	251. (b)	252. (b)
253. (b)	254. (c)	255. (d)	256. (b)	257. (c)	258. (d)	259. (b)
260. (c)	261. (b)	262. (d)	263. (b)	264. (b)	265. (a)	266. (c)
267. (b)	268. (b)	269. (c)	270. (b)	271. (c)	272. (b)	273. (b)
274. (a)	275. (a)	276. (d)	277. (b)	278. (a)	279. (c)	280. (a)
281. (d)	282. (d)	283. (c)	284. (a)	285. (c)	286. (b)	287. (b)
288. (a)	289. (d)	290. (c)	291. (b)	292. (c)	293. (a)	294. (d)
295. (b)	296. (c)	297. (c)	298. (a)	299. (b)	300. (b)	301. (b)
302. (a)	303. (b)	304. (c)	305. (a)	306. (b)	307. (c)	308. (c)
309. (d)	310. (c)	311. (b)	312. (c)	313. (c)	314. (b)	315. (c)
316. (d)	317. (b)	318. (b)	319. (a)	320. (a)	321. (b)	322. (d)
323. (a)	324. (a)	325. (a)	326. (a)	327. (a)	328. (c)	329. (a)
330. (b)	331. (b)	332. (d)	333. (a)	334. (b)	335. (c)	336. (c)
337. (c)	338. (a)	339. (c)	340. (c)	341. (a)	342. (c)	343. (a)
344. (c)	345. (b)	346. (a)	347. (b)	348. (c)	349. (d)	350. (a)
351. (b)	352. (d)	353. (d)	354. (a)	355. (c)	356. (b)	357. (a)

15

SIGNAL TRANSDUCTION

CHOOSE THE CORRECT ANSWER

1. Glyconeogenesis is regulated by
   (a) cortisol (b) GH
   (c) adrenalin and glycogen (d) all of these
2. Glycogenolysis is the process of conversion of
   (a) glycogen to glucose (b) glucose to glycogen
   (c) glycogen to glucose 6-phosphate (d) none of these
3. α-adrenergic signals of hypothalamic hypophysial unit suppresses
   (a) PSH (b) LH (c) PRL (d) GH release
4. Parathyroid gland acts through α -adrenergic receptors through which of the following?
   (a) cGMP (b) Ca^+ (c) cAMP (d) Na–K ATPase
5. Which of the following is/are trophic hormone(s)?
   (a) LH and FSH (b) TSH, ACTH, LH
   (c) TSH, ACTH, FSH, LH (d) all of these
6. Gr RH induces the synthesis and secretion of
   (a) FSH only (b) LH only (c) PRL only (d) FSH and LH
7. ACTH under the influence of corticotrophin-releasing hormone stimulates the synthesis of
   (a) cortisol (b) glucocorticoids
   (c) mineralocorticoids (d) cortisol and androgens

8. Feedback regulation is found for which of the following hormones?
   (a) TSH (b) ACTS (c) FSH and LH (d) all of these

9. What is the sequence of estrogen synthesis?
   (a) acetate ⟶ cholesterol ⟶ pregnenolone ⟶ progesterone ⟶ androgens
   (b) acetate ⟶ cholesterol ⟶ progesterone ⟶ estrogens
   (c) cholesterol ⟶ pregnenolone ⟶ progesterone ⟶ androgens ⟶ estrogens
   (d) none of these

10. PSH, LH, hCG and TSH contain
    (a) α-chain (b) β-chain
    (c) π- chain (d) both α and β chain

11. Hormone ⟶ receptor ⟶ membrane enzymes ⟶ cAMP ⟶ phosphorylase.
    In the above sequence which one is the most potential?
    (a) cAMP (b) membrane enzyme
    (c) hormone (d) receptor

12. Galactorrhea is a symptom caused by the deficiency of
    (a) FSH (b) LH (c) GH (d) PRL

13. $\text{ATP} \xrightarrow[\text{cyclase}]{\text{Adenylate}} \text{cAMP} \xrightarrow[\text{kinase}]{\text{Phosphodiester}} \text{AMP}$ Which one is more responsible for cell physiology?
    (a) ATP (b) cAMP (c) AMP (d) ATP + cAMP

14. Pituitary adenohypophysis contains which of the following hormone groups?
    (a) MSH, ACTH, FSH, LH, TSH, HCG
    (b) PRL, GH, HCG, TSH, FSH, LH, ACTH
    (c) MSH, FSH, LH, TSH, HCG
    (d) ACTH, FSH, LH, GH, MSH

15. Which of the groups of hormones are glycoprotein in nature?
    (a) ACTH, FSH., LH, PRL (b) LH, HCG, PRL, TSH
    (c) FSH, LH, TSH, HCG (d) all of them

16. The site for action of LH is
    (a) ovary (b) follicular cells (c) corpus luteum (d) egg

17. FSH is got by increasing intracellular cAMP in
    (a) testicular cells (b) ovarian cells
    (c) thecal cells (d) theca and granulosa cells

18. Which hormone is an indicator for pregnancy?
    (a) FSH in blood (b) LH in blood (c) hCG in blood (d) hCG in urine
19. Ovulation in vertebrates is induced by the injection of
    (a) gonadotropin hormones (b) FSH
    (c) LH (d) hCG
20. Which of the following hormones is responsible for the secretion of FSH and LH?
    (a) progesterone (b) testosterone
    (c) estrogen (d) estrogen + androgen
21. Myxoedema is a symptom caused due to
    (a) hypothyrodism (b) hyperthyrodism (c) hypo GH (d) hypo G + H
22. Secondary sexual characteristic are regulated by
    (a) pituitary hormones
    (b) testis or ovarian hormones
    (c) adrenal hormones
    (d) cumulative effects of all of these endocrine hormones
23. Cushing's syndrome is the abnormality of
    (a) adrenal gland tumour (b) pituitary tumour
    (c) thyroid tumour (d) pituitary adenohypophysis tumour
24. Autoimmune thyroiditis disorder is also known as
    (a) goitre (b) Grave's disease
    (c) Hashimoto disease (d) metabolic disorder
25. Which of the following group of hormones does adrenal gland secrete?
    (a) ACTH, MSH, cortisol, estrogens
    (b) adrenalin, noradrenalin, gluco-minerallo-corticoids
    (c) epinephrine, norepinephrine, glucocorticoids, minerallocorticoids, estrogens and androgen
    (d) none of these
26. Klinefelter's syndrome is caused by
    (a) pituitary hormones (b) gonadotropic hormones
    (c) thyroid hormones (d) FSH and LH
27. Hilar cells are located in
    (a) testes (b) ovary (c) thyroid gland (d) adrenal gland
28. FSH induces the follicular growth by secretion of
    (a) estrogens (b) testosterone (c) androgen (d) estradiol-17β

29. LH level is highest in the ovary during
    (a) follicular growth
    (b) oocyte maturation
    (c) ovulation
    (d) pregnancy
30. Which are the hormones that regulate lactation?
    (a) oxytocin
    (b) PRL
    (c) oxytocin + PRL
    (d) oxytocin + vasopressin + PRL
31. Pancreas secrete hormones, which are
    (a) insulin from β cells, glucagon from α cells
    (b) insulin from β cells, glucagon from β cells, somatostatin from α cells
    (c) glucagon from α cells, insulin from β cells and somatostatin
    (d) all of these
32. The most important effect of pheromones is on
    (a) behavioural pattern
    (b) reproductive pattern
    (c) co-ordination of behaviour or physiology
    (d) socialism
33. Most of the pheromones are secreted by
    (a) endocrine glands
    (b) mesocrine gland
    (c) glandular cells
    (d) saliva, sweating and urine
34. Tendrils coil around the substratum due to
    (a) geotropism (b) hydrotropism (c) thigmotropism (d) both (b) and (c)
35. Reproductive pheromones are responsible for
    (a) mimicry (b) colouration (c) alarming (d) copulation
36. Auxanometer is needed for
    (a) study of plant movement
    (b) measurement of plant growth
    (c) geotropic study of roots
    (d) calculation of plant nutrition in relation to growth
37. If total growth is plotted against time, the curve obtained is
    (a) S-shaped (b) parabolic (c) T-shaped (d) U-shaped
38. When growth is going on at a slow rate in the beginning, this phase is called
    (a) log phase
    (b) lag phase
    (c) senescence phase
    (d) linear phase
39. Leaf fall occurs when content of
    (a) gibberellin decreases
    (b) auxin decreases
    (c) auxin increases
    (d) abscisic acid decreases

40. Apical dominance is caused by
    (a) GA (b) IAA (c) ABA (d) florigen
41. Which of the following is a precursor of IAA synthesis in plants?
    (a) tyrosine (b) acetyl-CoA (c) tryptophan (d) phenylalanine
42. The growth regulator which delays ageing of plant organs is
    (a) auxin (b) cytokinin (c) gibberellin (d) 2,4-D
43. Gibberellins were first isolated from
    (a) an alga (b) a fungus (c) an angiosperm (d) bacteria
44. Bolting of genetic dwarfs is caused by
    (a) gibberellins (b) traumatin (c) ABA (d) cytokinins
45. Which of the following shows hormonal activity and is used as a weedicide?
    (a) zenatin (b) florigen (c) 2,4,5-t (d) ethylene
46. Photoreceptor pigment in plants for photomorphogenetic reactions is
    (a) phytochrome (b) kinetin (c) florigen (d) vernalin
47. IAA was isolated by
    (a) J.C. Bose (b) F.W. Went (c) Kogl *et al.* (d) Skoog and Miller
48. In the absence of light, increase in which of these takes place?
    (a) uptake of minerals (b) uptake of water
    (c) elongation of internodes (d) ascent of sap
49. Herbicide DCMU kills small weeds by
    (a) destroying chlorophyll (b) inhibiting PS-II
    (c) inhibiting PS-I (d) blocking protein synthesis
50. The gaseous fruit ripening plant growth regulator is
    (a) ethylene (b) kinetin (c) ABA (d) gibberellin
51. A detached leaf can be made to remain green for long by keeping it in
    (a) gibberellins (b) cytokinins (c) auxins (d) ABA
52. Which is known as " stress hormone"?
    (a) IAA (b) GA (c) zeatin (d) ABA
53. Phytochrome is involved in
    (a) flowering (b) seed germination
    (c) organogenesis (d) all of the above

54. The highest concentration of auxin exists in
   (a) leaves (b) growing tips (c) petioles (d) lower part of plant organs

55. Bananas can be prevented from over-ripening by keeping them
   (a) at room temperature (b) in refrigerator
   (c) in ABA (d) in ascorbic acid solution

56. Physiologically active phytochrome is
   (a) P_{730} (b) P_{660} (c) P_{700} (d) P_{680}

57. Effect of duration of light on the growth and flowering of plants is called
   (a) vernalization (b) photoperiodism
   (c) morphogenesis (d) photomorphogenesis

58. Abscisic acid controls
   (a) stem elongation (b) cell expansion and cell wall plasticity
   (c) cell division (d) leaf fall and dormancy

59. Which of the following is a phytohormone acting like growth inhibitor?
   (a) IA (b) NAA (c) ethylene (d) gibberellic acid-3

60. Induction of flowering by treating with low temperature is called
   (a) photoperiodism (b) temperature induction
   (c) vernalization (d) enhanced flowering

61. Zinc is essential for
   (a) auxin biosynthesis (b) gibberellin synthesis
   (c) chlorophyll synthesis (d) beta oxidation of fat

62. When the dark period (night) is interrupted by a short exposure to light, a short-day plant like sugar cane will
   (a) flower immediately resulting in loss of stem girth
   (b) not flower resulting in increase in stem girth
   (c) give more vigorous flowering resulting in decrease in sugar product
   (d) behave like a long-day plant and flower after a brief period

63. Sugar cane will accumulate more sugar if it is provided with
   (a) 2,4,5-T (b) 2,4-D (c) NAA (d) IBA

64. The hormone which can change the sex of a unisexual plant is
   (a) cytokinin (b) ABA (c) ethylene (d) auxin

65. The hormone which can substitute low-temperature treatment is
   (a) auxin (b) gibberellin (c) cytokinin (d) abscisic acid

66. The hormone which promotes maleness in flowering plants is
    (a) GA (b) IAA (c) kinetin (d) ABA
67. Pollen tube growth in the style of the carpel is
    (a) thigmotropic (b) chemotactic (c) chemonastic (d) chemotropic
68. An example of movement of variation is shown by
    (a) *Desmodium gyrans* (b) *Drosera*
    (c) *Mimosa pudica* (d) all of the above
69. Secondary roots are
    (a) positively geotropic (b) plagiogeotropic
    (c) plagiophototropic (d) ageotropic
70. Sprouting of potato can be prevented by
    (a) gibberellins (b) IAA
    (c) cytokinins (d) maleic hydrazide and IBA or NAA
71. Premature falling of leaves and fruits can be prevented by the application of
    (a) auxins (b) gibberellins (c) cytokinins (d) all of the above
72. Photoperiodism is supposed to be due to the synthesis of
    (a) cytokinins (b) florigen (c) gibberellins (d) auxins
73. Gibberellins are known to induce flowering more effectively in the case of
    (a) long-day plants (b) short-day plants
    (c) day neutral plants (d) all plants
74. Which of the following is an established long-day plant?
    (a) soybean (b) tobacco (c) wheat (d) xanthium
75. Which of the following is a bioassay of IAA?
    (a) avena curvature test (b) callus test
    (c) tobacco pith test (d) meristematic growth
76. The phase of growth where new cells are produced by cell divisions is called
    (a) formative phase (b) protoplasmic phase
    (c) phase of differentiation (d) phase of elongation
77. When grasses are mown, it causes
    (a) excessive lateral growth
    (b) control of excessive growth
    (c) control of flowering and resulting in vegetative growth
    (d) none of these

78. The universal natural auxin of plants is
    (a) ABA (b) IAA (c) NAA (d) IBA
79. In nurseries, the rooting hormone used is
    (a) IAA (b) 2,4-D (c) benzoic acid (d) IBA
80. Movement of male gametes towards archegonia in bryophytes and pteridophytes is an example of
    (a) chemotropism (b) chemonasty
    (c) chemotactic movement (d) autonomic movement of locomotion
81. Nyctinasty is
    (a) opening and closing of flowers due to differential growth on the two sides of the sepals and petals induced by temperature or light
    (b) closing up of leaves or petals during evening due to loss of turgor in cells
    (c) autonomic sleeping movement in leaves of some plants
    (d) movement of variation induced by shock or touch
82. IAA was first isolated from
    (a) corn germ oil (b) gibberella (c) human urine (d) *Rhizopus*
83. Which phytochrome overcomes dwarfism?
    (a) auxins (b) GA (c) CK (d) both (b) and (c)
84. Chemotropic movement is shown by
    (a) male gametes of bryophytes (b) male gametes of pteridophytes
    (c) pollen tube (d) all of the above
85. Cocklebur is
    (a) SDP (b) LDP (c) SLDP (d) LSDP
86. Went demonstrated that when the tip of *Avena* coleoptile was removed,
    (a) coleoptile growth stopped (b) coleoptile bends towards source of light
    (c) coleoptile bends away from source of light (d) coleoptile grows lengthwise
87. Under following condition

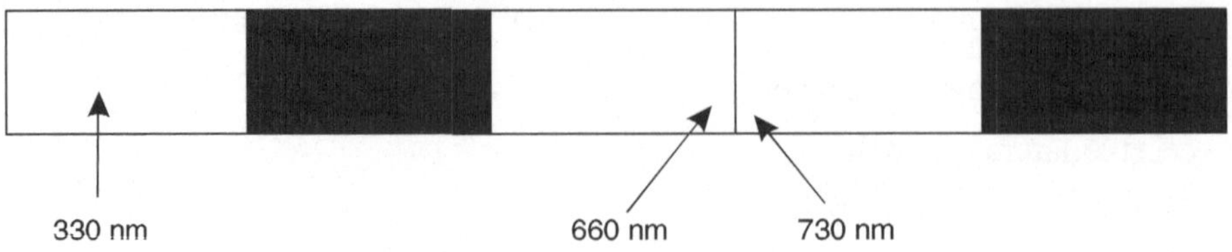

(a) LDP will flower but SDP will show vegetative growth
(b) SDP will flower but LDP will show vegetative growth
(c) both SDP and LDP will flower
(d) both SDP and LDP will show only vegetative growth

88. Which hormone was isolated and crystallized from milky endosperm of corn by Letham?
(a) zeatin (b) kinetin (c) heteroauxin (d) auxin-b

89. In *Gingko biloba*, seed dormancy is due to
(a) germination inhibitors (b) immature embryo
(c) very hard seed coat (d) seed coat being impermeable to water

90. By exogenous application of which hormone can biennials be induced to behave as annuals?
(a) NAA (b) IBA (c) IPA (d) GA

91. Scarification is a method for
(a) inducing flowering in a plant (b) breaking seed dormancy
(c) devernalization of plant (d) none of these

92. Progressive senescence of older leaves and lateral organs which may occur at any time of the year is called
(a) sequential senescence (b) simultaneous senescence
(c) synchronous senescence (d) both (b) and (c)

93. Which auxin is used to prevent the sprouting of potato tuber?
(a) NAA (b) IBA
(c) 2, 4-D (d) 2, 3, 6-trichlorobenzoic acid

94. Which of these statements is not correct?
(a) The exogenous supply of auxins increases formation of male flowers.
(b) Plants having excurrent habit have greater apical dominance.
(c) Auxins show polar transport.
(d) Foliar spray with NAA causes flowering in litchi.

95. 'Pomalin', particularly effective in increasing apple size, is a combination of
(a) auxins and GA (b) auxins and CK (c) CK and GA (d) ethylene

96. Who was the first to declare ethylene as a phytohormone?
(a) Crocker (b) Miller (c) Addicot (d) Letham

97. Which hormone promotes transverse growth and inhibits longitudinal growth?
(a) IPA (b) kinetin (c) ethylene (d) ABA

98. Thigmonasty is
    (a) paratonic movement of growth
    (b) paratonic movement of variation
    (c) paratonic movement of locomotion
    (d) autonomic movement of growth

99. In Acacia, the leaflets spread horizontally during the day and fold together and droop down during night. This type of movement is called
    (a) phototactic
    (b) phototropic
    (c) nyctinasty
    (d) autonomic movement

100. During mutations
    (a) the stem exhibits a kind of nodding movement in two directions
    (b) more growth occurs on the upper side of sepals and petals
    (c) less growth occurs on lower side of sepals and petals
    (d) cytoplasm moves around a vacuole in a cell

101. The plant organs (secondary roots) which are placed obliquely to the force of gravity are called?
    (a) ortinogeotropic (b) plagiotropic (c) diageotropic (d) apogeotropic

102. Which of the following is a biological activity of cytokinin?
    (a) expansion of excised radish cotyledons
    (b) delay in senescence
    (c) root growth inhibition test
    (d) more shooting

103. Which of the following phase has constantly dividing cells and is restricted to the apical meristem?
    (a) formative phase
    (b) phase of elongation
    (c) phase of maturation
    (d) phase of differentiation

104. Which temperature coagulates and damages the protoplasm and hinders growth?
    (a) 20–25°C (b) 28–30°C (c) 35–40°C (d) above 45°C

105. Natural CK 'Zeatin' was isolated from
    (a) autoclaved herring sperm DNA
    (b) corn kernels
    (c) coconut milk
    (d) both (b) and (c)

106. Find out the incorrect match.
    (a) Cosmos—SDP
    (b) Sugar beet—LDP
    (c) Tomato—Day Neutral plant
    (d) Cucumber—SDP

107. Precursor of IAA is
    (a) methionine (b) tryptophan (c) mevalonic acid (d) leucine

108. Which light is favourable for seed germination?
    (a) red (b) far red (c) blue (d) yellow

109. Which of the following is a negative photoblastic seed?
   (a) tobacco (b) onion (c) lettuce (d) mistletoe

110. Who proposed the existence of the hypothetical substance, vernalin?
   (a) Hendricks (b) Borthwick (c) Melchers (d) Chailakhyan

111. Accumulation of which form of phytochrome inhibits flowering in short-day plants?
   (a) P_r (b) P_{fr} (c) P_{660} (d) both (a) and (b)

112. Which hormone prevents premature fruit drop in apples?
   (a) IAA (b) GA_3 (c) IPA (d) CK

113. Epigeal germination is shown by
   (a) castor (b) mango (c) pea (d) gram

114. In which plant does reproduction occur in the second season?
   (a) wheat (b) radish (c) convolvulus (d) fragaria

115. Which is not a natural plant hormone?
   (a) GA_2 (b) GA_3 (c) IAA (d) 2,4-D

116. The optimum temperature for seed germination is
   (a) 10–15°C (b) 15–30°C (c) 30–40°C (d) 40–42°C

117. Which hormone is used for defoliation of forest trees?
   (a) TIBA (b) phosphon-D (c) AMO-618 (d) 2,4-D

118. Which of the following is a photomorphogenetic process?
   (a) seed germination (b) flowering
   (c) cholorophyll formation (d) all of the above

119. Abscisic acid causes closure of stomata because it causes
   (a) rapid movement of water in the guard cells
   (b) slow movement of water out of the guard cells
   (c) rapid movement of K^+ out of the guard cells
   (d) rapid movement of K^+ into the guard cells

120. The movement of auxins is
   (a) polar and basipetal (b) polar and acropetal
   (c) non-polar (d) not fixed

121. Tentacles of *Drosera* bend in response to the stimulus of touch by insects. This type of movement is called
   (a) thigmotropic (b) haptotropic (c) thigmonastic (d) nyctinasty

122. Zinc is required for the synthesis of
   (a) auxin (b) GA_3 (c) CK (d) GA_7
123. Reversal of apical dominance and a bushy growing habit of plant is due to
   (a) deficiency of calcium (b) excess of auxin in shoot
   (c) deficiency of potassium (d) low concentration of cytokinin
124. A point in the growth curve at which the exponential phase ends is said to be
   (a) inflexion (b) decreation phase
   (c) stationary phase (d) senescent phase
125. A hormone that has direct implication on water economy in plants is
   (a) auxin (b) gibberellin (c) ethylene (d) abscisic acid
126. A plant hormone that promotes phloem transport is
   (a) auxin (b) gibberellin (c) cytokinin (d) ethylene
127. A hormone that changes a plant's monopodial growing habit to bushy habit is
   (a) cytokinin (b) auxin (c) ABA (d) gibberellin
128. Which of the following is not a cytokinin?
   (a) benzoic acid (b) benzimidazole
   (c) benzyl adenine (d) 6-fufuryl aminopurine
129. Latex extraction from rubber plant is enchanced by use of a hormone
   (a) auxin (b) gibberellin (c) cytokinin (d) ethylene
130. Physiologically active form of phytochrome is the one that
   (a) absorbs red light (b) absorbs far red light
   (c) absorbs blue light (d) both (a) and (b)
131. Fruit traders use a chemical to prevent 'in transit' ripening of bananas. This chemical is
   (a) cytokinin (b) ethylene (c) abscisic acid (d) ascorbic acid
132. The phytohormone that acts as antheridiogen factor in ferns is
   (a) IPA (b) IAA (c) GA (d) TIBA
133. A growth pigment that regulates flowering in most plants is
   (a) phytochrome (b) florigen (c) auxin (d) cytokinin
134. Action spectrum of chloroplast exhibits a parallelism to absorption spectrum of
   (a) chlorophyll *a* (b) chlorophyll *b*
   (c) both (a) and (b) (d) chlorophylls and carotenoids

135. The specific role of growth hormones at cellular level is in
(a) enhanced release of Ca^{++} by ER
(b) enhanced intake of Ca^{++} through plasma membrane
(c) enhanced elasticity of cell wall
(d) all of the above

136. Photonastic movement is observed in
(a) nymphaea (b) silene (c) tulip (d) victoria

137. Find the odd one (w.r.t. photoperiodic response).
(a) four O' clock plant (b) spinach
(c) chrysanthemum (d) sugar beet

138. Photoperiodic stimulus in plants is perceived by
(a) stem apex (b) root (c) leaf (d) flowers

139. Tomato seeds show vivipary in their second generation due to
(a) ABA (b) ferulic acid (c) ethylene (d) absence of oxygen

140. Find the odd one (w.r.t. germination)
(a) pea (b) maize (c) *Sonneratia* (d) *Vicia faba*

141. *Mimosa* is very quick to respond to a shock stimulus. How much time does it take to start a reaction?
(a) 1 minute (b) 1 second (c) 1/10 second (d) 1/1000 second

142. Chemotropic movement is shown by
(a) pollen tube (b) male gametes of bryophytes
(c) coiling of young leaves (d) drooping of *Mimosa* leaves

143. Wheat plant shows
(a) whole plant senescence (b) sequential senescence
(c) simultaneous senescence (d) synchronous senescence

144. Which type of movement is shown by the insectivorous plant, Drosera, during capturing of prey?
(a) haptonastic (b) haptotropic (c) thigmotropic (d) diageotropic

145. Which of the following reduces transpiration by reducing stomatal opening for a period of few weeks without affecting other metabolic processes?
(a) PMA (b) waxes (c) silicon emulsions (d) CCC

146. Unicellular alga, *Chlamydomonas,* shows
(a) autonomic movement of locomotion (b) paratonic movement of locomotion
(c) both (a) and (b) above (d) movement of variation

147. A method of breaking dormancy, in which moist seeds are subjected for variable periods to low or high temperature in the presence of oxygen, is called
   (a) scarification (b) impaction (c) stratification (d) any of these

148. Vernalization was first studied in Europe on ______ varieties of ______.
   (a) summer, grasses (b) winter, cereals
   (c) summer, fruits (d) winter, vegetables

149. Nutation is autonomic growth movement during which
   (a) more growth occurs on one side and slow growth on the other side
   (b) more growth occurs at the base of thalamus
   (c) stem shows spiral coiling around a support
   (d) root grows towards source of water

150. Which of the following is a photomorphogenetic process?
   (a) flowering (b) chlorophyll formation
   (c) seed germination (d) all of the above

151. A short-day plant will flower under which condition?

(a) 

(b) 

(c) 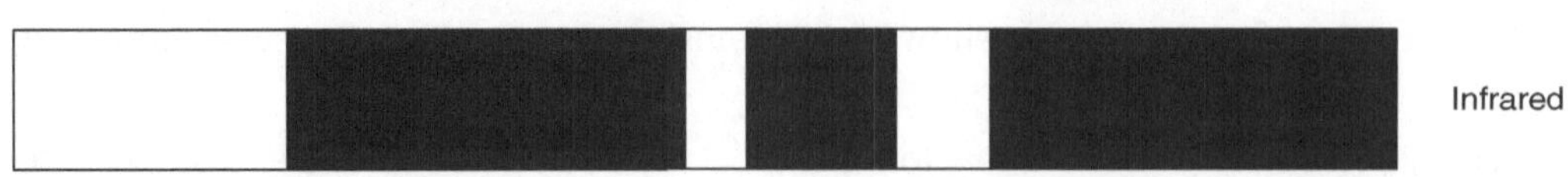

(d) 

152. The following diagram is showing

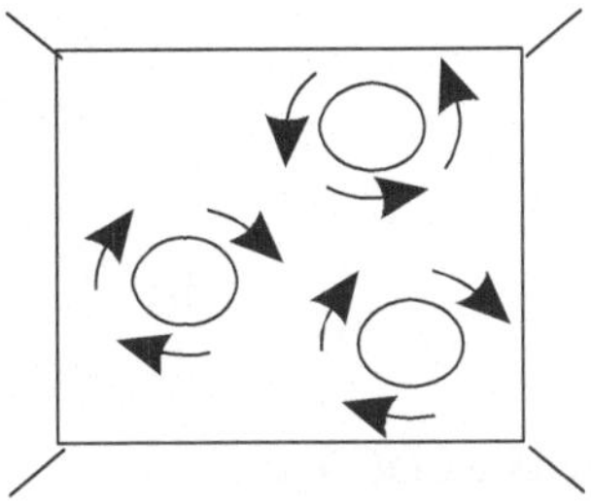

(a) heliotropism
(b) circulation
(c) rotation
(d) paratonic movement of locomotion

153. Went demonstrated that when the tip of oat coleoptile was removed,

(a) coleoptile bends towards source of light
(b) coleoptile bends away from source of light
(c) coleoptile grows lengthwise
(d) growth is stopped

154. Flowering hormone florigen is produced in ______ and is translocated through ______

(a) floral bud, xylem
(b) foral bud, pholoem
(c) leaves, pholem
(d) vegetative bud, intercellular spaces

155. Which of the following plant flowers after a period of vegetative growth, regardless of the photoperiod?

(a) radish
(b) tomato
(c) barley
(d) rice

156. Match the following

1. Kinetin	i. Gibbane skeleton
2. GA_3	ii. Root hair intiation
3. Ethylene	iii. Purine derivative
4. ABA	iv. Parthenocarpy in Rosa

(a) 1-iv, 2-iii, 3-ii, 4-i
(b) 1-iii, 2-iv, 3-ii, 4-i
(c) 1-iii, 2-i, 3-ii, 4-iv
(d) 1-ii, 2-i, 3-iv, 4-iii

157. If a mica plate is inserted on the shaded side of a coleoptile tip, what will happen?

(a) coleoptile tip will bend towards light
(b) coleoptile tip will show nutation
(c) it results in circum nutation
(d) no curvature will be formed

158. Ethylene and MH have antagonistic function towards

(a) promoting flowering in SDP
(b) increase in shelf life of cut vegetables
(c) promoting apical dominance
(d) breaking dormancy of storage organs

159. A photoblastic seed is given following treatment. In which of these cases will germination of seeds be maximum?

(a) seeds + R (b) seeds + R + Fr
(c) seeds + R + Fr + R (d) seeds + R + Fr + R + Fr

160. Seed dormancy in pulses is due to

(a) immature embryo (b) hard seed coat
(c) germination inhibitors (d) dormant embryo

161. Which of the following hormone is isolated and crystallized from milky endosperm of corn by Letham?

(a) kinetin (b) zeatin (c) gibberellin A_3 (d) dormin

162. What is correct for simultaneous senescence?

(a) It is progressive senescence of older leaves which may occur at any time of the year.
(b) It occurs in all annuals.
(c) It occurs in monocarpic plants.
(d) It is controlled by environmental factors rather than internal factors.

163. Naphthalene acetic acid is used for

(a) inducing rooting in cutting (b) inducing bud dormancy
(c) substitution of cold treatment (d) increasing shelf life of cut flowers

164. Which of the following is correct for phytochrome conversion reaction?

(a) 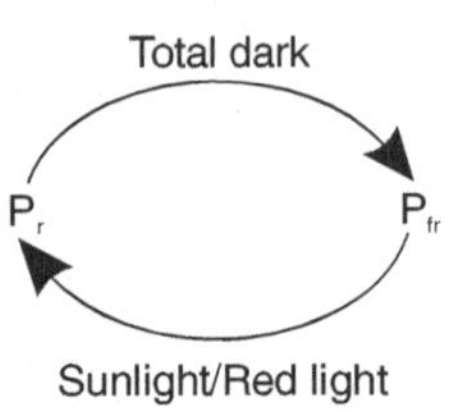

(b) 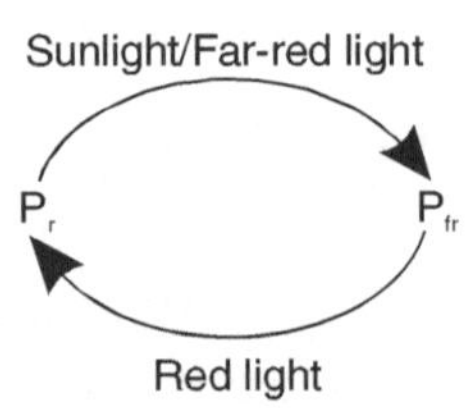

(c) 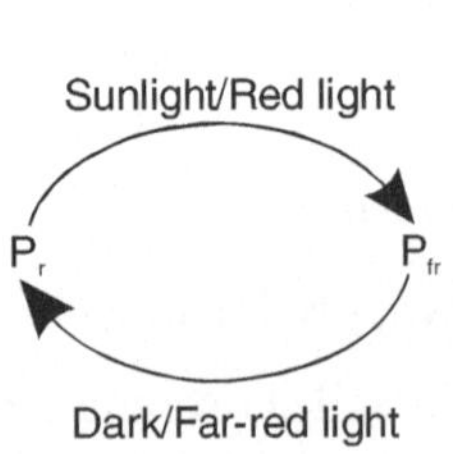

(d) 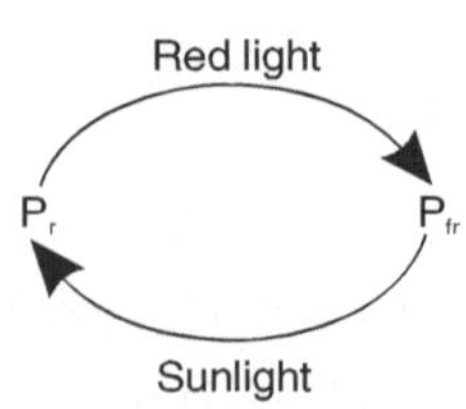

165. Abscisic acid causes closure of stomata because it causes

(a) rapid movement of water in the guard cells
(b) slow movement of water out of the guard cells
(c) rapid movement of K^+ out of the guard cells
(d) slow movement of K^+ out of the guard cells

190. TIBA is a/an
   (a) synthetic auxin (b) natural auxin
   (c) antiauxin (d) antigibberellin

191. By exogenous application of which hormone can biennials be induced to behave as annuals?
   (a) auxins (b) gibberellins (c) cytokinins (d) ethylene

192. Which kind of geotropic behaviour is shown by lateral branches of stem?
   (a) +ve orthogeotropic (b) –ve orthogeotropic
   (c) plageotropic (d) apogeotropic

193. According to statolith theory, the gravitational stimulus is perceived by
   (a) auxins (b) large starch grains
   (c) small starch grains (d) CK

194. Vernalization stimulus is perceived by
   (a) meristems (b) embryos
   (c) leaves (d) none of these

195. *Glycine max* is a/an
   (a) SDP (b) LDP (c) day neutral plant (d) SLDP

196. Match the following

1. Indole-3-acetic acid	i. Antigibberellin
2. 6-furfuryl aminopurine	ii. Natural auxin
3. NPA	iii. Synthetic gibberellin
4. AMO-1618	iv. Antiauxin
	v. Synthetic cytokinin

   (a) 1-v, 2-iv, 3-ii, 4-iii (b) 1-ii, 2-v, 3-iv, 4-i
   (c) 1-ii, 2-v, 3-i, 4-iv (d) 1-ii, 2-i, 3-iii, 4-v

197. Bud dormancy can be broken by
   (a) alternating low and high temperature (b) mechanical scarification
   (c) chemical scarification (d) light

198. Match the following

1. GA + Anthesins	i. True florigen
2. 2, 4-D + 2, 4, 5-T	ii. Agent orange
3. GA + CK	iii. Pomalin
4. ABA + kinetin	iv. Germination controller

   (a) 1-ii, 2-iii, 3-i, 4-iv (b) 1-i, 2-ii, 3-iii, 4-iv
   (c) 1-iv, 2-iii, 3-ii, 4-i (d) 1-iv, 2-i, 3-iii, 4-ii

199. Natural auxins are synthesized from
   (a) methionine (b) glycine (c) lysine (d) tryptophan

200. Exposure of light inhibits the process of germination in the seeds of which the following plants?
   (a) tobacco (b) capsella (c) allium (d) lactuca

201. Match the following

1. Ethylene	i. Pigment
2. Cytokinin	ii. Growth inhibitor
3. Phytochrome	iii. Flowering hormone
4. Florigen	iv. Phytohormone

   (a) 1-iv, 2-iii, 3-ii, 4-i (b) 1-iii, 2-iv, 3-ii, 4-i
   (c) 1-ii, 2-iv, 3-i, 4-iii (d) 1-ii, 2-i, 3-iii, 4-iv

202. Flowers open due to
   (a) environmental changes (b) epinasty
   (c) seismonasty (d) increase in turgor pressure

203. A biennial plant can be made annual by treating the plant with a plant hormone called
   (a) auxin (b) zeatin (c) 6-FAP (d) GA

204. A long-day plant can flower under short-day condition if it is
   (a) allowed to possess a certain number of leaves
   (b) allowed to attain a particular height
   (c) treated with florigen
   (d) treated with GA

205. Winter wheat is a/an
   (a) annual plant (b) polycarpic plant (c) C_4 plant (d) biennial plant

206. Gibberellins were first isolated from a plant source, i.e., immature seeds of
   (a) *Gibberella fujikuroi* (b) *Phaseolus coccineus*
   (c) *Avena sativa* (d) *Zea mays*

207. Phenomenon of phototropism is verified by
   (a) Korper–kappe theory (b) Went–cholodny theory
   (c) avena curvature test (d) apical cell theory

208. Triple response is a bioassay for
   (a) zeatin (b) ethylene (c) auxin (d) abscisic acid

209. Richmond Lang effect is associated with the growth hormone called
   (a) cytokinin (b) gibberellic acid
   (c) ethylene (d) indole 3-acetic acid

210. Opening of flowers of Crocus and Tulip is an example of
   (a) nyctinastic movements
   (b) thermonastic movement
   (c) photonastic movements
   (d) haptonastic movement

211. The day-neutral plant among the following is
   (a) *Cestrum mocturum*
   (b) *Secale cereale*
   (c) *Cucumis sativus*
   (d) *Xanthium pensylvanicum*

212. Critical photoperiod for Henbane plant is
   (a) 15.45 h (b) 11 h (c) 10 h (d) 12 h

213. On which of the following plants has no work been done in plant physiology?
   (a) *Phalaris canariensis*
   (b) *Avena sativa*
   (c) *Saccharum officinarum*
   (d) *Solanum tuberosum*

214. Pick the odd one out.
   (a) NAA (b) 2,4-D (c) TIBA (d) PAA

215. Ethylene was recognized as a growth regulator by
   (a) Addicot and Lynch
   (b) Gallston and Davis
   (c) Slatyr and Taylor
   (d) Skoog and Miller

216. High concentration of cytokinin in nutrient media induced ______ in callus mass.
   (a) adventitious root (b) shoot buds (c) gall formation (d) tumour formation

217. Phototropism phenomenon can be demonstrated using
   (a) clinostat
   (b) luxmeter
   (c) heliotropic chamber
   (d) photocalorimeter

218. Rotation and circulation are a kind of
   (a) autonomic movements
   (b) diurnal rhythm
   (c) paratonic movements
   (d) circadian rhythm

219. It is more appropriate to use the term ______ for a sensitive plant's drooping of leaves.
   (a) thigmotropism (b) haptonastic (c) hyponasty (d) seismotropic

220. Most pronounced avena curvature was observed under the effect of
   (a) auxanolonic acid (b) auxintriolic acid (c) heteroauxin (d) none of these

221. Opening of flowers is called
   (a) anthology (b) anthesis (c) horticulure (d) olericulture

222. Graviception is a property exhibited by
   (a) sand crystals (b) statoliths (c) drupers (d) none of these

223. Flowering in some biennial plants can be induced by
    (a) etiolation (b) bolting (c) cauliflory (d) pruning

224. The hormone required for flower formation is
    (a) vernalin (b) anthesin (c) florigen (d) phytochrome

225. Traumatropism occurs in ______ of plants.
    (a) growing regions (b) senescing regions
    (c) wounded regions (d) zone of maturation

226. Which of the following is a bioassay for thylene?
    (a) pea curvature test (b) triple pea test
    (c) chlorophyll retention test (d) barley endosperm test

227. Wound hormone is
    (a) jasmonic acid (b) traumatic acid (c) morphactins (d) lunularic acid

228. Why is chilling treatment necessary to accelerate flowering in biennial plants?
    (a) to produce florigen for flowering
    (b) to defunct inhibitors which inhibit flowering
    (c) to activate meristematic activity
    (d) to activate enzymes to promote flowering

229. Florigen was discovered by
    (a) Cajlachjan (b) Butler (c) Melchers (d) Garner

230. Prosopis is known for being
    (a) a phreatophyte (b) a drought-enduring xerophyte
    (c) a drought-resisting xerophyte (d) both (a) and (b)

231. Which of the following, added to a medium, will support not only growth of tobacco callus but also induce the formation of roots on it?
    (a) kinetin (b) indole acetic acid (c) kinetin + IAA (d) kinetin + IAA

232. What does a higher carbohydrate/nitrogen ratio do to a plant?
    (a) less vegetative growth (b) inhibits flowering
    (c) promotes flowering and fruiting (d) both (a) and (c)

233. Select the incorrect match
    (a) auxin—stimulates sprouting of potato (b) gibberellin—promotes flowering in LDP
    (c) ethylene—promotes lateral growth (d) cytokinin—prevents senescence

234. The movement of the peduncle of sunflower during daytime is an example of
    (a) phototaxis (b) photonasty (c) phototropism (d) both (b) and (c)

235. Growth of plants differs from growth of animals in being
    (a) localized (b) diffused
    (c) localized and diffused (d) limited

236. Growth in a root by observing the actual markings was first studied in India by
    (a) Sachs (b) Strasburger (c) Went (d) Paal

237. The estabilishment of localized qualitative differences in biochemicals, metabolic processes and structural organization in cells is known as
    (a) growth (b) log phase of growth
    (c) differentiation (d) development

238. The growing zones of root and shoot are respectively
    (a) 5 cm and 5–30 cm (b) 4.5 cm and 4–20 cm
    (c) 2 cm and 3–18.5 cm (d) 1 cm and 2–30 cm

239. Auxanometer is required for
    (a) studying rate of transpiration (b) measuring rate of respiration
    (c) finding out rate of photosynthesis (d) calculating rate of growth in length

240. The phase of growth where new cells are formed by the division of meristematic cells is known as
    (a) formative phase (b) phase of elongation
    (c) phase of differentiation (d) phase of mitosis

241. The rate of respiration of a cell in formative phase is
    (a) higher (b) lower
    (c) equal to differentiating cells (d) equal to elongating cells

242. The classical experiments in growth were performed by
    (a) Darwin and Lamarck (b) Lamarck and Boysen-Jensen
    (c) Boysen-Jensen and Darwin (d) de Vries and Paal

243. Enlargement of cells in the phase of elongation is in ______ direction.
    (a) linear (b) longitudinal
    (c) nodal (d) any

244. Cell enlargement is due to
    (a) florigen (b) traumatin (c) gibberellin (d) auxin

245. The phase of growth where synthesis of more of protoplasm, formation of a central vacuole and addition and extension of wall material takes place is known as
    (a) phase of cell formation (b) phase of elongation
    (c) phase of differentiation (d) primary growth

246. The differentiation which continues through all phases of growth is known as
   (a) structural differentiation (b) physiological differentiation
   (c) growth differentiation (d) total differentiation
247. The differentiation which is restricted to the later stages of growth of a cell is known as
   (a) physiological differentiation (b) structural differentiation
   (c) later differentiation (d) post differentiation
248. The respiratory activity in "cell maturation" phase as compared to other phases is
   (a) equal (b) less
   (c) more (d) dependent upon temperature
249. Maximum growth in root occurs
   (a) at its apex (b) in the root hair zone
   (c) behind the apex (d) in the presence of soil
250. When growth is going on at a slow rate, this phase is called as
   (a) lag period (b) log period
   (c) period of diminishing growth (d) linear phase of growth
251. In plants, optimum growth occurs in
   (a) blue light (b) red light (c) white light (d) green light
252. Maximum rate of growth is achieved during
   (a) lag period (b) log period (c) steady state (d) senescent phase
253. Cell differentiation is accompanied during period of
   (a) lag phase (b) log phase
   (c) diminishing growth (d) senescence
254. The period of steady state is never reached in the organs of
   (a) determinate structure (b) indeterminate structure
   (c) reproductive structure (d) all of these
255. When the rate of maximum growth is maintained for sometime, it is known as
   (a) J-shaped phase of growth (b) linear phase of growth
   (c) S-shaped phase of growth (d) all of these
256. The total period of growth is termed as
   (a) total period of growth (b) grand period of growth
   (c) graceful period of growth (d) magnified period of growth
257. Who studied the growth systematically for the first time?
   (a) Krebs (b) Sachs (c) Lysenko (d) Blackman

258. If total growth is plotted against time the curve obtained is called
   (a) S-shaped curve (b) G-shaped curve
   (c) T-shaped curve (d) U-shaped curve

259. Double sigmoid curve is found in the growth of
   (a) fruits (b) flowers
   (c) inflorescence (d) all organs showing growth

260. In plant, light is required for
   (a) photosynthesis (b) controlling distribution of hormones
   (c) controlling growth and movements (d) all of these

261. Maximum growth rate recorded in the stipe equivalent to 5 mm/min has been recorded in
   (a) *Dictyota dichotoma* (b) *Dendrocalamus giganteous*
   (c) *Dicanthium* (d) *Dictyophora phalloidea*

262. Two seeds are germinated, A in light and B in dark. Which of these is correct with reference to growth in height of plant?
   (a) A > B (b) B > A
   (c) A = B (d) No growth in either case

263. Plastids of etiolated plants possess
   (a) phycobilins (b) carotenoids and xanthophyll
   (c) chlorophylloid and carotenoids (d) chlorophyll and carotenes

264. According to Blackman (1919) the magnitude of relative growth has been named
   (a) efficiency index (b) growth index
   (c) grand period of growth (d) phase of differentiation

265. Mobilization of stored food in germinating seeds is triggered by
   (a) IAA (b) GA (c) MH (d) florigen

266. Viability of seed can be tested by
   (a) tetrazolium dye (b) germination test
   (c) both (a) and (b) (d) none of these

267. A condition when seed fails to germinate because of absence of favourable environmental conditions is called
   (a) dormancy (b) quiescence (c) viability (d) abscission

268. The seeds which are not influenced by the presence or absence of light are called
   (a) photoblastic (b) nonphotoblastic
   (c) light hard seeds (d) light-sensitive seeds

269. The seeds which are affected by the presence of light at the time of germination are called
   (a) non-photoblastic (b) photoblastic
   (c) light-sensitive seeds (d) none of these

270. The seeds of tomato cannot germinate in the presence of light and hence are called
   (a) negatively photoblastic (b) nonphotoblastic
   (c) photoblastic (d) light-sensitive seeds

271. Microbiotics are the seeds which can remain viable for
   (a) less than a month (b) less than a week
   (c) less than a year (d) less than three years

272. Mesobiotic seeds are those which can remain viable for
   (a) a month (b) a year (c) two years (d) 3–15 years

273. The seeds which can remain viable for more than 15 years are called
   (a) microbiotic (b) mesobiotic (c) macrobiotic (d) none of these

274. The germination where hypocotyl grows actively bringing the seed above the soil is called
   (a) epigeal (b) hypogeal (c) semi-epigeal (d) semi-hypogeal

275. Germination is favourable during
   (a) rabi (b) kharif (c) in all seasons (d) none of these

276. The process where the embryo of the seed continues to grow while the latter is attached to the parent plant is called
   (a) epigeal germination (b) hypogeal germination
   (c) vivipary (d) none of these

277. Vivipary is known to be present in
   (a) rhizophora (b) grasses (c) bamboos (d) hydrophytes

278. Evergreen trees remain green throughout the year on account of
   (a) absence of leaf fall (b) leaves falling in small number at intervals
   (c) supply of moisture throughout the year (d) cold climate

279. Correct sequence of cellular growth stage is
   (a) division–differentiation–elongation (b) division–elongation–differentiation
   (c) differentiation–division–elongation (d) elongation–differentiation–division

280. A sigmoid curve is obtained for the growth of
   (a) an organ (b) an organism (c) a population (d) all of these

281. Excess amount of CO_2

(a) retards growth
(b) accelerates growth
(c) affects growth slowly
(d) does not affect growth

282. Legume seeds exhibit dormancy due to

(a) undeveloped embryo
(b) absence of cytokinins
(c) hard seed coat
(d) absence of GA_3

283. A rhythmic phenomenon that occurs every 24 hours is a

(a) circadian rhythum
(b) biological clock
(c) photoperiodism
(d) daily phase cycle

284. If the tip of a seedling is cut off, growth as well as bending ceases because it hampers

(a) respiration
(b) photosynthesis
(c) perception of light stimulus
(d) transpiration

285. What does a higher C/N ratio do in a plant?

(a) improves vegetative growth
(b) inhibits flowering and fruiting
(c) promotes flowering
(d) both (a) and (b)

286. In an old leaf preparing to fall, the colour change is due to

(a) conversion of chloroplasts to leucoplasts
(b) chromoplasts being converted to chloroplasts
(c) chloroplasts being converted to chromoplasts
(d) chromoplasts being converted to elioplast

287. During leaf fall, the conversion of chlorophyll into chromoplasts is due to

(a) degradation of carotenoids
(b) synthesis of carotenoids
(c) degradation of chlorophyll and dominance of carotenoids
(d) predominance of prochlorophyll over carotenoids

288. The direct and dominating factor of growth is

(a) light (b) temperature (c) soil (d) wind

289. Crescograph, an instrument devised by J.C. Bose, was used to

(a) investigate sleep movement of bean leaves
(b) study the role of phytoclosure in seed germination
(c) record the progress of plant growth accurately
(d) distinguish between SDP and LDP

290. Auxanometer is used to measure
   (a) primary growth
   (b) secondary growth
   (c) growth of leaves
   (d) effect of environment on growth

291. Tomato seeds fail to germinate due to a germination inhibitor in the pulp of fruit. It is
   (a) abscisic acid (b) ferulic acid (c) coumarin (d) thiourea

292. Plant growth is
   (a) limited (b) unlimited (c) unlocalized (d) diffused

293. Phytotron is a device by which
   (a) electrons are bombarded
   (b) protons are liberated
   (c) plants are grown in controlled environment
   (d) mutations are produced in plants

294. A green plant will turn yellow if placed in the dark. This is called
   (a) phototropism (b) dormancy (c) chlorosis (d) etiolation

295. The law of compound interest for growth does not justify one of the following factors.
   (a) Growth is dependent upon the intial size.
   (b) Growth is dependent upon the rate of growth.
   (c) Growth is dependent upon the time interval.
   (d) Growth is dependent upon the location of meristem.

296. By which method is testa made permeable to water?
   (a) stratification (b) denudation (c) vernalization (d) scarification

297. The activity of α-amylase in the endosperm of barley germinating seeds is induced by
   (a) ethylene (b) cytokinin (c) IAA (d) gibberellin

298. The usual shape of the growth curve is
   (a) linear
   (b) inverted bell-shaped
   (c) sigmoidal
   (d) zig-zag

299. Which light is necessary for the germination of photoblastic seeds?
   (a) red (b) green (c) far-red (d) blue

300. Auxanometer is used for measuring
   (a) primary growth (b) secondary growth (c) transpiration (d) phototropism

301. Movement of auxin is largely
   (a) acropetal (b) basipetal (c) both (a) and (b) (d) centripetal

302. The hormone responsible for apical dominance is
   (a) IAA (b) GA (c) ABA (d) florigen

303. If a growing plant is decapitated, then
   (a) axillary buds are inactivated (b) axillary buds are activated
   (c) leaves become yellow and fall off (d) growth stops

304. Growth substance that stimulates nodule formation in leguminous plants is
   (a) NAA (b) IAA (c) IBA (d) ABA

305. Avena curvature test was first done by
   (a) J.C. Bose (b) F.W. Went (c) Thimann (d) Skoog

306. The phenomenon of apical dominance can be overcome by exogenous application of
   (a) auxins (b) gibberellins (c) cytokinis (d) ethylene

307. Bakanae disease is associated with discovery of
   (a) 2, 4-D (b) IAA (c) GA (d) ABA

308. Bolting does not require
   (a) short days (b) long days
   (c) internode elongation in rosette plants (d) cold nights
   (e) gibberellin application

309. Gibberellin was first extracted from a species of the fungus
   (a) *Fusarium* (b) *Gelidim* (c) *Gracelaria* (d) *Aspergillus*

310. Genetic dwarfness can be controlled by treating plants with
   (a) cytokinin (b) gibberellic acid (c) auxin (d) antigibberellin

311. The hormone that breaks dormancy of potato tuber is
   (a) IAA (b) ABA (c) zeatin (d) gibberellin

312. The hormone primarily connected with cell division is
   (a) IAA (b) NAA (c) cytokinin/zeatin (d) gibberellic acid

313. Which of the following is a cytokinin?
   (a) phytochrome (b) leucine (c) ethylene (d) zeatin

314. Cytokinin synthesis is maximum in
   (a) roots (b) leaves (c) shoot tip (d) fruit

315. The gaseous fruit ripening phytohormone is
   (a) ethylene (b) kinetin (c) GA (d) ABA

316. Which is not a phytohormone?
   (a) IAA (b) GA (c) phytochrome (d) florigen

317. Ethylene gas
(a) is saturated hydrocarbon
(b) slows down ripening of apples
(c) retards ripening of tomatoes
(d) speeds up maturation of fruits

318. The hormone responsible for ageing is
(a) GA
(b) IAA
(c) ABA
(d) cytokinin

319. Which is a stress hormone preventing seeds during drought?
(a) benzyl aminopurine
(b) dichlorophenoxy acetic acid
(c) ethylene
(d) abscisic acid

320. Hormones differ from enzymes in that they are
(a) found in both living and dead cells
(b) found only in plants
(c) used up in the metabolism
(d) not used up in the process

321. The existence of growth-regulating chemicals in plants was first suggested by
(a) Darwin
(b) Went
(c) Sachs
(d) Paal

322. A hormone used to induce root formation in horticulture is
(a) IBA
(b) NAA
(c) ABA
(d) 2, 4-D

323. Parthenocarpic fruits can be produced by the application of which of the following auxins?
(a) indole acetic acid (IAA)
(b) indole butyric acid (IBA)
(c) 2-4 dichlorophenoxyacetic acid (2,4-D)
(d) all of the above

324. What happens on the removal of the apical bud in a plant
(a) makes the plant bushy
(b) availability of more auxins to lateral buds
(c) formation of lateral branches
(d) all of the above

325. A substance formed at one place and that expresses its effect at a distant place is called as
(a) hormone
(b) enzyme
(c) pheromone
(d) vitamin

326. High concentration of auxin is found in
(a) root apex
(b) shoot apex
(c) petiole base
(d) node

327. Differentiation of shoot is controlled by
(a) high gibberellin : auxin ratio
(b) high gibberellin : cytokinin ratio
(c) high auxin : cytokinin ratio
(d) high cytokinin : auxin ratio

328. The deficiency of the element zinc in plants affects the biosynthesis of
(a) ethylene
(b) abscisic acid
(c) auxin
(d) cytokinin

329. A hormone promoting flowering in LDP and also controling sex expression is
(a) cytokinins
(b) auxins
(c) ethylene
(d) gibberellins

330. Sometimes, sprouting of potato tubers occurs in storage. It is due to
   (a) cytokinins (b) auxins (c) gibberellins (d) coumarins

331. Which hormones help in breaking the dormancy of seed?
   I. ABA II. CK III. GA IV. IAA
   (a) I, III, II (b) I, II, IV (c) II, III, IV (d) I, II, III

332. Bakanal disease in Japan was due to a fungus known as
   (a) *Gibberella fujikori* (b) *Fusarium moniliforme*
   (c) Both (a) and (b) (d) *Plasmopara viticola*

333. Apical dominance is not affected by
   (a) IAA (b) gibberellins
   (c) IBA (d) indole acetaldehyde

334. Seedless water melons are obtained by
   (a) triploidy (b) GA treatment
   (c) haploidy (d) vegetative propagation

335. Which of the following prevents premature fall of fruit?
   (a) NAA (b) ethylene (c) GA_3 (d) zeatin

336. Cytokinin synthesis is maximum in
   (a) roots (b) leaves (c) fruits (d) stems

337. From which of the following was kinetin isolated?
   (a) coconut milk (b) pineapple (c) soybean (d) groundnut

338. Which of the following types of phytohormones resembles the nucleic acids in some structural aspects?
   (a) cytokinin (b) auxins (c) gibberellins (d) abscisic acid

339. Coconut milk factor/hormone found in fluid endosperm of coconut is
   (a) abscisic acid (b) cytokinin (c) an auxin (d) a gibberellin

340. A plant hormone used for inducing morphogenesis in tissue culture is
   (a) abscisic acid (b) gibberellins (c) cytokinins (d) ethylene

341. Male hormone in plant is
   (a) auxin (IAA) (b) gibberellin (GA) (c) cytokinin (CK) (d) ethylene

342. During seed germination the enzyme diastase converts
   (a) starch to glycogen (b) starch to simple sugars
   (c) starch to maltose (d) reserve food to sugars and fats

343. Which of the following types of hormones stimulates synthesis of α-amylase in the endosperm of geminating seeds of barley?
(a) IAA (b) ABA (c) GA (d) ETH (ethylene)

344. Water stress, closure of stomata and seed dormancy are related to
(a) auxin (b) cytokinin (c) gibberellin (d) abscisin

345. Which one among the following chemicals is used for causing defoliation of forest trees and as weedicide?
(a) AMO 1618 (b) phosphon D (c) malic hydrazide (d) 2, 4-D

346. Which combination is used in fruit ripening?
(a) 80% CH_4 + 20% O_2 (b) 80% C_2H_4 + 20% CO_2
(c) 80% CO_2 + 20% CH_4 (d) 80% C_2H_4 + 20% O_2

347. What is agent orange?
(a) a hazardous chemical used in luminous paints
(b) colour used in fluorescent lamps
(c) a weedicide containing dioxin
(d) a biodegradable insecticide

348. Which of the following cannot be isolated from plants?
(a) vitamin B_{12} (b) vitamin C (c) niacin (d) riboflavin

349. The pigment responsible for photomorphogenetic response is
(a) phytochrome (b) chlorophyll (c) cytochrome (d) xanthophyll

350. What is true about phytochrome?
(a) P_r absorbs red light and becomes P_{fr} (b) P_r abosorbs yellow light and becomes P_{fr}
(c) P_{fr} abosorbs yellow light and becomes P_r (d) P_{fr} absorbs red light and becomes P_r

351. Phytochrome is found in
(a) alagae/thallophytes (b) vascular cryptogams
(c) fungi (d) flowering plants/angiosperms

352. Pick up the correct explanation
(a) xanthium—long-day plant (b) sunflower—short-day plant
(c) wheat—short-day plant (d) tomato—day-neutral plant

353. Flowering in short-day plants is induced by
(a) short days and interrupted long nights (b) short days and uninterrupted long nights
(c) short nights (d) long days with interrupted nights

354. In short-day plants, flowering is inhibited by
   (a) interruption of dark by white or red light
   (b) dark interruption by far-red light
   (c) dark interruption by red light followed by far-red light
   (d) none of these

355. Phytochrome is involved in picking stimulus in
   (a) phototropism (b) photorespiration (c) photoperiodism (d) photosynthesis

356. Phytochrome in plants is sensitive to
   (a) green light (b) blue light
   (c) red light (d) red and far-red light

357. What will be the effect on phytochrome in a plant subjected to continuous red light?
   (a) increased synthesis of phytochrome
   (b) level of phytochrome decreases
   (c) phytochrome is destroyed
   (d) destruction and synthesis of phytochrome remains in equilibrium

358. The pigment that absorbs red and far-red light in plants is
   (a) cytochrome (b) xanthophyll (c) phytochrome (d) carotene

359. Which one prevents photo-oxidation and pigment destruction?
   (a) phytochrome (b) phytohormone (c) phycocyanin (d) phycoerythrin

360. Plants respond to photoperiods due to the presence of
   (a) phytohormones (b) phytochrome (c) enzymes (d) stomata

361. The term phytochrome was used by
   (a) Garner and Allard (b) Borthwick and Bendricks
   (c) Borthwick (d) Moore

362. Phytochrome becomes biologically active in
   (a) green light (b) blue light (c) red light (d) none of these

363. Which of the following statements about phytochrome is true?
   (a) P_r absorbs yellow light and becomes P_{fr} (b) P_r absorbs red light and becomes P_{fr}
   (c) P_{fr} absorbs red light and becomes P_r (d) P_{fr} absorbs yellow light and becomes P_r

364. An LDP is
   (a) tobacco (b) *Glycine max* (soyabean)
   (c) spinach (d) *Mirabilis* (4 O'clock plant)

365. If a tree flowers thrice a year in October, January, and July in northern India, the plant is
   (a) photo- and thermo-sensitive
   (b) photo- and thermo-insensitive
   (c) photo-sensitive and thermo-insensitive
   (d) thermosensitive and photo-insensitive

366. Flash of light in dark inhibits flowering in
   (a) SDP (b) LDP (c) LSDP (d) DNP

367. In SDP, flowering is controlled by which of the following conditions?
   (a) dark condition is interrupted by white/red light
   (b) dark condition in interrupted by far red light
   (c) dark condition is interrupted by white light followed by far red light
   (d) in short day plant flowering cannot be interrupted

368. Flowering in SDP is induced by
   (a) short day and interrupted long night (b) short night
   (c) long day and interrupted night (d) none of these

369. Which is the correct combination for photoperiodism?
   (a) tomato—DNP (b) xanthium—LDP
   (c) sunflower—SDP (d) wheat—SDP

370. A long-day plant flowers only when it is exposed to
   (a) red light (b) light more than critical day length
   (c) light equal to cirtical day length (d) light less than critical day length

371. Seasonal leaf fall is due to stopping of the following hormone.
   (a) auxin (b) cytokinin (c) gibberellin (d) ethylene

372. "Skototropic movements" are those induced by
   (a) light (b) touch (c) night (d) heat

373. Peristome teeth of moss show
   (a) hydochasy (b) xerochasy (c) hydrotropism (d) chemotropism

374. Tendrils and climbers show which type of movement?
   (a) taxis (b) tropism (c) nasty (d) epitaxis

375. The response of plants to light was first of all discovered by
   (a) Lamarck (b) Darwin (c) Thimann (d) Went

376. The leaf of *Mimosa* (touch-me-not) on touching becomes flaccid and finally drop down due to
(a) movement of water into intercellular spaces in pulvinus and pulvinules
(b) water movement from leaflet bases
(c) change in water concentration with changing osmatic pressure
(d) presence of nervous system

377. Geotropic response is percieved by
(a) mature roots (b) root cap
(c) root hairs (d) elongating cells of quiescent centre

378. Movement of sunflower peduncle towards the direction of sun is called
(a) photonasty (b) phototropism (c) nyctinasty (d) seismonasty

379. Opening of the sunflower is said to be
(a) seismonastic (b) photonastic (c) thermonastic (d) photoperiodism

380. Closure of lamina in venus fly trap (Dionaea) after an insect happens to fly over it is due to
(a) thigmotropism (b) haptonasty (c) chemotropism (d) chemonasty

381. Plant movement is response to diffuse stimulus of light is called
(a) phototropism (b) photolysis (c) phototaxis (d) photonasty

382. Movement of sunflower peduncle from east to west towards the direction of sun is due to
(a) photonasty (b) phototropism (c) nyctinasty (d) seismonasty

383. In Oxalis, the flowers open in the morning and close in the evening because of
(a) photonasty (b) photrotopism (c) phototaxis (d) nyctinasty

384. Tendrils exhibit twining due to
(a) thigmotropism (b) seismonasty (c) heliotropism (d) diageotropism

385. Phototropic and geotropic movements are linked to
(a) gibberellins (b) enzymes (c) auxin (d) cytokinins

ANSWERS

1. (d)	2. (c)	3. (c)	4. (c)	5. (c)	6. (c)	7. (d)
8. (d)	9. (d)	10. (a)	11. (c)	12. (d)	13. (d)	14. (b)
15. (c)	16. (c)	17. (d)	18. (d)	19. (a)	20. (c)	21. (a)
22. (d)	23. (a)	24. (c)	25. (c)	26. (d)	27. (b)	28. (d)
29. (c)	30. (c)	31. (c)	32. (c)	33. (d)	34. (d)	35. (d)
36. (b)	37. (a)	38. (b)	39. (b)	40. (b)	41. (c)	42. (b)
43. (b)	44. (a)	45. (c)	46. (a)	47. (c)	48. (c)	49. (b)
50. (a)	51. (b)	52. (d)	53. (d)	54. (b)	55. (d)	56. (a)
57. (b)	58. (d)	59. (c)	60. (c)	61. (a)	62. (b)	63. (b)
64. (d)	65. (b)	66. (a)	67. (d)	68. (d)	69. (b)	70. (d)
71. (a)	72. (b)	73. (a)	74. (c)	75. (a)	76. (a)	77. (a)
78. (b)	79. (d)	80. (c)	81. (b)	82. (a)	83. (b)	84. (b)
85. (a)	86. (a)	87. (b)	88. (a)	89. (b)	90. (d)	91. (b)
92. (a)	93. (a)	94. (a)	95. (c)	96. (a)	97. (c)	98. (b)
99. (c)	100. (a)	101. (b)	102. (d)	103. (a)	104. (d)	105. (d)
106. (d)	107. (b)	108. (a)	109. (b)	110. (c)	111. (b)	112. (a)
113. (a)	114. (b)	115. (d)	116. (b)	117. (d)	118. (d)	119. (c)
120. (a)	121. (c)	122. (a)	123. (c)	124. (a)	125. (d)	126. (c)
127. (a)	128. (a)	129. (d)	130. (b)	131. (d)	132. (a)	133. (a)
134. (a)	135. (d)	136. (c)	137. (a)	138. (c)	139. (b)	140. (c)
141. (c)	142. (a)	143. (a)	144. (a)	145. (a)	146. (c)	147. (c)
148. (b)	149. (a)	150. (d)	151. (a)	152. (b)	153. (d)	154. (d)
155. (b)	156. (c)	157. (d)	158. (d)	159. (c)	160. (b)	161. (b)
162. (d)	163. (a)	164. (c)	165. (c)	166. (a)	167. (a)	168. (a)
169. (a)	170. (d)	171. (a)	172. (b)	173. (d)	174. (d)	175. (b)
176. (d)	177. (a)	178. (a)	179. (d)	180. (d)	181. (c)	182. (a)
183. (b)	184. (a)	185. (b)	186. (c)	187. (b)	188. (b)	189. (d)
190. (c)	191. (b)	192. (c)	193. (b)	194. (a)	195. (d)	196. (b)
197. (a)	198. (b)	199. (a)	200. (b)	201. (c)	202. (b)	203. (d)
204. (d)	205. (d)	206. (b)	207. (b)	208. (b)	209. (a)	210. (b)
211. (c)	212. (b)	213. (d)	214. (c)	215. (b)	216. (b)	217. (c)
218. (a)	219. (b)	220. (c)	221. (b)	222. (b)	223. (b)	224. (b)
225. (c)	226. (b)	227. (b)	228. (b)	229. (a)	230. (d)	231. (a)
232. (d)	233. (a)	234. (c)	235. (a)	236. (b)	237. (c)	238. (d)
239. (d)	240. (a)	241. (a)	242. (c)	243. (a)	244. (d)	245. (b)
246. (b)	247. (b)	248. (b)	249. (c)	250. (a)	251. (c)	252. (b)
253. (c)	254. (b)	255. (b)	256. (b)	257. (b)	258. (a)	259. (a)
260. (d)	261. (d)	262. (b)	263. (c)	264. (a)	265. (b)	266. (c)

267.	(b)	268.	(b)	269.	(b)	270.	(a)	271.	(d)	272.	(d)	273.	(c)
274.	(a)	275.	(c)	276.	(c)	277.	(a)	278.	(b)	279.	(a)	280.	(d)
281.	(a)	282.	(c)	283.	(a)	284.	(c)	285.	(b)	286.	(c)	287.	(c)
288.	(a)	289.	(c)	290.	(a)	291.	(b)	292.	(b)	293.	(c)	294.	(d)
295.	(d)	296.	(d)	297.	(d)	298.	(c)	299.	(a)	300.	(a)	301.	(b)
302.	(a)	303.	(b)	304.	(b)	305.	(b)	306.	(c)	307.	(c)	308.	(a)
309.	(a)	310.	(b)	311.	(d)	312.	(c)	313.	(d)	314.	(a)	315.	(a)
316.	(c)	317.	(d)	318.	(c)	319.	(d)	320.	(c)	321.	(a)	322.	(a)
323.	(b)	324.	(d)	325.	(a)	326.	(b)	327.	(d)	328.	(c)	329.	(d)
330.	(c)	331.	(d)	332.	(c)	333.	(b)	334.	(b)	335.	(a)	336.	(c)
337.	(a)	338.	(a)	339.	(b)	340.	(c)	341.	(b)	342.	(b)	343.	(c)
344.	(d)	345.	(b)	346.	(b)	347.	(c)	348.	(a)	349.	(a)	350.	(a)
351.	(d)	352.	(d)	353.	(b)	354.	(a)	355.	(c)	356.	(d)	357.	(d)
358.	(c)	359.	(a)	360.	(b)	361.	(b)	362.	(c)	363.	(b)	364.	(c)
365.	(b)	366.	(a)	367.	(a)	368.	(d)	369.	(a)	370.	(b)	371.	(a)
372.	(c)	373.	(b)	374.	(b)	375.	(b)	376.	(a)	377.	(b)	378.	(b)
379.	(b)	380.	(b)	381.	(c)	382.	(b)	383.	(a)	384.	(d)	385.	(c)

16

MICROSCOPY

CHOOSE THE CORRECT ANSWER

1. Feulgen reaction is a technique that has been developed by Feulgen and Rossenbeck to study or to stain
   (a) DNA (b) protein (c) PNA (d) lipids
2. A phase contrast microscope is used
   (a) for increasing power
   (b) to study cells and their components in a living state using a phase difference in the light rays
   (c) to localize cell structure stained with fluorochromes
   (d) for obtaining X-ray diffraction pattern of DNA
3. Which of the following statements is correct with respect to microsomes?
   (a) Microsomes as discrete entities are found only in the nucleolus and are revealed by electron microscopy pictures.
   (b) Microsomes as discrete entities are not found in the intact cell but are the result of the fragmentation of most of the cytoplasmic membranous components.
   (c) Microsomes as discrete entities are found always attached to the outer surface of rough ER
   (d) Microsomes as discrete entities are found in all living cells uniformly distributed throughout the cell.
4. Magnification of a "compound microscope" does not depend upon
   (a) focal length of objective (b) focal length of eyepiece
   (c) tube length of microscope (d) numerical aperture of objective

5. The maximum magnification possible with an electron microscope is
   (a) 200 times (b) 20,000 times (c) 2000 times (d) 200,000 times

6. The higher resolution in an electron microscope is due to
   (a) low wavelength of ultraviolet radiation as a light source
   (b) high numerical aperture of a series of glass lenses
   (c) powerful electromagnetic lenses
   (d) very low wave length (.005 mm) of electron beam used as light source

7. 1 micron is
   (a) 1/100 of a mm (b) 1/1000 of a mm
   (c) 1/10,000 of a mm (d) 1/1,00,000 of a mm

8. To determine the ultrastructure of a cell organelle, the most likely method to be used would be
   (a) autoradiography (b) phase-contrast microscopy
   (c) microdissection (d) electron microscopy

9. A fluorescent microscope is used
   (a) for increasing resolving power
   (b) for obtaining X-ray diffraction pattern of DNA
   (c) to study cells and their components in a living state using a phase difference in the light rays
   (d) to localize cell structure stained with fluorochromes.

10. One of the important techniques used to study the synthesis of molecules and to trace the metabolic pathway or events in cells is
    (a) chromatography (b) autoradiography
    (c) microspectrophotometry (d) calorimetry

11. Which of the following types of microscopes use ultraviolet rays of higher wavelengths?
    (a) ultraviolet microscope (b) polarizing microscope
    (c) electron microscope (d) fluorescent microscope

12. X-ray crystallography is used to study
    (a) arrangement of proteins (b) composition of protein and nucleic acids
    (c) three-dimensional structure of proteins (d) structure of lipids

13. Feulgen stain, a specific stain for DNA, contains
    (a) cotton blue (b) acridine (c) neutral red (d) basic fuchsin

14. Separation of chromatids during metaphase or behaviour of chromosomes during mitosis and meiosis can be best studied by
    (a) phase contrast microscopy (b) scanning electron microscopy
    (c) X-ray microscopy (d) cell fraction method

15. In contrast microscopy, the source of light used is
    (a) infrared light
    (b) beams of high-speed electrons
    (c) ultraviolet light
    (d) ordinary day light
16. To destroy the cell boundary, the tissues are homogenized in
    (a) 1.0 M salt solution
    (b) 1.0 M sucrose solution
    (c) 0.25 M sucrose solution
    (d) 0.25 M NaCl solution
17. Feulgen reaction of DNA is due to
    (a) phosphoric acid
    (b) removal of RNA but not DNA
    (c) phosphoric acid, carbohydrates and N_2 bases
    (d) aldehyde produced by acid hydrolysis
18. Which of the following is a vital dye?
    (a) Janus green (b) methylene blue (c) neutral red (d) all the above
19. Cellulosic cell walls can be specifically stained by
    (a) methylene blue (b) Sudan IV (c) zinc chloriodide (d) phloroglucinol
20. Cells or organelles can be separated by the method of
    (a) autoradiography
    (b) thin-section microtomy
    (c) differential and gradient centrifugation
    (d) X-ray diffraction
21. The resolving power of an electron microscope is
    (a) 1Å (b) 10 Å (c) 100 Å (d) 1000 Å
22. The technique other than microscopy used for the study of the cell is
    (a) plasmolysis
    (b) autoradiography
    (c) obliteration
    (d) none of the above
23. Electron microscope is made up of
    (a) polarizer and analyser filters
    (b) electromagnetic lenses
    (c) objective and ocular lenses
    (d) fluorochromes
24. Electron microscope was discovered by
    (a) R. Brown (b) Zernike (c) Wilson (d) Knoll and Ruska
25. Resolving power of microscope means capacity to
    (a) magnify the image
    (b) distinguish between two organelles
    (c) distinguish between two objects
    (d) distinguish between two close points

26. The resolving power of human eye is
    (a) 250 μ (b) 100 μ (c) 1000 μ (d) 10 μ
27. The maximum magnification possible using light microscope is
    (a) 200 times (b) 1000 times (c) 2000 times (d) 20000 times
28. The chromosome network cannot be stained in
    (a) haematoxylin (b) carmine (c) Feulgen stain (d) crystal violet
29. If cells are broken up and sedimented by centrifugation, the new structural fraction obtained is
    (a) lysosome (b) ribosome (c) microsome (d) centrosome
30. Resolving power of a light microscope is
    (a) 0.1 μ (b) 0.2 μ (c) 0.3 μ (d) 0.4 μ
31. Which of the following is a technique for the quantitative estimation of substances in a cell?
    (a) X-ray crystallography (b) microspectrophotometry
    (c) autoradiography (d) chromatography
32. Which of the follwoing would most probably be used to observe unstained living tissues?
    (a) fluorescence microscope (b) electron microscope
    (c) ultracentrifuge (d) phase contrast microscope
33. In SEM the source of illumination is
    (a) infrared light (b) ordinary day light
    (c) beam of electrons (d) ultraviolet light
34. Janus green is a stain suitable for staining
    (a) mitochondria (b) nuclei (c) vacuole (d) cytoplasm
35. Who among the following scientists developed the cytochemical technique for the presence of DNA?
    (a) Johannsen (b) Feulgen and Rossenbeck
    (c) McClintock (d) Hammerling
36. A living cell can be studied using
    (a) electron microscope (b) phase contrast microscope
    (c) compound microscope (d) interference microscope
37. A living cell can be observed under
    (a) light microscope (b) fluorescent microscope
    (c) phase contrast microscope (d) electron microscope
38. How many times is the resolution of light microscope greater than that of human eye?
    (a) 100 times (b) 250 times (c) 500 times (d) 1000 times

39. The molecular organization of cell components can be revealed by
    (a) phase contrast microscope
    (b) electron microscope
    (c) X-ray diffraction technique
    (d) all the above

40. Because of the limitations in resolving power in light microscope the magnification is not useful when the size is more than
    (a) 500 times object size
    (b) 100 times object size
    (c) 1000 times object size
    (d) 1500 times object size

41. Wavelength of electrons is
    (a) less than light
    (b) more than light
    (c) equal to light
    (d) like the X-rays (.055 mm)

42. The electron microscope differs from the ordinary light microscope in having high resolving power due to the use of
    (a) fast electrons which have a shorter wavelength than ordinary light rays, as the source of illumination
    (b) electrons which have a longer wavelength than the ordinary light, as the source of illumination.
    (c) UV rays as the source of illumination
    (d) gamma rays as the source of illumination

43. Angstrom, the unit of measurement in electron microscopy is equal to
    (a) 0.01 μm
    (b) 0.01 μm
    (c) 0.0001 μm
    (d) 0.00001 μm

44. Electron microscope has an advantage over light microscope in
    (a) having greater magnification power
    (b) that it requires no light
    (c) giving depth focus
    (d) not advantageous at all

45. Resolving power of a compound microscope is influenced by
    (a) the numerical aperature of eye lens
    (b) the numerical aperture of objective lens
    (c) the numerical aperture of both
    (d) none of the above

46. Which of the following is generally used to observe unstained living tissues (or a culture of living cells)?
    (a) polarization microscope
    (b) phase contrast microscope
    (c) radioautogram
    (d) ultracentrifuge

47. Resolving power of an electron microscope is
    (a) about 10 times that of light microscope
    (b) about 50 times that of light microscope
    (c) more than 100 times that of light microscope
    (d) more than 100 times that of visible microscope

48. Which of the following microscopes has the highest resolving power?
    (a) electron microscope
    (b) UV microscope
    (c) fluorescent microscope
    (d) polarizing microscope

49. Transparent living cells can be properly studied under the
    (a) phase contrast microscope
    (b) electron microscope
    (c) ordinary optical microscope
    (d) X-ray microscope

50. The wavelength of radiations used in electron microscopes is
    (a) more than visible light
    (b) equal to visible light
    (c) less than visible light
    (d) like the X-rays

51. Vital stains are employed in the study of
    (a) preserved tissues
    (b) fixed tissues
    (c) frozen tissues
    (d) living cells

52. Technique by which subcellular components are separated on the basis of their physical properties is
    (a) autoradiography
    (b) electrophoresis
    (c) cell fractionation
    (d) chromatography

53. If a biochemical analysis of mitochondria is to be done, the best procedure would be to
    (a) plasmolyse the cells, filter it and take the debris
    (b) grind the cell and filter out the mixture and take the debris
    (c) select cells which have a large number of mitochondria
    (d) subject the cells of cell fraction (centrifuge) and obtain mitochondria

54. Pure fraction of cellular components can be obtained by or the isolation of individual organelles from cell homogenates is achieved through
    (a) microspectrophotometry
    (b) differential centrifugation
    (c) X-ray diffraction
    (d) chromatography

55. Autoradiography technique was used to
    (a) find out the absorption maximum of chlorophyll pigments
    (b) study photorespiration in certain plants
    (c) trace the path of carbon in photosynthesis
    (d) establish the CO_2 evolved during respiration

56. In recent years, considerable attention is being given to DNA–protein interaction, because expression of genes is regulated through binding of specific proteins on regulatory DNA sequences. The pattern of protein binding on DNA can be studied by
    (a) ultracentrifugation
    (b) X-ray crystallography
    (c) light microscope
    (d) electron microscope

57. The transmission electron microscope has a resolution of
    (a) 2000 nanometres (b) 200 nanometres
    (c) 2 nanometres (d) 0.2 nanometres

58. We require this instrument to see chromosomes in the living cell.
    (a) ordinary lens (b) electron microscope
    (c) phase contrast microscope (d) light microscope

59. For studying ribosomes, which of the following is used?
    (a) electron microscope (b) X-ray crystallography
    (c) ultracentrifugation (d) observing another microscope

60. Three-dimensional structure of protein can be elucidated by using
    (a) spectrophotometer (b) electron microscope
    (c) X-ray crystallography (d) SEM microscope

61. Using acridine orange stain (after acetylation) on a tissue section, it is possible to obtain
    (a) green fluorescence for DNA and red for RNA
    (b) red fluorescence for DNA and green for RNA
    (c) red fluorescence for both DNA and RNA
    (d) green fluorescence for both DNA and RNA

62. Feulgen and Rossenbeck reaction is employed for the detection of
    (a) DNA alone (b) RNA alone (c) DNA and RNA (d) proteins

63. In transmission electron microscope (TEM), the objectives for viewing ultrastructure are
    (a) electromagnetically controlled (b) electrostatically controlled
    (c) electronically controlled (d) made up of quartz

64. Electron microscope differs from light microscope in having
    (a) an eyepiece lens
    (b) condensing lens system
    (c) a high resolving power
    (d) an objective to form and magnify the image

65. If detailed structure of surface of pollen grain is required to be examined, the instrument required is
    (a) scanning electron microscope (b) light microscope
    (c) phase contrast microscope (d) none of these

66. Polarization microscope is unique as it tells the observer
    (a) the orientation of molecules
    (b) the three-dimensional structure of the compound
    (c) both (a) and (b)
    (d) none of the above

67. Millon's reagent in cytochemical techniques is used for locating
    (a) DNA (b) proteins (c) lipids (d) sucrose

68. DNA fingerprinting has proved to be useful in forensic science. It involves the use of
    (a) minisatellite (b) ribosomal RNA (c) cDNA (d) bacterial DNA

69. RFLP (restriction fragment length polymorphism) study is a technique for
    (a) transferring genes from unrelated species
    (b) isolating single genes
    (c) isolating single gene products
    (d) identifying genetic (DNA) homologies

70. Electron microscope revealed the presence of
    (a) chloroplast (b) ribosome (c) mitochondria (d) lysosome

71. The basis for DNA fingerprinting is
    (a) knowledge of human karyotype
    (b) occurrence of Restriction Fragment Length Polymorphism (RFLP)
    (c) availability of cloned DNA
    (d) phenotypic differences between individuals

72. DNA probes are used in humans for
    (a) disease diagnosis (b) disease control
    (c) disease resistance (d) disease tolerance

73. The process in which the amount of DNA, RNA and protein can be known at a time is called
    (a) tissue culture (b) cellular fractioning
    (c) phase contrast microscopy (d) autoradiography

74. Electroencephalogram (EEG) is used in the diagnosis of
    (a) AIDS (b) tuberculosis (c) epilepsy (d) brain

75. The magnetic resonance generated by the nuclei of hydrogen atoms of biological tissues subjected to an external magnetic field is the basis of
    (a) NMR (Nuclear Magnetic Resonance) (b) ECG (electrocardiograph)
    (c) CT scanning (d) all of the above

76. The unit of measurement in transmission electron microscopy is
    (a) 0.001 μm (b) 0.00001 μm (c) 0.01 μm (d) 0.0001 μm

77. Molecules containing radioactive isotopes are used in
    (a) chromatography (b) autoradiography
    (c) electrophoresis (d) spectroscopy

78. Feulgen positive reaction is the test for
    (a) RNA (b) proteins (c) lipids (d) DNA

79. The resolving power of a light microscope is
    (a) 3000 Å (b) 0.05 Å (c) 10 Å (d) 275 Å

80. Tomographic electron microscopy is used to analyse
    (a) nanoparticles (b) microparticles (c) milliparticles (d) femtoparticles

81. The resolving power of transmission electron microscope (TEM) is greater than light compound microscope by
    (a) 10 times (b) 100 times (c) 1000 times (d) 1000000 times

82. The resolving power of an optical instrument is the ability to distinguish two closely placed objects as two distinct objects. It depends on
    (a) numerical aperture (b) wavelength of light
    (c) both (a) and (b) (d) none of these

83. EEG (Electroencephalogram) is an investigation technique used for diagnosis of
    (a) brain infection (b) epilepsy and other brain disorders
    (c) effects of drugs on the brain (d) sleep disorders

84. Zernicke invented
    (a) polarizing microscope (b) fluorescent microscope
    (c) phase contrast microscope (d) interference microscope

85. Iodine test is used to detect
    (a) typhoid (b) malaria (c) fats (d) carbohydrates

86. Feulgen reaction is employed to study
    (a) DNA (b) RNA (c) protein (d) fats

87. Kepler and Galileo developed
    (a) oil immersion (b) field glasses
    (c) telescope (d) classroom microscope

88. In electron microscope, high-speed electrons are used having a wavelength of about
    (a) 0.01Å (b) 0.05Å (c) 0.15 Å (d) 0.25 Å
89. Types of lenses in a compound microscope are
    (a) 3 (b) 4 (c) 2 (d) 1
90. Resolving limit of our eye is
    (a) 1" (b) 1' (c) $\frac{1}{60}'$ (d) $\frac{1}{60}''$

ANSWERS

1. (a)	2. (b)	3. (b)	4. (d)	5. (d)	6. (d)	7. (b)
8. (d)	9. (d)	10. (b)	11. (d)	12. (c)	13. (d)	14. (a)
15. (b)	16. (c)	17. (d)	18. (d)	19. (c)	20. (c)	21. (b)
22. (a)	23. (b)	24. (d)	25. (d)	26. (b)	27. (c)	28. (b)
29. (c)	30. (c)	31. (b)	32. (d)	33. (c)	34. (a)	35. (b)
36. (b)	37. (a)	38. (c)	39. (c)	40. (a)	41. (d)	42. (a)
43. (c)	44. (a)	45. (b)	46. (b)	47. (c)	48. (a)	49. (a)
50. (d)	51. (d)	52. (c)	53. (d)	54. (b)	55. (c)	56. (b)
57. (c)	58. (c)	59. (a)	60. (c)	61. (a)	62. (a)	63. (a)
64. (a)	65. (a)	66. (a)	67. (b)	68. (c)	69. (d)	70. (b)
71. (b)	72. (a)	73. (d)	74. (c)	75. (a)	76. (d)	77. (b)
78. (d)	79. (a)	80. (a)	81. (b)	82. (c)	83. (b)	84. (c)
85. (d)	86. (a)	87. (d)	88. (b)	89. (c)	90. (d)	

17

ANALYTICAL TECHNIQUES

Choose the Correct Answer

1. In partition chromatography, the broadening of peaks can be reduced by
   (a) very small starting genes
   (b) rapid separation
   (c) slow separation and very small starting genes
   (d) rapid separation and very small starting genes
2. For the separation of non-polar material, the typical stationary phase is
   (a) benzene (b) alcohol (c) amides (d) water
3. For the separation of polar material, the typical stationary phase is
   (a) benzene (b) amides (c) amines (d) water
4. For the separation of non-polar material, typical mobile phase is
   (a) alcohols or amides (b) water
   (c) benzene (d) all of these
5. In the separation of polar materials, the typical mobile phase is
   (a) alcohol or benzene (b) water
   (c) benzene (d) all of these
6. In adsorption chromatography,
   (a) the fraction of solutes bound to the stationary phase is constant
   (b) the fraction bound is not constant but decreases with decreasing concentration
   (c) both (a) and (b)
   (d) none of these

7. In adsorption chromatography, for the separation of small organic molecules and proteins, the adsorbent used is
   (a) alumina (b) silica gel (c) hydroxyapatite (d) all of these
8. For separation of sterols and amino acid by adsorption chromatography, the material used is
   (a) hydroxyapatite (b) calcium phosphate gel
   (c) silica gel (d) alumina
9. For separation of nucleic acids by adsorption chromatography, the material used is
   (a) alumina (b) silica gel
   (c) hydroxyapatite (d) activated carbon
10. Adsorption of molecules is more strong in
    (a) adsorption chromatography in general (b) ion-exchange chromatography
    (c) paper chromatography (d) thin layer chromatography
11. The total volume of material, both solid and liquid, in the column is known as
    (a) void volume (b) bed volume (c) elution volume (d) liquid volume
12. The volume of the mobile phase in the column is known as
    (a) bed volume (b) void volume (c) elution volume (d) liquid volume
13. Channelling in adsorption chromatography is the result of
    (a) homogeneous bed in the column (b) inhomogeneity in the column
    (c) presence of gas in the elution medium (d) elution solvent being water
14. During adsorption chromatography, a single substance separating in multiple peaks is due to
    (a) molecules being degraded during passage (b) channelling
    (c) retention (d) rise in temperature
15. What is the requirement for the mesh size of the support in columns?
    (a) larger the mesh number, smaller the particle (b) larger the mesh number, larger the particle
    (c) both (a) and (b) (d) none of these
16. In thin layer chromatography, the spot size is reduced as compared to paper chromatography because
    (a) time taken for separation is small
    (b) material used in thin layer is pulverised
    (c) material used in thin layer is fibrous
    (d) there is associated capillarity in material used
17. In GLC, if the liquid phase is adsorbed to a solid support, the process is known as
    (a) capillary operation (b) open-tubular operation
    (c) packed-column operation (d) none of these

18. In GLC, if the liquid is adsorbed to the inner surface of a tube or column, the system is known as
    (a) capillary operation
    (b) open-tubular operation
    (c) packed-column operation
    (d) both (a) and (b)
19. If the composition of a gas changes, the resistance changes and the electrical resistance of a wire is temperature-dependent. This principle is used in
    (a) thermal-conductivity cell
    (b) argon-ionization detector
    (c) flame-ionization detector
    (d) none of these
20. Hydrogen gas is burned with air in the presence of the column effluent. This process is used in
    (a) thermal-conductivtiy cell
    (b) argon-ionization detector
    (c) flame-ionization detector
    (d) all of these
21. Sephadex cannot be prepared in a form that can accept molecules whose molecular weight is greater than 5×10^6. This is solved by
    (a) decreasing the length of cross-linking molecules
    (b) increasing the length of cross-linking molecules
    (c) prolonged swelling
    (d) using dry beads
22. A linear polymer of D-galactose with 3, 6-acrylamide is
    (a) dextran
    (b) agarose
    (c) acrylamide
    (d) none of these
23. Polyacrylamide gels are prepared by cross-linking acrylamide with
    (a) 3, 6-anhydro-1-galactose
    (b) guanidinium chloride
    (c) N-N′-methylene bisacrylamide
    (d) TEMED
24. Most preparative work require the use of gels of which of the following nature?
    (a) superfine
    (b) fine
    (c) coarse
    (d) both (a) and (b)
25. Maximum number of grams of a particular molecule that can penetrate the pores of a particular weight of gel is known as
    (a) capacity of a gel
    (b) strength of a gel
    (c) flexibility of a gel
    (d) rigidity of a gel
26. The chromatographic behaviour of almost all substances in gels is
    (a) independent of temperature and pH
    (b) independent of ionic strength and buffer composition
    (c) both (a) and (b) above
    (d) none of these

27. Gel chromatography is advantageous because
    (a) conditions for maximum stability can be maintained
    (b) virtually there is no adsorption and less gene spreading
    (c) elution volume is related in a simple manner to molecular weight
    (d) all of these

28. In ion-exchange chromatography, the ion exchanger contains charged groups which are
    (a) non-covalently linked (b) linked through hydrogen bonds
    (c) covalently linked (d) linked via van der Waals forces

29. A typical strongly acidic cation exchanger is a
    (a) sulphonic acid group (b) phenolic hydroxyl group
    (c) quaternary amino group (d) carboxylic group

30. If an H^+ is attached to a cation exchanger, the exchanger is said to be in
    (a) acidic form (b) basic form (c) neutral form (d) none of these

31. If an ion exchanger has covalently charged group as positive then it is said to be
    (a) cation exchanger (b) anion exchanger
    (c) both (a) and (b) (d) none of these

32. The ability of an ion exchanger to take up exchangeable ions is known as
    (a) total capacity (b) available capacity
    (c) ground capacity (d) measurable capacity

33. The available capacity of an ion-exchanger is dependent upon
    (a) pH (b) ionic strength (c) porosity of matrix (d) all of these

34. If a protein is stable at pH value above the isoelectric point, then the exchanger required is
    (a) cation (b) anion (c) both (a) and (b) (d) none of these

35. If a protein is stable at pH value below the isoelectric point, then the exchanger required is
    (a) cation (b) anion (c) both (a) and (b) (d) none of these

36. For extremely high resolution (as in HPLC), the support usually is of mesh size
    (a) 50–100 (b) 100–200 (c) 200–400 (d) 7400

37. In paper chromatography, when the stationary phase and mobile phase are both water, then separation is because of
    (a) adsorption
    (b) partitioning between water molecules of similar nature
    (c) partitioning between water molecules, because the physical structure of bound water is very different from that of free water
    (d) none of these

38. If a weakly ionized substance that requires a very low or high pH for ionization is chromatographed, the exchanger used is
    (a) strong (b) weak (c) medium (d) neutral

39. Which of the following statements is true?
    (a) Weak exchanges show greater resolution of substances if charge difference is small.
    (b) Macromolecule having strong charge is best eluted on weak exchangers.
    (c) Weak ion exchangers are best for separation of molecules with a high charge.
    (d) All of these.

40. Cellulose ion exchangers have proved to be the best for purifying large molecules such as proteins and polynucleotides because
    (a) matrix is fibrous (b) all functional groups are on the surface
    (c) both (a) and (b) (d) none of these

41. Both pH and ionic strength of eluting buffer are increased for
    (a) an anion exchanger (b) a cation exchanger
    (c) both (a) and (b) (d) none of these

42. ______ reacts very strongly with the amino group and is extremely useful in coupling enzymes, nucleic acid, antibodies, antigens, proteins, etc.
    (a) cyanogen bromide (b) activated charcoal
    (c) agarose (d) acrylamide

43. An ion exchanger is used to separate single-stranded polynucleotide. A gradient of increasing salt concentration at pH 2.0 is used for elution. Which elutes first?
    (a) triadenylic acid (b) pentaadenylic acid
    (c) hexaadenylic acid (d) heptaadenylic acid

44. An enzyme with mol. weight 56,000 is used in catalytic amount to cleave a protein of mol. weight 65,000 into fragments of mol. weight 43,000 and 22,000. Gel chromatography is used to assess the result. In which order will the proteins be eluted?
    (a) 65,000–56,000–43,000–22,000 (b) 65,000–43,000–22,000
    (c) 22,000–43,000–65,000 (d) 22,000–43,000–56,000–65,000

45. A globular protein aggregates to form either a tetrahedral tetramer or a linear tetramer. The mixture is chromatographed on sephadex gel. The elute which forms first is
    (a) tetrahedral (b) linear (c) both (a) and (b) (d) none of these

46. The R_f value of substance A and B are 0.34 and 0.68 when chromatographed on paper. What is the ratio of the distance proved after three hours, assuming that neither substance has run off the paper?
    (a) A : B:: 2 : 1 (b) A : B :: 1 : 2 (c) A : B :: 1 : 1 (d) A : B :: 2 : 2

47. Suppose that an enzyme is dissociated into four identical subunits and that you want to test enzymatic activity of the individual subunits, and you also know that there are no tetramers remaining in the sample. What chromatographic system would you choose to free the monomers from tetramers?
    (a) paper chromatography (b) thin layer chromatography
    (c) gel-chromatography (d) affinity chromatography

48. In attempting to purify a protein, it is found that the protein is tightly bound to DNA. When DNase is added to destroy DNA, it is found that the DNase binds to the complex without digesting DNA. To dissociate DNA–protein complex, 2 M NaCl is added, the concentration at which ion exchanges retard either DNA or protein. What will you use?
    (a) affinity chromatography (b) paper chromatography
    (c) gel chromatography (d) ion-exchange chromatography

49. In comparing a molecule of DNA that is circular with one that is linear, both with mol. weight 6×10^6, which would elute first for an agarose column?
    (a) linear (b) circular (c) both (a) and (b) (d) none of these

50. A procedure has been devised for purifying a particular enzyme using gel chromatography. In an effort to increase the amount of material to be handled, you use a sample whose concentration of protein is ten times that normally used. What is the nature of elution?
    (a) The enzyme elutes with void volume because it aggregates.
    (b) The enzyme elutes very slowly because of high concentration.
    (c) Enzyme is inactivated.
    (d) None of these.

ANSWERS

1. (d)	2. (a)	3. (b)	4. (a)	5. (b)	6. (b)	7. (a)
8. (c)	9. (c)	10. (b)	11. (b)	12. (a)	13. (b)	14. (b)
15. (a)	16. (b)	17. (c)	18. (d)	19. (a)	20. (c)	21. (b)
22. (b)	23. (c)	24. (b)	25. (a)	26. (c)	27. (d)	28. (c)
29. (a)	30. (a)	31. (b)	32. (a)	33. (d)	34. (b)	35. (a)
36. (d)	37. (c)	38. (a)	39. (d)	40. (a)	41. (b)	42. (a)
43. (a)	44. (b)	45. (b)	46. (b)	47. (c)	48. (c)	49. (a)
50. (a)						

18

PHOTOSYNTHESIS

CHOOSE THE CORRECT ANSWER

1. The process of ATP formation in chloroplasts is known as

   (a) photophosphorylation (b) photosynthesis
   (c) phosphorylation (d) none of these

2. PS-I is reactive in

   (a) red light (b) far-red light (c) both (a) and (b) (d) none of these

3. Pigment system II involves

   (a) chlorophyll *a*-680 (b) chlorophyll *a*-660
   (c) chlorophyll *a*-670 (d) all of these

4. In PS-II system, which functions as antenna?

   (a) chlorophyll *a*-660 (b) chlorophyll *a*-670
   (c) chlorophyll *a*-678 (d) all of these

5. PS-II system is active in

   (a) red light (b) far-red light (c) both (a) and (b) (d) none of these

6. In red drop effect, the curve drops dramatically in the region

   (a) at 680 light (b) above 680 nm (c) below 680 nm (d) none of these

7. Which of the following is a Hill oxidant?

   (a) benzoquinone (b) $NADP^+$
   (c) ferricyanide (d) all of the above

8. How many electrons are needed to reduce 6 molecules of CO_2?
   (a) 36 (b) 48 (c) 24 (d) 84
9. RuBisCO
   (a) is a bifunctional enzyme
   (b) has molecular weight of 5,57,000
   (c) is composed of 8 large subunits and 8 small subunits
   (d) all of the above
10. Compensation point is the value of a factor where there is a
   (a) beginning of photosynthesis
   (b) little photosynthesis
   (c) photosynthesis equal to the rate of respiration
   (d) neither photosynthesis nor respiration
11. The first stable product of Hatch and Slack cycle is
   (a) PGA (b) OAA (c) RuBP (d) malate
12. Photosynthetically active radiation (PAR) is
   (a) 390–700 nm (b) 3900–7600 Å
   (c) 400–760 nm (d) 4000–7000Å
13. Which one of the following pigment lacks phytol tail?
   (a) $C_{55}H_{72}O_5N_4Mg$ (b) $C_{35}H_{32}O_5N_4Mg$
   (c) $C_{55}H_{70}O_6N_4Mg$ (d) $C_{54}H_{70}O_6N_4Mg$
14. Q_{10} value of photochemical and chemical reaction is respectively
   (a) 1 : 1 (b) 2 : 1 (c) 2 : 3 (d) 1 : 2
15. Find the odd one (w.r.t. Z scheme)
   (a) NAD (b) plastocyanin (c) ferredoxin (d) plastoquinone
16. First unstable product of C_3 pathway is
   (a) phosphoglyceraldehyde (b) phosphoglyceric acid
   (c) dihydroxyacetone (d) β-oxo-acid
17. Dicken cycle is conditioned by
   (a) high O_2 (b) high temperature and bright light
   (c) lower CO_2 (d) all of the above
18. Which one the following is required during photolysis of water?
   (a) Mn, Mg, Ca (b) Cl, Mn, Mo (c) Ca, Cl, Mn (d) all of the above

19. Threshold value of CO_2 in C_4 plant is

(a) 0–10 ppm (b) 10–50 ppm (c) 50–150 ppm (d) 150–200 ppm

20. Chloroplasts have peripheral reticulum and starch grains in

(a) mesophyll cells of C_3 plants (b) mesophyll cells of C_2 plants
(c) mesophyll cells of C_4 plants (d) bundle sheath cells of C_4 plants

21. ATP is

(a) adenine triphosphate (b) adenosine tetraphosphate
(c) adenosine triphosphate (d) adenine tetraphosphase

22. The light compensation point in a heliophyte is

(a) 100–400 foot candles (b) 50–500 foot candles
(c) 5–100 foot candles (d) 500–1000 foot candles

23. How many ATP molecules are required by C_4 pathway to form one molecule of glucose?

(a) 18 (b) 16 (c) 30 (d) 32

24. Photorespiration results in the loss of carbon to the tune of

(a) 10% (b) 25% (c) 50% (d) 75%

25. Match List-I and II and choose the correct option from the assigned codes

List I	**List II**
A. *Alloteropsis semi-alata*	1. C_4 cycle in mesophyll cells
B. *Atriplex spongiosa*	2. C_3 cycle in mesophyll cells
C. *Euphorbia corollata*	3. C_3 cycle in bundle sheath cell
D. *E. maculata*	4. C_4 and C_3 ecotypes

Codes

	A	B	C	D
(a)	4	1	2	3
(b)	1	2	3	4
(c)	2	4	1	3
(d)	4	2	3	1

26. Hill's reaction concludes that

(a) Oxygen is released from Hill's oxidant.
(b) Electrons for the reduction of CO_2 are obtained from water.
(c) $NADPH_2$ is formed as hydrogen acceptor.
(d) Carbon dioxide is essential for photosynthesis.

27. At high O_2 concentration, the rate of photosynthesis decreases due to
    (a) Warburg effect (b) Pasteur effect
    (c) Emerson effect (d) Richmond Lang effect
28. Photo-oxidation of water (in photosynthesis) occurs in association of
    (a) cytochrome B_6 (b) plastocyanin (c) PS-II (d) PS-I
29. Assimilatory power is
    (a) reduction of RuBP (b) oxidation of H_2O
    (c) reduction of CO_2 (d) synthesis of malic acid
30. Oxidative decarboxylation is a step in the process of
    (a) photorespiration (b) photosynthesis
    (c) respiration (d) photoperiodism
31. Pick the odd one out (with respect to use in plant physiology).
    (a) C^{11} (b) C^{14} (c) O^{18} (d) S^{35}
32. If four O_2 molecules are released in a photochemical reaction, how much of assimilatory power will be generated?
    (a) 10 $NADPH_2$ + 12 ATPs (b) 10 $NADPH_2$ + 8 ATPs
    (c) 8 $NADPH_2$ + 12 ATPs (d) 8 $NADH_2$ + 12 ATPs
33. Red drop is created by wavelength corresponding to
    (a) visible light (b) ultraviolet radiation
    (c) infrared light (d) blue part of spectrum
34. Dimorphism of chloroplasts is seen in
    (a) Osmund and Holtum plants (b) Hatch and Slack plants
    (c) Calvin plants (d) lower plants
35. Find out the minimum number of quanto needed to generate 3 molecules of oxygen in photosynthesis.
    (a) 12 (b) 24 (c) 30 (d) 18
36. Z-scheme of photosynthesis has been given by
    (a) Calvin and Benson (b) Hill and Bendall
    (c) Arnon and Whatley (d) Emerson and Arnold
37. Photosynthesis is carried out by the roots of some
    (a) halophytes (b) epiphytes (c) heliophytes (d) sciophytes
38. Credit of discovery of photosynthesis goes to
    (a) Barnes (b) Priestly (c) Ingenhousz (d) Van Helmont

39. Chlorophylls absorb visible light region of which of the following wavelengths?
    (a) 400–500 nm (b) 600–800 nm
    (c) 400–500 nm and 600–700 nm (d) 600–400 nm
40. Maximum O_2 evolution occurs from
    (a) forests (b) marine phytoplankton
    (c) crops (d) land mass
41. Phosphorescence is the radiation emitted from excited chlorophyll molecule at the rate of
    (a) 10^{-12} sec (b) 10^{-9} sec (c) 10^{-2} sec (d) 10^{-8} sec
42. At enhanced oxygen concentrations, the rate of photosynthesis decreases, and the phenomenon is called
    (a) Pasteur effect (b) Richmond-Lang effect
    (c) Warburg effect (d) Emerson effect
43. Light harvesting complex (LHC) refers to
    (a) quanta (b) quantasomes (c) complete grana (d) reaction centre
44. When concentration of O_2 is high, RUBisCO functions as
    (a) oxygenase (b) oxidase (c) peroxidase (d) hydrolase
45. During photosynthesis in sugar cane, conversion of oxaloacetic acid into malic acid occurs in
    (a) mesophyll cells (b) bundle sheath cells
    (c) both (a) and (b) (d) guard cells
46. What is the role of PS-I in photosynthesis?
    (a) synthesis of ADP (b) reduction of NADP to $NADPH_2$
    (c) photolysis of water (d) release of O_2 from H_2O
47. Find the wrong pair.
    (a) Englemann—Action spectrum
    (b) Chl. *c*—$C_{35}H_{32}O_5N_4Mg$
    (c) Quantum requirement—12% in plant cell of photosynthesis
    (d) ATP requirement—2 ATP per C_4 cycle
48. During noncyclic photophosphorylation, electrons are transferred from PQ to PC through
    (a) Cyt. b_6 (b) Cyt. b_{559} (c) Cyt. f (d) Fd
49. Chlorophyll *a* is called primary photosynthetic pigment because
    (a) it is responsible for emission of electrons during cyclic photophosphorylation
    (b) it is responsible for emission of electrons during non-cyclic photophosphorylation
    (c) both (a) and (b)
    (d) it is blue-green in colour and is soluble in ether

50. Smoke reduces photosynthesis by
    (a) blocking hill reaction
    (b) blocking dark reaction
    (c) both (a) and (b)
    (d) reducing chlorophyll and carotenoid synthesis
51. During synthesis of one molecule of glucose in sugar cane, how many ATP molecules are used?
    (a) 30 (b) 40 (c) 36 (d) 18
52. Which is not a C_4 plant?
    (a) *Saccharum officinarum* (b) *Euphorbia maculata*
    (c) *Alloteropsis* (d) none of the above
53. Low CO_2 compensation point is observed in
    (a) C_2 plant (b) C_3 plant (c) C_4 plant (d) all plants
54. During C_4 cycle, malic acid is decarboxylated to form pyruvic acid and CO_2 in
    (a) palisade parenchyma cells of mesophyll (b) spongy parenchyma cells of mesophyll
    (c) bundle sheath cells (d) both in spongy and parenchyma cells
55. Which cycle is operated in pineapple for CO_2 fixation?
    (a) C_4 cycle (b) C_3 cycle (c) C_2 cycle (d) CAM cycle
56. Who proposed that green plants purify air in the presence of light?
    (a) Priestley (b) Ingenhousz (c) Warburg (d) Van Helmont
57. Protochorophyll differs from chlorophyll in lacking
    (a) 4 hydrogen atoms in one of its pyrrole rings
    (b) 4 hydrogen atoms in two of its pyrrole rings
    (c) 2 hydrogen atoms in one of its pyrrole rings
    (d) 2 hydrogen atoms in two of its pyrrole rings
58. Which one of the following statements about cytochrome P-450 is wrong?
    (a) It has an important role in metabolism. (b) It contains iron.
    (c) It is an enzyme involved in oxidation. (d) It is a coloured cell.
59. Which one of the following prokaryotic algae possesses chlorophyll *b*?
    (a) *Anacystis* (b) *Oscillatoria* (c) *Prochloron* (d) *Segtonema*
60. The wavelength of light most absorbed during photosynthesis is
    (a) 400 nm (b) 500 nm (c) 550 nm (d) 660 nm
61. The organelle associated with photorespiration is called as
    (a) lysosome (b) peroxisome (c) glyoxisome (d) mesosome

62. Which one of the following elements is required for photosynthetic oxygen evolution?
    (a) copper (b) iron (c) manganese (d) zinc

63. NADPH is generated through
    (a) photosystem I (b) photosystem II
    (c) anaerobic respiration (d) glycolysis

64. The CO_2 content in the atmosphere is
    (a) 0.034% (b) 0.34% (c) 3.34% (d) 4%

65. The core metal of chlorophyll is
    (a) Fe (b) Mg (c) Ni (d) Cu

66. Warburg effect is decreased rate of photosynthesis at
    (a) low concentration of CO_2 (b) high concentration of CO_2
    (c) higher concentration of CO_2 (d) none of the above

67. Photosynthetic evolution of oxygen requires
    (a) coenzyme (b) Fe enzyme (c) Mn enzyme (d) Zn enzyme

68. The C_4 plants are different from C_3 plants with reference to the
    (a) substrate that accepts CO_2 in carbon assimilation
    (b) type of end product of photosynthesis
    (c) type of pigments involved in photosynthesis
    (d) none of the above

69. Which of the following is the first product of photorespiration?
    (a) glycolic acid (b) serine (c) phosphoglycolate (d) glycine

70. C_4 plants have higher net photosynthetic rate because
    (a) they have no photorespiration
    (b) they have Kranz type of anatomy
    (c) they have PEP as CO_2 acceptor
    (d) they can photosynthesize in low light intensity

71. Photo-oxidation refers to
    (a) photolysis of water
    (b) photorespiration
    (c) light-induced breakdown of substance in the presence of oxygen
    (d) all of the above

72. Action spectrum of photosynthesis was described in 1883 by
    (a) Calvin (b) Hill (c) Arnon (d) Englemann

73. The electron energy transfer between accessory pigments and chlorophyll molecules takes place by

(a) transduction (b) radiation (c) resonance (d) fluorescence

74. The two enzymes responsible for primary carboxylation in C_3 and C_4 plants respectively are

(a) RuBP carboxylase and pyruvate carboxylase
(b) PEP carboxylase and RuBP carboxylase
(c) RuBP carboxylase and PEP carboxylase
(d) PEP carboxylase and pyruvate carboxylase

75. RuBisCO (RuBP-caboxylase oxygenase) is a protein which comprises

(a) 5% of total chloroplast portion (b) 11% of total chloroplast portion
(c) 16% of total chloroplast portion (d) 20% of total chloroplast portion

76. Role of light in photosynthesis was discovered by

(a) Priestley (b) Von Helmont (c) Ingenhousz (d) de Saussure

77. Which element is essential for photolysis of water?

(a) nitrogen (b) chlorine (c) carbon (d) oxygen

78. In C_4 plants, initial CO_2 fixation takes place in the chloroplasts of

(a) guard cells (b) spongy mesophyll
(c) palisade tissue (d) bundle sheath

79. Photorespiration takes place only in

(a) green parts of the plant (b) all the living cells of the plant
(c) mitochondria (d) root

80. Assimilatory powers produced during photosynthesis are

(a) H_2O and O_2 (b) RuDP and RuMP
(c) $C_6H_{12}O_6$ and PGAL (d) ATP and $NADPH_2$

81. C_4 plants differ from C_3 plants with respect to

(a) number of ATP molecules consumed (b) first product
(c) substrate which accepts carbon dioxide (d) all of the above

82. Which of the following is present at 3rd carbon atom of II pyrrole ring of chlorophyll *a* molecule?

(a) aldehyde group (b) carboxyl group
(c) methyl group (d) Mg

83. Which of the following occurs in C_3 cycle?

(a) reductive carboxylation (b) oxidative carboxylation
(c) photophosphorylation (d) oxidative phosphorylation

84. Which of the following wavelengths is active in view of photosynthesis?
    (a) 400–500 nm (b) 400–700 nm (c) 200–450 nm (d) 510–600 nm

85. Photorespiration is favoured by
    (a) high O_2 and low CO_2 (b) low O_2 and high O_2
    (c) low temperature and high O_2 (d) low light and high O_2

86. Which of the following statements is correct about C_4 type of phyotosynthesis?
    (a) C_4 cycle is an independent cycle.
    (b) C_4 cycle is adjunct to Calvin cycle.
    (c) RuBP carboxylase has higher affinity for CO_2.
    (d) Carboxylase is present in the bundle sheath cells.

87. The isotopes that have proved useful in researches on photosynthesis are
    (a) O^{18} and O^{14} (b) C^{11} and P^{32} (c) P^{32} and O^{18} (d) C^{14} and O^{18}

88. Non-cyclic photophosphorylation involves
    (a) pigment system II only (b) pigment system I only
    (c) both pigment systems I and II (d) none of these

89. Photophosphorylation means
    (a) formation of ATP from ADP (b) formation of NADP
    (c) formation of ADP from ATP (d) formation of PGA

90. Which one of the following is needed in both photosynthesis and respiration?
    (a) sunlight (b) chlorophyll (c) glucose (d) cytochrome

91. The first CO_2 acceptor in C_4 plants is
    (a) PEP (b) RuDP (c) PGA (d) OAA

92. Chlorophyll is soluble in
    (a) water (b) organic solvents
    (c) water and organic solvents (d) no solvent

93. Calvin cycle occurs in
    (a) chloroplast (b) cytoplasm (c) mitochondria (d) glyoxysomes

94. Photolysis of water during photosynthesis occurs with the help of
    (a) PS-I (b) PS-II (c) cytochrome (d) ferredoxin

95. In place of water, some anaerobic photosynthetic bacteria use other hydrogen donors, e.g., H_2S because
    (a) they cannot tolerate oxygen
    (b) other donors are freely available in the growth medium
    (c) other donors need less energy input
    (d) other donors yield more glucose molecules

96. Which of the following is a photosynthetic, free-living, aerobic, N_2 fixing bacteria?
    (a) *Anabaena* (b) *Rhizobium* (c) *Azospirillum* (d) *Azotobacter*

97. The process of photosynthesis takes place in
    (a) the cells of leaves (b) the chloroplasts
    (c) the leaves (d) the entire plant

98. Photosynthesis is a process by which plants prepare food materials. Which of the following statements is correct?
    (a) Carbohydrates are formed using CO_2 and H_2O in the presence of sunlight with the help of chlorophyll.
    (b) Carbohydrates are formed with CO_2 and H_2O with the help of chlorophyll.
    (c) Carbohydrates are formed by the use of CO_2 and H_2O.
    (d) Carbohydrates are formed using CO_2 in the presence of sunlight.

99. In which of the following plant types, are both PEP carboxylase and RuBP carboxylase present in the chloroplasts?
    (a) CAM (b) C_4 (c) C_3 (d) CAM and C_4

100. Thylakoid bodies in the grana possess the photosynthetic units that are called
    (a) glyoxysomes (b) quantiosomes (c) polysomes (d) ribosomes

101. The chlorophyll molecule has
    (a) phosphate head and porphyrin tail (b) porphyrin head and phytol tail
    (c) phytol head and porphyrin tail (d) none of the above

102. Hill reaction in photosynthesis refers to
    (a) absorption of C_2 (b) CO_2 fixation (c) photoreduction (d) splitting of water

103. The alga which is used in researches related to photosynthesis is
    (a) *Chlorella* (b) *Fucus* (c) *Acetabularia* (d) *Chlamydomonas*

104. The step of photosynthesis that requires direct light is
    (a) excitation of chlorophyll (b) fixation of CO_2
    (c) transfer of H_2 to NAD (d) regeneration of chlorophyll

105. Which of the following is not correct for C_4 plants?
   (a) These are less efficient for photosynthesis.
   (b) These are less efficient for photorespiration.
   (c) They have Kranz type of anatomy in leaves.
   (d) These have PEP as CO_2 acceptor.

106. The presumed size of chlorophyll molecules is
   (a) Head: 15 × 15 Å, Tail: 20 Å
   (b) Head:10 × 12 Å, Tail: 15 Å
   (c) Head: 20 × 20 Å, Tail: 25 Å
   (d) Head: 15 × 15 Å, Tail: 25 Å

107. The carbon dioxide acceptor in C_3 plants is
   (a) phosphoenol pyruvate (PEP)
   (b) ribulose 1, 5-dishosphate (RuBP)
   (c) phosphoglyceric acid (PGA)
   (d) ribulose monophosphate (RuMP)

108. Which of the following is least effective in photosynthesis?
   (a) sunlight (b) red light (c) blue light (d) green light

109. Which colour of light gives maximum absorption peak of chlorophyll?
   (a) red (b) blue (c) green (d) yellow

110. Formation of ATP in photosynthesis and respiration is an oxidation process, which utilizes the energy from
   (a) cytochromes (b) ferredoxin (c) electrons (d) carbon dioxide

111. Which photosynthetic bacteria possess both PS-I and PS-II?
   (a) purple sulphur bacteria
   (b) cyanobacteria
   (c) purple nonsulphur bacteria
   (d) green sulphur bacteria

112. The assimilatory power formed in light reaction of photosynthesis is
   (a) ATP (b) ATP + $NADPH_2$ (c) $NADPH_2$ (d) ADP + $NADH_2$

113. Which of the following is sensitive to longer wavelengths of light?
   (a) PS-II (b) PS-I (c) phosphorylation (d) photolysis

114. A very efficient converter of solar energy with net productivity of 2–4 kg/m^2 or more is
   (a) wheat (b) sugarcane (c) rice (d) Bajra

ANSWERS

1. (a)	2. (c)	3. (d)	4. (d)	5. (a)	6. (b)	7. (d)
8. (c)	9. (d)	10. (c)	11. (d)	12. (d)	13. (b)	14. (d)
15. (c)	16. (d)	17. (d)	18. (c)	19. (a)	20. (d)	21. (c)
22. (a)	23. (b)	24. (b)	25. (a)	26. (b)	27. (a)	28. (c)
29. (a)	30. (a)	31. (b)	32. (c)	33. (a)	34. (b)	35. (c)
36. (b)	37. (b)	38. (c)	39. (c)	40. (b)	41. (b)	42. (c)
43. (b)	44. (b)	45. (a)	46. (b)	47. (c)	48. (c)	49. (c)
50. (a)	51. (a)	52. (d)	53. (c)	54. (c)	55. (d)	56. (b)
57. (c)	58. (d)	59. (c)	60. (d)	61. (b)	62. (c)	63. (b)
64. (a)	65. (b)	66. (c)	67. (c)	68. (a)	69. (c)	70. (c)
71. (d)	72. (d)	73. (c)	74. (c)	75. (c)	76. (c)	77. (b)
78. (b)	79. (a)	80. (d)	81. (d)	82. (c)	83. (a)	84. (b)
85. (a)	86. (b)	87. (d)	88. (c)	89. (a)	90. (d)	91. (a)
92. (b)	93. (a)	94. (b)	95. (a)	96. (a)	97. (b)	98. (a)
99. (a)	100. (b)	101. (a)	102. (d)	103. (a)	104. (a)	105. (a)
106. (a)	107. (b)	108. (d)	109. (b)	110. (c)	111. (b)	112. (b)
113. (b)	114. (b)					

19

IMMUNE SYSTEM

Choose the Correct Answer

1. Lecithin is decomposed to
   (a) choline (b) neurine (c) both (a) and (b) (d) none of these
2. Urate stones are formed due to
   (a) hyperuricemia (b) gout (c) both (a) and (b) (d) none of these
3. Renal calculi formation is due to urine
   (a) infection (b) stagnation (c) both (a) and (b) (d) none of these
4. Mixed gall stones are composed of
   (a) cholesterol (b) bile pigments (c) both (a) and (b) (d) none of these
5. The portion of antigenic molecules which are involved in actual binding with antibody-combining sites are termed
   (a) antigenic determinants (b) antigen receptors
   (c) antibodies (d) none of these
6. Choleic acid formed by deoxycholic acid assists the absorption of many important
   (a) insoluble compounds (b) soluble compounds
   (c) both (a) and (b) (d) none of these
7. The combining sites on the surface of the lymphocytes are antibody-like molecules called
   (a) antigen receptors (b) antigenic determinants
   (c) antibodies (d) none of these

8. Many substances such as fatty acids, phenols, naphthalene, etc. combine with deoxycholic acid to form
   (a) choleic acid (b) acetic acid
   (c) phenol (d) trichloroethyl alcohol
9. During digestion, the gall bladder contracts by the stimulation of the hormone cholecystokinin which is produced by the
   (a) small intestine (b) large intestine (c) pancreas (d) none of these
10. Bile in the gall bladder is composed of
   (a) bicarbonate (b) water (c) chloride (d) all of these
11. Substances which can give an immune response when introduced into an animal are called
   (a) antigens (b) antibodies (c) polymers (d) none of these
12. Which of the following when entering the duodenum and upper jejunum stimulates the secretion of enterocrinin which induces the flow of intestinal juice?
   (a) fats (b) proteins (c) carbohydrates (d) all of these
13. Pancreatic lipase is specific for the hydrolysis of primary ester linkages which occurs in position ________ of triglycerol.
   (a) 1 and 3 (b) 2 and 4 (c) 1 and 2 (d) 2 and 3
14. Pancreatic amylase is an
   (a) α-amylase (b) endoamylase (c) both (a) and (b) (d) none of these
15. The normal concentration of IgG is
   (a) 710 to 1530 mg/dl (b) 710 to 1230 mg/dl
   (c) 710 to 1030 mg/dl (d) none of these
16. Carboxypedptidase is a ________ containing enzyme.
   (a) zinc (b) silver (c) calcium (d) phosphorus
17. Pancreozymin stimulates the pancreas to produce a viscous fluid low in ________ but high in enzyme content.
   (a) bicarbonate (b) chloride (c) calcium (d) mucin
18. The secretion of pancreatic juice is controlled by
   (a) neural means (b) hormonal means
   (c) both (a) and (b) (d) none of these
19. Paracasein is acted upon by
   (a) pepsin (b) coproteoses (c) peptones (d) all of these

20. Pepsin converts native proteins into
    (a) proteoses (b) peptones (c) pepsin (d) both (a) and (b)
21. Hydrochloric acid stimulates duodenum to liberate
    (a) secretin (b) pepsin (c) both (a) and (b) (d) none of these
22. Carbonic anhydrase catalyses the formation of H_2CO_3 from
    (a) H_2O (b) CO_2 (c) both (a) and (b) (d) none of these
23. Saliva facilitates swallowing by the lubricating action of the
    (a) glycoprotein (b) peptones (c) pepsin (d) none of these
24. Which of the following mechanisms causes the initiation of gastric secretion?
    (a) nervous (b) reflex (c) both (a) and (b) (d) none of these
25. Salivary amylase is activated by
    (a) chloride ion (b) sulphide ion (c) phosphate (d) none of these
26. No antibody activity is associated with
    (a) IgD (b) IgE (c) IgG (d) IgM
27. Renin causes the ______ of milk.
    (a) coagulation (b) oxidation (c) reduction (d) none of these
28. Toxic substance are formed in the body by ______ action in the large intestine.
    (a) bacterial (b) viral (c) reduction (d) none of these
29. The toxic substance that is formed in the body during the degradation of RBC is
    (a) bilirubin (b) glucuronic acid (c) both (a) and (b) (d) none of these
30. Detoxification processes take place in the
    (a) liver (b) kidney (c) heart (d) all of these
31. Alcohols and adehydes are detoxified by which of the following processes.
    (a) oxidation (b) reduction (c) coagulation (d) all of these
32. Chloral is oxidized to
    (a) trichloracetic acid (b) *p*-acetylaminophenol
    (c) trichlorethyl alcohol (d) none of these
33. Aniline is oxidized to
    (a) *p*-aminophenol (b) trichloroacetic acid
    (c) choleic acid (d) none of these

34. Acetanilide is oxidized to
    (a) *p*-aminophenol (b) trichloroacetic acid
    (c) trichloroethyl alcohol (d) none of these

35. Benzene is oxidized to
    (a) phenol (b) acetic acid (c) alcohol (d) none of these

36. Chloral on reduction produces
    (a) trichloroethylacohol (b) *p*-acetylaminophenol
    (c) trichloroacetic acid (d) none of these

37. Picric acid on reduction yields
    (a) picramic acid (b) acetic acid (c) bicarbonates (d) none of these

38. Atropine on hydrolysis forms
    (a) tropic acid (b) tropine (c) bicarbonates (d) none of these

39. Aspirin is hydrolysed to
    (a) salicylic acid (b) acetic acid (c) both (a) and (b) (d) none of these

40. Procaine on hydrolysis yields
    (a) *p*-aminobenzoic acid (b) diethylaminoethanol
    (c) both (a) and (b) (d) none of these

41. The process by which the chemical to be detoxified combines with another chemical formed by the body is called
    (a) conjugation (b) coagulation (c) reduction (d) none of these

42. Bilirubin diglucuronide is formed from
    (a) bilirubin (b) glucuronic acid (c) both (a) and (b) (d) none of these

43. Indoxyl sulphate is the product fromed by the combination of protein and
    (a) indoxyl (b) glycine (c) both (a) and (b) (d) none of these

44. Which of the following is formed from hippuric acid?
    (a) benozoic acid (b) glycine (c) both (a) and (b) (d) none of these

45. Nicotinuric acid is formed from
    (a) nicotinic acid (b) glycine (c) both (a) and (b) (d) none of these

46. Bromobenzene and cysteine form
    (a) *p*-bromophenoyl (b) meracapturic acid (c) both (a) and (b) (d) none of these

47. N-methyl pyridine is formed from
    (a) pyridine (b) active methionine (c) both (a) and (b) (d) none of these

48. Cyanide and sodium thiosulphate form
    (a) thiocyanate (b) sodium sulphate
    (c) both (a) and (b) (d) none of these

49. Immunoelectrophoresis consists of both electrophoretic separation and immune precipitation of
    (a) proteins (b) lipids (c) alcohols (d) carbohydrates

50. Half of the patients with allergic diseases have increased serum
    (a) IgE levels (b) IgD levels (c) IgG levels (d) IgM levels

51. The B lymphocytes are responsible for humoral immunity which is expressed by the production of specific circulating plasma proteins termed
    (a) antibodies (b) thiocyanate (c) polymers (d) all of these

52. How many chains of polypeptide does the basic unit of all immunoglobulin molecules consist of?
    (a) four (b) two (c) both (a) and (b) (d) none of these

53. Immunoglobulins composed of more than one basic monomeric unit are termed
    (a) polymers (b) anibodies (c) both (a) and (b) (d) none of these

54. IgM is a
    (a) pentamer (b) trimer (c) both (a) and (b) (d) none of these

55. The polypeptide chains are folded 3-dimensionally with ______ bonds to form areas called domains.
    (a) disulphide (b) bicarbonate (c) polypeptide (d) none of these

56. The part of the antibody molecule which combines with antigens is formed by a few amino acids in the F region of
    (a) H chains (b) L chains (c) both (a) and (b) (d) none of these

57. The major types of L chains in man are
    (a) kappa (b) lambda (c) both (a) and (b) (d) none of these

58. About 70 per cent of the human immunoglobulin molecules carry kappa light chains and 30 per cent carry ______ light chain.
    (a) kappa (b) lambda (c) both (a) and (b) (d) none of these

59. In IgG, the H chain is termed as gamma chain; and in IgE, it is termed as
    (a) epsilon chain (b) L chain (c) H chain (d) none of these

60. The half of the light (L) chain towards the carboxy terminus is termed as the constant region, while the amino terminal half is the variable region of the
    (a) light chain (b) heavy chain (c) epsilon (d) none of these

61. The portion of the immunoglobulin molecule which binds the specific antigen is formed by the amino terminal portions of
    (a) H chain (b) L chain (c) both (a) and (b) (d) none of these

62. The domains of the protein chains form globular regions with
    (a) secondary structure (b) tertiary structure
    (c) primary structure (d) both (a) and (b)

63. IgG comprises ______ per cent of the gamma globulins and contains 2–4 per cent carbohydrates?
    (a) 80 (b) 70 (c) 60 (d) 50

64. What per cent of carbohydrate does IgA contain?
    (a) 5 to 10 (b) 2 to 4 (c) 4 to 6 (d) 10 to 15

65. IgM contains 576 amino acids and has a mass of
    (a) 9, 50, 000 daltons (b) 8, 50, 000 daltons
    (c) 7, 50, 000 daltons (d) none of these

66. IgM is the major immunoglobulin expressed on the surface of B cells and its carbohydrate content is
    (a) 10 to 12% (b) 5 to 10% (c) 4 to 6% (d) 2 to 4%

67. IgE is present in the serum in very low concentration as a single basic unit with ______ ε chains.
    (a) heavy (b) light (c) both (a) and (b) (d) none of these

ANSWERS

1. (c)	2. (c)	3. (c)	4. (c)	5. (a)	6. (a)	7. (a)
8. (a)	9. (a)	10. (d)	11. (a)	12. (d)	13. (a)	14. (c)
15. (a)	16. (a)	17. (a)	18. (c)	19. (a)	20. (d)	21. (a)
22. (c)	23. (a)	24. (c)	25. (a)	26. (a)	27. (a)	28. (a)
29. (a)	30. (a)	31. (a)	32. (a)	33. (a)	34. (a)	35. (a)
36. (a)	37. (a)	38. (c)	39. (c)	40. (c)	41. (a)	42. (c)
43. (c)	44. (c)	45. (c)	46. (c)	47. (c)	48. (c)	49. (a)
50. (a)	51. (a)	52. (c)	53. (a)	54. (a)	55. (a)	56. (c)
57. (c)	58. (b)	59. (a)	60. (a)	61. (c)	62. (d)	63. (a)
64. (a)	65. (a)	66. (a)	67. (a)			

20

CANCER

I. Choose the Correct Answer

1. Many cancers are associated with the abnormal production of
   (a) carbohydrates (b) fats (c) proteins (d) minerals
2. The tumour markers are
   (a) enzymes (b) hormones (c) proteins (d) all of the above
3. Cancer is caused by which of the following groups
   (a) radiant energy (b) chemical compounds
   (c) viruses (d) all of the above
4. DNA of gene is damaged by
   (a) ultraviolet rays (b) X-rays (c) γ-rays (d) all of the above
5. Ultraviolet radiation may cause the formation of
   (a) purine dimers (b) pyrimidine dimers
   (c) nucleoside dimers (d) nucleotide dimers
6. Carcinogenicity caused by radiant energy causes damage to
   (a) DNA (b) RNA (c) mRNA (d) tRNA
7. Chemicals can cause human cancers up to
   (a) 60% (b) 70% (c) 80% (d) 90%
8. The carcinogenic substances may be
   (a) organic molecules (b) inorganic molecules
   (c) metals (d) all of the above

9. The ultimate carcinogen attacks the nucleophilic group in
   (a) DNA (b) RNA (c) protein (d) all of the above
10. The specific mono-oxygenase responsible for the metabolism of polycyclic aromatic hydrocarbons is termed as
   (a) decarboxylase (b) cytochrome oxidase
   (c) cytochrome P450 (d) deaminase
11. Promoters can cause
   (a) promotion (b) initiation (c) acceleration (d) phagocytosis
12. The premier in carcinogenesis is
   (a) DNA (b) RNA
   (c) rRNA (d) none of the above
13. Many tumour cells exhibit abnormal
   (a) Golgi bodies (b) RNA
   (c) chromosomes (d) all of the above
14. Burkitt's lymphoma is a fast-growing cancer caused by
   (a) monocytes (b) B-lymphocytes (c) neutrophils (d) esinophils

II. Fill in the Blanks

1. Oncogenic viruses contain either ______ or ______ as their genome.
2. The ______ antigens are bound tightly to DNA.
3. The genome of the retrovirus contains ______ genes.
4. The integrated double-stranded cDNA is called a ______.
5. Translocation is a type of chromosomal change seen in ______ cells.
6. The lower activity of ______ can cause stimulation of the activity of adenylate cyclase.
7. Metastasis is a ______ process to analyse in humans.
8. Changes in the ______ chains of cell glycoproteins may result in ______.
9. Markers are most often detached from ______ stages of cancer rather than ______ stages of cancer.
10. Apurinic sites may form by ______ of corresponding bases.

11. The process by which procarcinogens are converted to active carcinogens is called ______.
12. Monooxygenases and transfereses cause the metabolism of ______ and other xenobiotics.
13. Most of the chemical carcinogens are ______.
14. In the growth phase of cancer, ______ treatment is very important.
15. The basic aim in the treatment of cancers is to use drugs that kill ______ cells but that are not toxic to the ______ cells.
16. Cancer cells which exhibit MDR have increased amounts of ______.
17. Humans of all ______ are victimized by cancer.
18. Gene ______ is involved in the action of most growth factors.
19. Once a cell becomes a tumour cell, its tendency for ______ is increased.
20. Some growth factors can affect the cells that ______ them.
21. Exposure of cells to PDGF can cause rapid activation of cellular ______.
22. ______, a gene that influences susceptibility to early-onset breast and ovarian cancer, has been isolated.
23. BRCA1 is a protein of ______ amino acids containing a ______ finger domain as its amino terminal region and is a ______ factor.
24. BRCA1 is associated with increased susceptibility to ______ cancer only and mapped to chromosome ______.
25. P^{53} is a nuclear______of molecular mass ______.
26. P^{53} acts as a transcriptional ______, regulating genes that participate in cell division.
27. P^{53} acts as ______ checkpoint control for ______ damage. If DNA is excessively damaged, the activity of P^{53} is ______ resulting in inhibition of cell division.
28. P^{53} hastens the ______of potentially dangerous cells which have the potential to become malignant.
29. The mutational ______ differs in different cancers.
30. Transversions are frequent in ______ cancer and ______ cancer.
31. Estrogens block effects of androgens in ______ tumours.
32. Methyl alkylates denature ______ and other molecules in______.
33. Tumour markers are detected in ______ stages rather than ______ stages of cancer.

34. The tumour marker calcitonin is used in ______ and alpha-fetoprotein is used as a marker in the cancer of the ______ and ______.

I. Answers

1. (c)	2. (d)	3. (d)	4. (d)	5. (b)	6. (a)	7. (c)
8. (c)	9. (d)	10. (b)	11. (b)	12. (a)	13. (b)	14. (b)

II. Answers

1. DNA or RNA
2. T
3. 4
4. provirus
5. cancer
6. GTPase
7. complex
8. oligosaccharide, metastasis
9. advanced, early
10. elimination
11. metabolic activation
12. procarcinogens
13. mutagens
14. chemotherapy
15. cancer, normal
16. proteins
17. ages,
18. activation
19. malignancy
20. synthesize
21. proto-oncogenes
22. BRCAI
23. 1863, zinc, transcription
24. breast, 13q, 12-13

25. phosphoprotein, 53 kDa
26. regulator, genes
27. G_1, DNA, increased
28. death
29. spectrum
30. lung, liver
31. prostatic
32. DNA, myeloma
33. advanced, early
34. medullary carcinoma, liver, germ cell.

GATE QUESTION PAPERS

GATE-1996

1. In each of the subquestions, one answer among the alternative, is correct. Choose the correct alternative and write in your answer book the letter a, b, c or d along with the corresponding subquestion number. $(6 \times 1 = 6)$

1.1 Which one of the following chromatographic techniques is most suitable for separating glucosamine from glucuronic acid?

(a) Affinity chromatography
(b) Molecular exclusion
(c) Ion-exchange
(d) Hydrophobic chromatography

1.2 The dipeptide Lys-Glu on electrophoresis at pH 6.0 would

(a) Move towards anode (+)
(b) Move towards cathode(–)
(c) Remain stationary
(d) Get degraded

1.3 The following compounds were separately dissolved in water to the concentration 10^{-3} mol L^{-1}

(i) 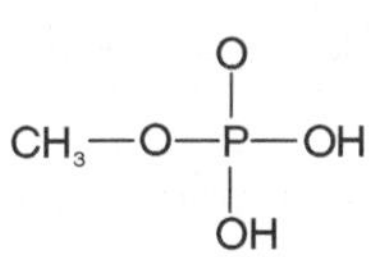

(ii) CH_3NH_2

(iii) CH_3CO_2H

(iv) 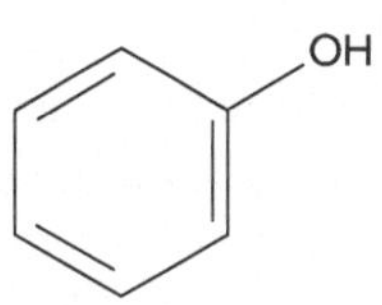

(v) 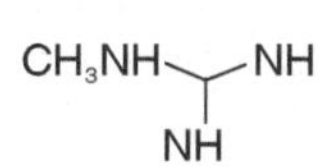

The pHs of the solutions are in the order

(a) i < iii < iv < v
(b) i < ii < iii < iv < v
(c) iii < iv < v < i < ii
(d) i < iv < v < iii < ii

1.4 Which one of the following is used for sequential amino terminal cleavage of peptides and proteins?

(a) Phenylglyoxal
(b) Phenylisothiocyanate
(c) 2, 4-Dinitrofluorobenzene
(d) Phenyl methyl sulfonyl fluoride

1.5 The minimum molecular weight of a pure heme protein containing 0.426% by weight of iron (at. wt. 56) is

(a) 13,100 (b) 60,000 (c) 26,300 (d) 100,000

1.6 The decapeptide Asp-Gly-Glu-A/la-Lys-Met-Leu-Arg-Phe-Val on trypsin treatment will give

(a) Two fragments (b) Five fragments (c) Four fragments (d) Three fragments

2. Match the entries in Column A against those in Column B and write the matching pairs in your answer book.

Column A	**Column B**
(a) The amino acid with maximum number of codons (6)	(i) Cys
(b) The characteristic amino acid in papain active site	(ii) TrP
(c) The amino acid precursor in heme biosynthesis	(iii) Ser
(d) The amino acid contributing the most to protein absorbance at 280 nm	(iv) Gly
	(v) Pro

3. In each of the subquestions, one answer among the alternatives is correct. Choose the correct alternative and write in your answerbook the letter a, b, c, or d along with the corresponding question number.

3.1 On doubling the enzyme concentration, the kinetic parameter that changes is

(a) K_m (b) V_{max} (c) K_{cat} (d) both V_{max} and K_{cat}

3.2 An enzymes does the following in catalysing a reaction

(a) Stabilizes the substrate (b) Decreases the equilibrium constant

(c) Increases the forward reaction rate (d) Hastens the approach to equilibrium

3.3 Two isozymes with identical turnover number have the K_m values 1 mM for isozyme I and 10 mM (for isozyme II). Which of the following statements is true with respect to the relative catalytic efficiencies of the isozymes?

(a) Isozyme II is superior to isozyme I (b) Isozyme I is superior to isozyme II

(c) Both are equally efficient (d) None of the above

3.4 Proline racemase and alanine racemase catalyse similar reactions but with distinctly different chemical mechanisms. One uses pyridoxal phosphate the other does not. The most likely reason is

(a) Proline is a helix breaker while alanine is not

(b) Proline is an imino acid while alanine is an amino acid

(c) Proline and alanine have different molecular weights

(d) Proline and alanine have different P_I values

3.5 The coenzyme involved in the biosynthesis of thymidine from uridine is

(a) Cyanocobalamine (b) S-adenosylmethionine

(c) Pantothenic acid (d) Folic acid

4. In each of the subquestions, one answer among the alternative is correct. Choose the correct alternative and write in your answerbook the letter a, b, c, or d along with the corresponding question number. (4 × 1 = 4)

4.1 The chemical bond energy mobilized during respiration is conserved in the form of

(a) Glucose 6-phosphate (b) UDP
(c) ATP (d) Pyrophosphate

4.2 The role of $NADP^+$ during photosynthesis is to

(a) Dislodge electrons from chlorophyll (b) Accept electrons from water
(c) Catalyse the combination of CO_2 and H_2O (d) Carry out photolysis

4.3 Death by cyanide poisoning is due to the inhibition of

(a) Cytochrome P_{450} (b) Cytochrome *c* oxidase
(c) Cytochrome *b* (d) Cytochrome *c* reductase

5. In each of the subquestions, one answer among the alternatives is correct. Choose the correct alternative and write in your answerbook the letter a, b, c, or d along with the corresponding question number (6 × 1 = 6)

5.1 The length of a 3000 kb linear B DNA fragment will be approximately

(a) 10 millimetre (b) 1 millimetre (c) 100 millimetre (d) 0.1 mellimetre

5.2 The following double-stranded DNA fragments have distinct T_m values

```
(i)   G G C A T A T T A G G C
      | | | | | | | | | | | |
      C C G T A T A A T C C G

(ii)  G C G G T A T T G G C G
      | | | | | | | | | | | |
      C G C C A T A A C C G C

(iii) T A T A G C G C A T T A
      | | | | | | | | | | | |
      A T A T C G C G T A A T
```

The T_m values are in the order

(a) i < ii < iii (b) iii < ii < i (c) ii < i < iii (d) iii < i < ii

5.3 The mRNA transcribed from the DNA sequence 5PACTTGATTc-OH3 is

(a) 5’pTGAACTAAG-OH3’ (b) 5’pUGAACUAAG-OH3’
(c) 5’pGAATCAAGT-OH3’ (d) 5’pGAAUCAAGU-OH4’

5.4 During translation, AUG codes for methionine at

(a) The end of a polypeptide chain
(b) The start of a polypeptide chain
(c) The start as well as the interior of a polypeptide chain
(d) None of the above

5.5 Following is the pictorial representation of the transcription of strand A of a DNA fragment

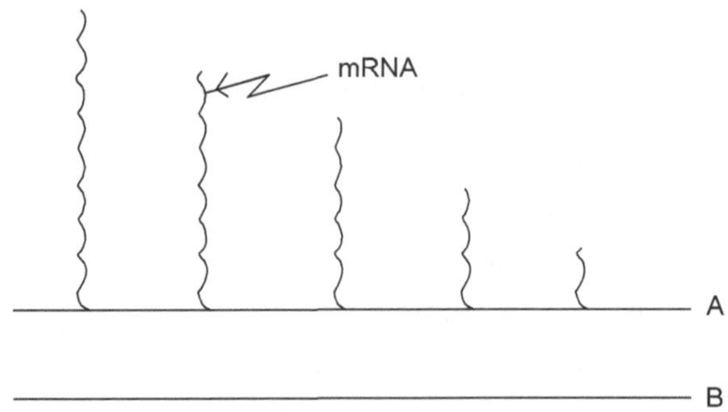

Which of the following assignment of the DNA strand polarities is correct?

(a) 5' ———————— 3'A
    3' ———————— 5'B

(b) 3' ———————— 5'A
    3' ———————— 5'B

(c) 3' ———————— 5'A
    5' ———————— 3'A

(d) 5' ———————— 3'A
    5' ———————— 3'A

5.6 In addition to oligonucleotide primers, the polymerase chain reaction (PCR) requires

(a) DNA ligase
(b) *Taq* polymerase
(c) Topoisomerase
(d) RNA polymerase

6. (i) List two most important factors that influence membrane fluidity (2)

(ii) Define with one example each (a) Facilitated transport, and (b) Active transport (2)

(iii) In a globular protein most apolar amino acid side chains are buried inside while most ionizable and polar side chains are surface-exposed. Why? (2)

(iv) Reduced and urea denatured lysozyme, when dialysed against water, regains almost full activity with all correct S—S bonds. What prevents the formation of incorrect S—S bonds? (1)

7. (i) 5×10^{-4} mole L^{-1} tyrosine solution in a path length of 1 cm gave an absorbance of 0.7. What is the absorbance when tyrosine concentration is 1.5×10^{-4} mol L^{-1} and the path length is 3 cm?

(ii) Calculate the ΔG° for proton transport from cytoplasm into a lysosome, given that Cytoplasmic pH = 7.0 lysosomal pH = 5.0 = 1.98×10^{-3} kcal/deg mol, $T = 27^\circ C$ (2)

(iii) Arginase, a trimeric enzyme of MW 40,000, contains three active sites. Under optimal conditions, 5 μg of pure emzyme hydrolysed 3.0 μ mol of arginine per min.

(a) What is the specific activity of the enzyme?
(b) What is the turnover number per active site?

8. (i) Patients suffering from phenylketonuriea are advised to avoid intake of food rich in phenylalanine. Explain the metabolic logic. (2)

(ii) There is a decrease in ethanol production when yeast cells are transferred from anaerobic environment. Explain the biochemical basis of the effect. (2)

(ii) In the following scheme representing glycogen breakdown upon the stimulation of liver cells by adrenalin, identify the elements A, B, C and D, or A', B', C' and D' (2)

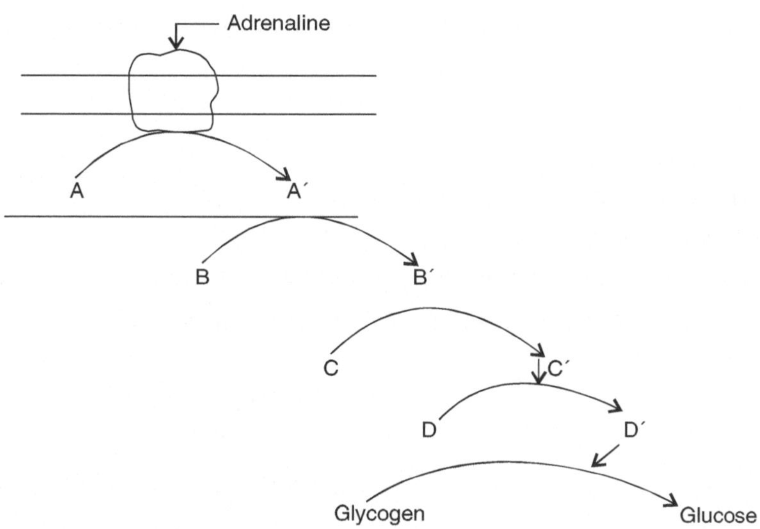

9. (i) Define each of the following:

(a) Immunogen (b) Hapten (c) Epitope (3)

(ii) Identify from the following list, the antibody species which is (a) Most efficient, and (b) Least efficient in complement fixation. (2)

(1) Fab (2) (Fab)' (3) IgG (4) IgM

(iii) Describe the logic in the use of anti–IgM antibodies to separate T and B cells. (1)

GATE-1997

1. For each subquestion given below, four alternatives are provided of which only one is correct. Write the correct answer in the answer book by writing a, b, c or d along with the corresponding subquestion number (10 × 1 = 10)

1.1 The Michaelis constant. K_m is
- (a) Numerically equal to 1/2 V_{max}
- (b) Dependent on enzyme concentration
- (c) Independent of pH
- (d) Numerically equal to the substrate concentration that gives half–maximal velocity

1.2 For determining the C-terminal amino acid of a polypeptide chain, the reagent that would be useful is
- (a) Trypsin
- (b) Carboxypeptidase
- (c) Phenyl isothiocyanate
- (d) 1 (N) HCI

1.3 For the reaction, fructose 6-P + $P_i \longrightarrow$ fructose 1,6-bisphosphate + H_2O (equilibrium constant K_{eq} = 0.001 at pH 7; T = 300 K; R = 2 cal/mol degree) the standard free energy change $\Delta G°$ is approximately equal to
- (a) + 4.10 kcal/mol
- (b) – 4.1 1kcal/mol
- (c) + 2.1 kcal/mol
- (d) – 2.1 kcal/mol

1.4 In contrast to the resting state, vigorously contracting muscle shows
- (a) An increased conversion of pyruvate to lactate
- (b) Decreased oxidation of pyruvate to CO_2 and water
- (c) A decreased NADH/NAD ratio
- (d) Decreased concentration of AMP

1.5 The catabolism of hemoglobin
- (a) Occurs in the red blood cells
- (b) Involves the oxidative cleavage of porphyrin ring
- (c) Results in the liberation of CO_2
- (d) Is the sole source of bilirubin

1.6 A ganglioside must contain in its structure
- (a) N-acetyl neuraminic acid (NANA), hexoses, sphingosine, long-chain fatty acid
- (b) NANA, a hexose, a fatty acid, sphingosine, phosphorylcholine
- (c) NANA, sphingosine, ethanolamine
- (d) NANA, hexoses, fatty acid, glycerol

1.7 Insulin does all of the following except

(a) Enhanced glucose transport into muscle (b) Enhanced glycogen formation by liver
(c) Increased lipolysis in adipose tissue (d) Gluconeogenesis inhibition in liver

1.8 The genetic code is

(a) Degenarate in that many triplets code for more than one amino acid
(b) Read in the direction of 3'⟶5'
(c) Generally referred to as universal, since it is nearly the same in all organisms
(d) Degenerate for all the amino acids

1.9 To clone a gene corresponding to a protein with partial amino acid sequence, Met–Trp–Cys–Trp (number of codons for Met = 1, Cys = 2, Trp = 1), the number of oligonucleotides that need to be designed to screen c-DNA library is

(a) 2 (b) 4 (c) 5 (d) 8

1.10 DNA sequencing by Sanger's method involves the use of

(a) Ribonucleotide (b) 3'-deoxyribonucleotide
(c) 2', 3'-dideoxyribonucleotide (d) Fluorodinitrobenzene

2. Match the entries in Column A with those in Column B and write the matching pairs in the answer book

Column A	**Column B**
(a) Oxidative decarboxylation	(i) Vitamin D
(b) Water-soluble antioxidant	(ii) Ascorbic acid
(c) Glutamate residue	(iii) Thiamine
(d) Vitamin	(iv) Folic acid
(e) Increased CS_2 uptake	(v) β-Carotene

3. In each of subquestion, one answer among the alternatives is correct. Write the correct answer in your answer book using letter a, b, c or d along with the corresponding subquestion number. (10 × 1 = 10)

3.1 Which of the following statements about the structure of B-DNA is incorrect?

(a) Within the double helix, there are 10 bases per turn of the helix.
(b) Separation of the two strands of the double helix requires untwisting of the helix.
(c) The double helix contains antiparallel chains.
(d) Molar amount of adenine plus thymine is equal to the molar amount of guanine plus cytosine.

3.2 Which of following is not a characteristic of tRNA?

(a) It contains a codon.
(b) It contains an anticodon.
(c) It can become attached covalently to an amino acid.
(d) It interacts with mRNA during transcription.

3.3 Which of the carbon atoms of pyruvate would be labeled during glycolysis of glucose having ^{14}C at C-1?

(a) Carboxylate carbon (b) Carbonyl carbon
(c) Methyl carbon (d) None of these

3.4 Which of the following is a constituent of chlorophyll?

(a) Plastoquinone (b) Fe^{2+}
(c) Fe^{3+} (d) Substituted tetrapyrrole

3.5 Which of the following statements about natural sterols is incorrect?

(a) Cholesterol is the most abundant sterol in animal tissue.
(b) All the carbon atoms of cholesterol are derived from acetyl CoA.
(c) β-sitosterol is the most abundant plant sterol.
(d) Dietrary β-sitosterol and cholesterol are absorbed to about the same extent in the intestine of normal humans.

3.6 Which of the following statements is correct?

(a) The α-helix can be composed of more than one polypeptide chain.
(b) α-sheets exist only in the antiparallel form.
(c) α-bends often contain praline.
(d) Motifs are a type of secondary structure.

3.7 Which one of the following statements concerning Ca^{2+} is correct?

(a) Intracellular concentration of Ca^{2+} is higher than that of the extracellular fluid.
(b) The concentration of cytosolic Ca^{2+} is transiently decreased by activation of phospholipase C.
(c) The effects of Ca^{2+} is most often mediated by calmodulin.
(d) Intracellular Ca^{2+} is mostly free.

3.8 Which of the following is not a membrane lipid?

(a) cholesterol (b) choline
(c) phosphoglycerides (d) cerebrosides

3.9 Which of the following enzymes is inhibited by methotrexate, an anticancer drug?

(a) Ribonucleotide reductase (b) Reverse transcriptase
(c) Dihydrofolate reductase (d) DNA polymerase

3.10 Which of the following statements about monoclonal antibodies is incorrect?

(a) They can be produced in large amount by growing many mice, each of which is immunized with the same antigen.
(b) They are useful in quantifying specific proteins in human blood.
(c) They can be used to purify scarce proteins by affinity chromatography.
(d) They can be used to prepare catalytic antibodies.

4. (i) List three desired characteristics of a vector, useful for DNA cloning. (3)

   (ii) Give a reason why genetic information is encoded in DNA rather than in RNA. (2)

5. (i) Although trypsin is a proteolytic enzyme, it does not digest the tissue in which it is produced. Explain why? (2)

   (ii) When paper electrophoresis at pH6.0 is performed on a mixture of alanine, arginine, glutamic acid, lysine and serine,

   (a) Which compound(s) would move towards the anode?

   (b) Which compound(s) would move towards the cathode?

   (c) Which compounds(s) would remain near the origin? (3)

6. (i) Calculate the yield in ATP molecules of the complete oxidation of palmitic acid ($C_{16}O$) (3)

   (ii) Explain why arsenate (ASO_4^3) acts as an uncoupling agent for oxidative phosphorylation. (2)

7. Write short notes on

   (i) Liposomes (lipid vesicles) (3)

   (ii) Ribozymes (2)

8. (i) Give reasons why T_m value of a linear double-stranded DNA in 0.3 M NaCI solution is decreased when

   (a) The length of DNA molecule is decreased

   (b) The concentration of NaCl is decreased

   (c) Urea is added to the solution

   (ii) When a monoclonal antibody preparation is treated with β-mercaptoethanol and then electrophoresed (PAGE), two bands appear. On the other hand, a single band appears without the treatment of β-mercaptoethanol. Explain why. (2)

GATE-1999

1. For each sub-question given below, four alternatives are provided of which only one is correct. Write the correct answer in the answer book by writing a, b, c or d along with the corresponding sub-question number $20 \times 1 = 20$

   1. Which of the following statements is incorrect?

      (a) Myoglobin consists of a single polypeptide chain in addition to a heme prosthetic group.

      (b) The red colour associated with the oxygenated forms of myoglobin and hemoglobin is due to heme.

      (c) Hemoglobin has a higher affinity for oxygen compared to myoglobin as seen from the oxygen binding curves.

      (d) The α and β chains of the hemoglobin tetramer are slightly shorter than the single chain of myoglobin.

   2. The molecular structure of a disaccharide is given below. Identify the compound.

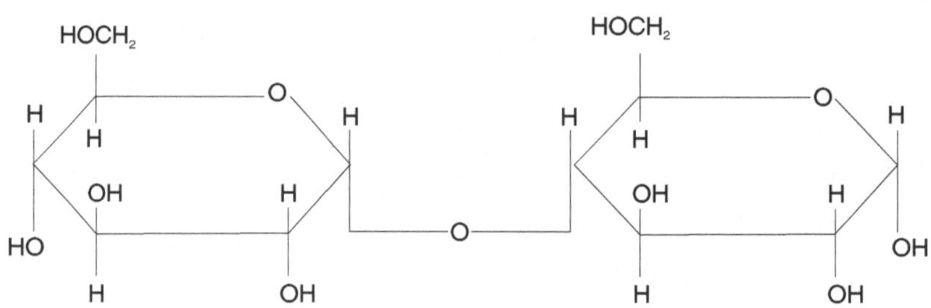

      (a) Lactose (b) Maltose (c) Sucrose (d) Cellobiose

   3. J chain is a glycopeptide chain associated with which of the following immunoglobulins?

      (a) IgA (b) IgG (c) IgM (d) IgE

   4. Which statement about ribonuclease is correct?

      (a) Ribonuclease is an exonuclease.

      (b) The enzyme-mediated cleavage occurs at the pyrimidine and purine nucleotide residues.

      (c) The hydrolysis product is a polynucleotide with a 5'-phosphate terminus.

      (d) The hydrolysis product is a polynucleotide with a 3'-phosphate terminus.

   5. Molecular weight of an unknown protein can be found out by

      (a) Electrophoresis (b) Ion-exchange chromatography

      (c) Affinity chromatography (d) None of the above techniques

   6. N-terminal amino acids are usually determined by Sanger's method using

      (a) Ninhydrin reagent (b) 2, 4-dinitrobenzene (c) Hydrazine (d) Concentrated nitric acid

7. Cleavage of the following peptide with Chymotrypsin generates

N_3N—Gly—Arg—Ser—Phe—Gly—Asn—Lys—Trp—G—Val—COO

(a) 2 fragments (b) 3 fragments (c) 4 fragments (d) No cleavage

8. The average molecular weight of an amino acid in a protein is

(a) 112 (b) 109 (c) 110 (d) 137

9. Which of the following peptides can be easily detected by absorbance of 280

(i) Leu—Tyr—Met—Ala—Glu (ii) Ser—Thr—Trp—Ile—Leu

(iii) Ac—Ala—Glu—Gln—Ser—Asp—Lys (iv) Thr—Tyr—Trp—Val—Ile

(a) (i), (ii) and (iv) (b) (i) and (iv) (c) (ii) and (iii) (d) (ii) and (iv)

10. Phospholipids containing cyclopropane rings have

(a) A higher T_m than saturated acyl chains of the same length
(b) Closer packing of the acyl chains in the gel phase
(c) Greater fluidity in membranes that they form a part of
(d) All of the above

11. The oxygen saturation curves for two oxygen-binding proteins are shown in the figure below. Each protein binds four oxygens. Which protein has a higher affinity for oxygen?

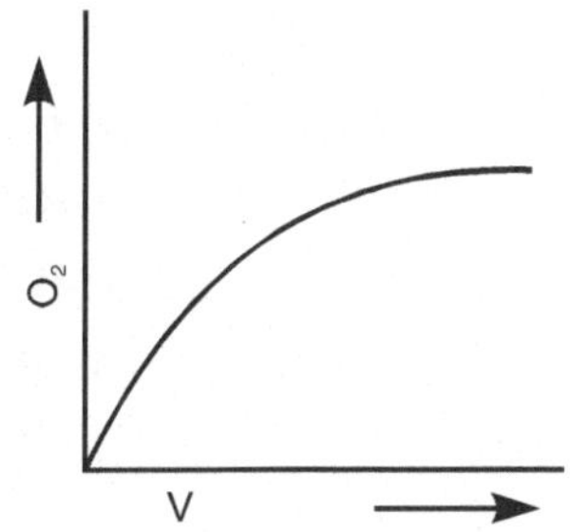

(a) Protein A (b) Protein B
(c) Both proteins A and B (d) None of the proteins

12. The formation of glucose in photochemical reactions involves the reduction of CO_2 with the production of O_2. For this process, which of the following agents is responsible?

(a) $FADH^+$ (b) $NADPH^+$ (c) ADP (d) None of the above

13. Although T-cell and B-cell membranes contain some shared receptors, the receptor specific for B-cell membrane is

(a) Pokeweed mitogen (PWM) (b) Phyto hemagglutinin (PHA)
(c) Concanavalin A (con A) (d) Sheep red blood cells (SRBCs)

14. Most antibodies are synthesized by the

(a) Central lymphoid organs (b) Peripheral lymphoid organs
(c) Primary lymphoid organs (d) Macrophages

15. Of the immunological techniques commonly used
    (a) Enzyme linked immunosorbent assay (ELISA) can be used to assay antibodies only
    (b) Immunoelectrophoresis can be used to determine the concentration of IgG because of its cathodic migration
    (c) The precipitation in immuno-diffusion cannot detect antigenic relationship
    (d) Radioimmunoassay (RIA) is used for quantitation of any substance that is immunogenic or haptenic.
16. Immunosuppressive measures are most effective when administered
    (a) Just prior to antigen exposure (b) One week before antigen exposure
    (c) At the time of antigen exposure (d) Following antigen exposure
17. Which statement about oxidative phosphorylation is not true
    (a) The five assemblies of proteins and cofactors involved include the electron-transferring complexes I-IV and ATP synthase.
    (b) The mobile carrier ubiquinone links complexes I and II with complex III and cytochrome C links complex III with complex IV.
    (c) For every two electrons transferred from NADH to Q by complex I, two protons are translocated to the intermembrane space.
    (d) FAD, FMN, iron-sulphur clusters and copper atoms are other co-factors that participate in electron.
18. Z-DNA is a third type of DNA helix with structure
    (a) CGATGC (b) ATGCCG (c) GCTAAT (d) CGCGCG
19. The TATA box plays a key role in assembling active transcription complexes by
    (a) Binding with TATA box-binding protein (b) Binding with DNA protein
    (c) Binding with RNA protein (d) It does not bind with any protein
20. The order of the predicted rate of diffusion across a lipid bilayer (highest to lowest) is
    (a) Toluene, galactose, phenylalanine, chloride ion
    (b) Chloride ion, phenylalanine, galactose, toluene
    (c) Galactose, phenylalanine, chloride ion, toluene
    (d) Toluene, phenylalanine, galactose, chloride ion

2. Match the entries in the Column A with those in Column B and write the matching pairs in the answer book.

Column A	**Column B**
a. Hexokinase	(i) isocitrate
b. Pyruvate dehydrogenase	(ii) 3-phosphoglycerate
c. Aconitase	(iii) Glucose 6-phosphate
d. Succinate dehydrogenase	(iv) Acetyl CoA
e. Ribulose diphosphate carboxylase	(v) Fumarate

3. (a) What are the immediate consequences of a limitation in the following factors on the rate of photosynthesis?

   (i) light and CO_2

   (ii) The rate of electron transport

   (iii) The activity of RuB is CO

   (b) The two ionizable – OII, groups of the phosphate group in ex-D-Ribose 5- phosphate have pKa values of 1.2 and 6.6. What is the structure of the predominant form at physiological pH?

4. (a) In aqueous solutions below pH 3.0, the synthetic polypeptide polyglutamate, forms regions of α -helices. Above pH 5.0 it assumes an extended conformation. Explain this observation.

   (b) In glycolysis, there are two reactions that require one ATP each and two reactions that produce one ATP each. This being the case, how can glycolysis of glucose to lactate lead to the net production of two ATP molecules per glucose molecule?

5. (a) The transformation of glucose to fructose 6-phosphate occurs as follows.

   Glucose + ATP $\longrightarrow$ Glucose 6-phosphate + ATP

   $\Delta G^{o} = 4.0$ kcal/mol

   Glucose 6-phosphate $\longrightarrow$ Fructose 6-phosphate

   $\Delta G^{o} = +4.0$ kcal/mol

   Calculate the standard free energy change and equilibrium constant for this transformation at 25°.

   (b) What are the main features of a zinc finger?

6. (a) *In vitro* sequencing reactions are commonly carried out at 65°C and usually employ a DNA polymerase isolated from bacteria that grow at high temperatures. What is the advantage of this?

   (b) What are the main features of the enzyme linked immunosorbent assay (ELISA) method?

7. (a) The chromosome of a certain bacterium is a circular, double-stranded DNA molecule of 5.2×10^6 base pairs. The rate of replication-fork movement is 1000 nucleotides per second. Calculate the time required to replicate the chromosome.

   (b) Suppose you want to express the β -globin gene using an expression vector. What factors are important for maximizing the efficiency of transcription and translation if the expression vector is.

   (i) a eukaryotic vector (ii) a prokaryotic vector (iii) a fusion-protein vector

GATE-2000

1. For each sub-question given below, four alternatives are provided of which only one is correct. Write the correct answer book by writing a, b, c or d along with the corresponding sub-question number.

1.1. The properties of water include

(a) The ability to form hydrophobic bonds with itself

(b) A disordered structure in the liquid state

(c) A low dielectric constant

(d) Being a dipole, with the negative end at the oxygen atom

1.2. Deficiency of which one of these enzymes cause defect in the storage of glycogen.

(a) Lactate dehydrogenase
(b) Glyceraldehyde 3-phosphate dehydrogenase
(c) Phosphorylase
(d) Glycogen synthetase

1.3. The turnover number of chymotrypsin is 100 s^{-1} and for DNA polymerase it is 15 s^{-1}. This means that

(a) Chymotrypsin binds to its substrates with higher affinity than does DNA polymerase.

(b) The velocity of the reaction is always greater than that of DNA polymerase.

(c) The velocities of reactions catalysed by both enzymes at saturating substrate level could. be made equal if 6.7 times more DNA polymerase than chymotrypsin were used.

(d) The velocities of chymotrypsin reaction at a particular enzyme concentration and saturating substrate levels is lower than that of DNA polymerase reaction under the same conditions.

1.4. Which of the following statements is true? Enzyme catalysis of a chemical reaction

(a) Increases the forward and reverse reaction rates

(b) Decreases ΔG° so that the reaction can proceed spontaneously

(c) Increases the energy of transition state

(d) Decreases the entropy of reaction

1.5. Which of the following enzymes can be irreversibly inactivated with diisopropyl fluoridate (DIPF)?

(a) Carboxypeptidase A
(b) Trypsin
(c) Lysozyme
(d) *Eco*RI endonuclease

1.6. Histones have very high percentage of arginine and lysine residues (15–30%). For this class of proteins, which of the following reagents would be a suitable choice for generating peptides in the determination of the amino acid sequence of the protein

(a) Cyanogen bromide (b) Thermolysin
(c) Trypsin (d) *N*-bromosuccinamide

1.7. The half-life of [^{32} p] phosphate is approximately 14 days. Forty-two days after the purchase of a batch of [^{32}p], the radioactivity present was 1.25 mCi. The radioactivity on the date of purchase would be

(a) 5 mCi (b) 0.625 mCi (c) 2.5 mCi (d) 10 mCi

1.8. A 100 ml. of 0.1 M sodium acetate solution was with 300 ml. of 0.3 M sodium acetate. The molarity of the final solution would be

(a) 330mM (b) 250mM (c) 200mM (d) 400mM

1.9. A double-stranded DNA and has 30% thymine. The percentage of cytosine is

(a) 30% (b) 20% (c) 70% (d) 15%

1.10. The following type of interaction is mainly responsible for aggregation of proteins in dilute solutions

(a) Hydrogen bonds (b) Hydrophobic interactions
(c) Disulfide bonds (d) Peptide bonds

1.11. The summarization of this chromophore by light is the first event in visual excitation

(a) Retinol (b) All-*trans* retinal (c) 11-*cis* retinal (d) Retionic acid

1.12. In the transaminases the cofactor is linked via Schiff base formed between the cofactors and the

(a) α-NH_2 group of N-terminal amino (b) β-NH_2 group of asparagines
(c) ε-NH_2 group of Iysine (d) None of the above

1.13. An enzyme protein forms 0.001% of the total soluble protein in a crude extract The degree of purification required to obtain a homogeneous enzyme is

(a) 1000-fold (b) 100-fold (c) 100,000-fold (d) 10-fold

1.14. A homogeneous protein of native molecular weight 100, 000 gave a single band of molecular weight 50,000 on SDS-PAGE in the presence of β–mercaptoethanol. N terminal analysis gave two amino acids alanine and leucine in equal proportions. Hence the

(a) Protein is a homodimer
(b) Protein is contaminated with another protein
(c) Protein has two polypeptides linked by disulfide bridges
(d) None of the above

1.15. Cholecalciferol (Vitamin D), in order to become biologically active, undergoes hydroxylation at two positions in the molecule. Hydroxylation at the 25th position is done in the liver whereas hydroxylation at the 1st position occurs in

(a) Skin (b) Pancreas (c) Intestine (d) Kidney

1.16. How many asymmetric carbon atoms are present in the compound whose structure is given below

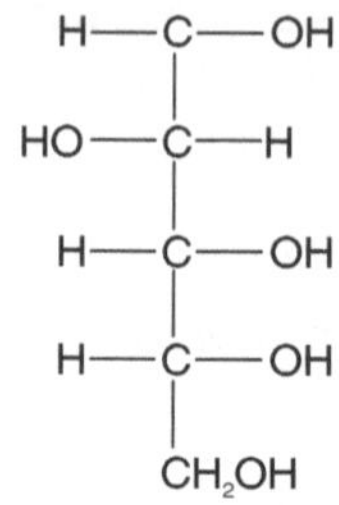

(a) 1 (b) 2 (c) 3 (d) 4

1.17. Diphtheria toxin inhibits protein synthesis by

(a) Causing formation of ADP–EF2 complex
(b) Release of peptidy l-t-RNA from the 'P' site
(c) Binding to factor EF-1
(d) Inhibiting peptide bond formation

1.18. Antigenic peptides are presented to T cells by

(a) TCR/CD3 (b) CD 28 (c) CTLA4 (d) MHC

1.19. Which of the following statements about cyclic photophosphorylation is not correct?

(a) It does not involve NADPH formation.
(b) It uses electrons supplied by photosystem II.
(c) It involves a substrate level phosphorylation.
(d) It does not generate oxygen.

1.20. Which of the following statements about biological membranes is not true?

(a) They contain carbohydrates that are covalently bound to proteins and lipids.
(b) They are large sheet-like structures with closed boundaries.
(c) They are symmetric because of the symmetric nature of lipid bilayers.
(d) They contain specific proteins that mediate their distinctive functions.

2. Match the entries in Column 1 with those in Column 2 and write matching pairs in the answer book.

Column-1	**Column-2**
(a) Van der Waals bonds	(i) Spatial arrangement of amino acids that are near each other in the linear sequence
(b) γ - Globulins	(ii) Glycosylation inhibitor
(c) Tertiary structure	(iii) Cooperativity of oxygen binding
(d) Hill coefficient (n_H)	(iv) Involves polarizable atoms
(e) Tunicamycin	(v) Immune protection

3. (a) In the Entner–Doudoroff pathway, 2-keto 3-deoxy 6-phosphogluconate is acted upon by an enzyme resulting in the formation of two 3-carbon compounds. Name the enzyme and the products formed

(b) Two first-order constants k_2 and k_3 and one second-order rate constant k_1, define K_m by the relationship between the two enzymes. By substituting the appropriate units for the rare constants in this expression, show that K_m must be expressed in terms of concentration.

4. (a) Arsenate, AsO_4^{3-}, is an uncoupling reagent for oxidative phosphorylation but unlike dinitrophenol, it does not transport protons across the mitochondrial membrane. How might arsenate function as an unocoupler?

(b) Removal of the thymus gland from a young animal not only compromises the production of cytotoxic T cells, but also decreases antibody production by B cells. Why is this the case?

5. (a) Sketch the appropriate plots if the following variables are plotted. Assume that single Michaelis-Menten kinetics apply.

(i) v versus [S] (ii) [S] versus time (iii) [ES] versus time

(b) When lactose is used as an inducer, there is a lag before the enzymes of the lactose operon are synthesized. With IPTG there is no lag. Explain.

6. (a) The structure of a dimer of murein present in bacterial cell wall is represented below.

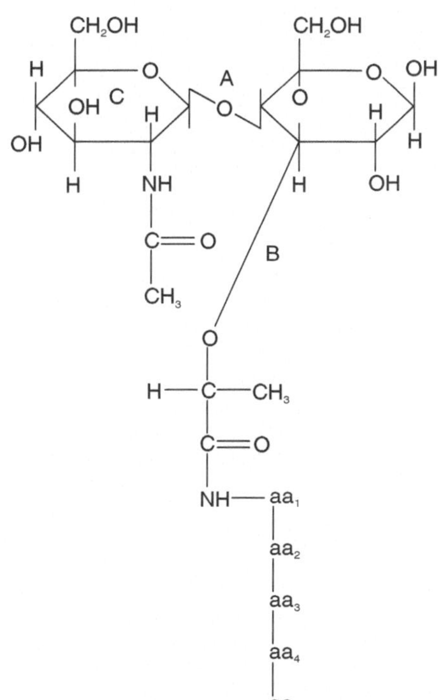

(i) Name the linkage at A (ii) Name the linkage at B (iii) Name the sugar at C

(b) The DNA in a bacterium is uniformly labeled with ^{15}N and the organism is shifted to a growth medium containing ^{14}N-labeled DNA precursors. After two generations of growth, the DNA is isolated and is subjected to density-gradient equilibrium sedimentation. What proportion of light-density DNA to intermediate-density DNA would you expect to find?

7. (a) If $^{18}O_2$ were added to C3 plants on a bright, sunny day, would you expect glycine subsequently isolated from the leaves to be labeled? Explain.

(b) Early experiments on the problem of protein folding suggested that the native three-dimensional structure of a protein was an automatic consequence of its primary structure. Cite an experimental evidence that shows that this is the case.

GATE-2001

1. For each sub-question given below, four alternatives are provided of which only one is appropriate Write the answer in the answer book by writing a, b, c, or d along with the corresponding sub-question number.

1.1 The pH of 1 mM HCl is

(a) 1 (b) 2 (c) 0 (d) 3

1.2 The length of an α helical section of a polypeptide chain of 20 residues would be.

(a) 30 Å (b) 20 Å (c) 5.4 Å (d) 3.6 Å

1.3. Which one of these reagents is best suited for identification of the amino terminal residue of a peptide of which you have less than 0.1 μg.

(a) Performic acid (b) Dansyl chloride
(c) Fluorodinitrobenzene (d) Cyanogen bromide

1.4. If the solution of double-stranded DNA is heated above its melting temperature its absorbance will,

(a) Decrease (b) Increase
(c) Remain unchanged (d) Initially increase and then decrease

1.5 β-amylase cleaves

(a) α (1-4) glycosidic bonds (b) β (1-4) glycosidic bonds
(c) α (1-6) glycosidic acid (d) β (1-6) glycosidic bonds

1.6 Gangliosides contain

(a) A ceramide structure (b) Glucose or galactose
(c) Sialic acid (d) All of the above

1.7 A southern transfer of *E. coli* DNA after complete digestion with E*co*RI was probed with labeled cDNA probe of a gene which occurs only once in the *E. coli* genome. If the gene contains one *Eco*RI cleavage site near its centre, the number of radioactive bands you are most likely to find on autoradiography would be

(a) 0 (b) 1 (c) 2 (d) 3

1.8. The largest energy reserve (kilocalories) in human is

(a) Blood glucose (b) Liver glycogen
(c) Muscle glycogen (d) Adipose tissue triacylglycerol

1.9. Both strands of DNA serve as templates concurrently in

(a) Replication (b) Mismatch repair (c) Excision repair (d) All of the above

1.10 In which of the following does the inner surface of a closed membrane or vesicle become the outer surface of a closed membrane or vesicle?

(a) Fusion of two intracellular vesicles
(b) Transfer of an endoplasmic reticular membrane into Golgi membrane via vesicle
(c) Exocytosis of a secretory vesicle?
(d) Division of a bacterial cell

1.11 Enzyme linked immunosorbent assay used to detect antigens or antibodies utilizes those enzymes that

(a) Have a high turnover rate
(b) Yield a stable colored product
(c) Are stable on conjugation to proteins
(d) All of the above

1.12 Which of the following horomones uses cyclic AMP as the second messenger

(a) Follicle stimulating hormone
(b) Insulin
(c) Progesterone
(d) Gonadotropin releasing hormone

1.13 IgM class of antibodies are more effective than IgG in bringing about complement-mediated lysis because

(a) IgM has higher affinity for complement components
(b) IgM has higher affinity for antigen binding
(c) IgM is a pentamer
(d) IgM is present in higher concentration

1.14 Which of the following statements is false

(a) MHC class II molecules are present on all nucleated cells
(b) MHC class II molecules are made up of two polypeptides
(c) MHC class I molecules are present on all nucleated cells
(d) T cell response is MHC-restricted

1.15 Which of these substrates is best suited for measuring RNA systhesis by RNA polymerase?

(a) g p^{32} ATP
(b) a p^{32} d ATP
(c) g p^{32}d ATP
(d) a p^{32} UTP

1.16 The tripeptide sequence encoded by the following polydeoxyribonucleotide sequence, 5'CCC AAA TAC 3' would be

(a) Pro ly stry
(b) Met phe gly
(c) Met gly lys
(d) Met phe lys

1.17 All of the following are true about transposons except

(a) Transposons move from one location to a different one within a chromosome.
(b) Both donors and target sites must be homologous.
(c) Transposons may activate a gene.
(d) Transposons may inactivate a gene.

1.18 The class of immunoglobulilns that can get transported across epithelial cell is

(a) IgG
(b) IgE
(c) IgA
(d) IgM

1.19 A muscle extract is dialysed exhaustively against 10 mM sodium phosphate buffer, pH 7.0. The extract does not contain any ATPase activity. If ATP is now added, the following cofactor that also must be added to convert glucose to glucose-6-phosphate is

(a) NAD (b) $FADH_2$ (c) Mg^2 (d) Fe^{3+}

1.20 Injection of dinitrophenol (DNP) into a rat causes an immediate increase in its body temperature, because

(a) DNP acts as an inhibitor of mitochondrial ATPase
(b) DNP blocks the electron transport chain
(c) DNP inhibits succinic dehydrogenase
(d) DNP uncouples electron transport from oxidative phosphorylation

1.21 A rat liver extract is treated with avidin. Which of the following conversion in gluconeogenesis will not occur?

(a) Glyceraldehyde 3-p⟶glucose 6-P (b) Malate⟶Oxaloacetate
(c) Pyruvate⟶Phosphoenopyruvate (d) Fumarate⟶Phosphoenolpyruvate

1.22 The direct effect of cyclic AMP in the protein kinase A pathway is to

(a) Activate adenylate cyclase
(b) Dissociate regulatory subunits form protein kinase A
(c) Phosphorylate protein kinase A
(d) Release hormones from target tissue

1.23 The active site amino acid residue that could be involved in a reaction catalysed by an enezyme with a pH optimum of 4 would be

(a) Arginine (b) Cysteine (c) Serine (d) Glutamate

1.24 The isoelectric pH of γ globulin, human serum albumin (HAS), ribonuclease (RNase), and hen egg lysozyme (HEL) are 6.6, 4.9, 7.8 and 11.0 respectively. The order in which these proteins will be eluted from a CM cellulose ion exchange column by an increasing salt gradient at pH 7.0 would be

(a) HAS, γ globulin, RNase and HEL (b) HEL, RNase, γ globulin, HSA
(c) HEL, γ globulin, RNase, HAS (d) HAS, RNase, HEL, γ globulin

1.25 A solution contains DNA polymerase along with Mg^{2+} salts induces formation of

(a) Single-stranded RNA base-paired with a primer with a free 3'OH group
(b) Single-stranded DNA base-paired with a primer with a free 5' OH group
(c) Single-stranded DNA base-paired with a primer with a free 3'OH group
(d) Single-stranded DNA base-paired with a free 3'OH group at each end

2. Give correct answers to the following

(a) Which of the following amino acids are likely to have their side chains on the inside of globular proteins in solution, Val, pro, phe, asp, lys, ile, his
(b) Which of the following ion exchangers would be suitable for purification of DNA?

(c) What is the coenzyme form of vitamin B6? Name the reactions in which this coenzyme is involved.

(d) In a preparation of mitochondia, oxidation of fatty acids is carried out in the presence of CoA, O_2 and ω_2.

3. Answer the following in one sentence each (DO not exceed) (1 × 5 = 5)

(a) What are RAG genes?

(b) What are recombination single sequences?

(c) How does simultaneous expression of IgM and IgD occur in mature B cells?

(d) Name any four members of the immunoglobulin superfamily other than immunoglobulin?

(e) Define monoclonal antibodies.

4. State true or false against each of the following (1 × 5 = 5)

(a) Methylation of lysine ε–amino group occurs in histones to modulate their interaction with DNA

(b) In all cases of diabetes mellitus, circulatory levels of insulin are low or negligible.

(c) While bone marrow is the muscle, glycogen functions as glucose reserve for maintenance of blood glucose concentration.

(d) Both in the liver and muscle, glycogen functions as a glucose reserve for maintenance of blood glucose concentration.

(e) Type I topoisomerases create transient single-stranded breaks in DNA, whereas type II topoisomerases make transient double-stranded breaks in DNA.

5. Match the entries in column 1 with the appropriate one in the column 2 and write matching pairs in the answer book

(a) Ionophore	(i) Cathepsin
(b) Oubain	(ii) Na^+K^+ATPase
(c) Malonate	(iii) Homogentisic acid oxidase
(d) Alkaptonuria	(iv) Phenylalanine hydroxylase
(e) Protein hydrolyzing enzymes	(v) Penicillin
	(vi) Gramicidine A
	(vii) Competitive inhibitor of succunic dehydrogenase

6. (a) The K_m of a certain enzyme is μ mole/min. However the V_{max} remained the same, even at a tenfold lower concentration of the substrate. Explain. (2)

(b) Restriction analysis of a fragment of dsDNA with II and III gave fragments of sizes 2.8 kb and 1.2 kb. The same DNA upon digestion with BamH1, yielded fragment sizes 1.8 kb, 1.0 kb, 0.9kb and 0.3 kb. Making use of this data draw a restrication map of the DNA fragment.

(c) Define isoenzymes.

GATE-2002

1. For each sub-question given below, four alternatives are given of which only one is appropriate Write the answer in the answer book by writing a, b, c, or d along with the corresponding sub-question number. $(1\times 25 = 25)$

1.1 Protein may be separated according to size by

(a) Reverse phase chromatography (b) Ion exchange chromatography
(c) Molecular exclusion chromatography (d) Isoelectric focusing

1.2 Which of the following has a quaternary structure?

(a) α-chymotrypsin (b) Hemoglobin
(c) Insulin (d) Myoglobin

1.3 Which of the following anti-tumor agent acts by impairing the de novo purine synthesis

(a) Cytosine arabinoside (b) 5-Fluorouracil
(c) Methotrexate (d) Hydroxyurea

1.4 Which of the following is a conservative substation?

(a) Val to Ile (b) Asp to Pro (c) Lys to Leu (d) Try to Ala

1.5 cAMP activates all except one of the following molecules,

(a) Glycogen phosphroylase (b) Hexokinase
(c) 6- Phosphofructokinase-1-kinase (d) Try to Ala

1.6 Which of the following statements is correct in case of glycoproteins?

(a) May contain 95% or more of carbohydrate
(b) Have the carbohydrate linked to the protein by either N- or O-glycosidic bonds
(c) Found only on cell membranes
(d) Always contain a serially repeating carbohydrate unit

1.7 Phospholipids are involved in all except one of the following,

(a) Cell-cell recognition (b) Single transduction
(c) Surfactant function on the lungs (d) Mediator of hypersensitivity

1.8 Insulin-dependent diabetes mellitus is associated with

(a) High levels of insulin (b) Severe weight gain
(c) Destruction of β-cells of pancreas (d) Mutation of insulin receptor

1.9 The Z-DNA helix

(a) Has fewer base pairs per turn than the B-DNA
(b) Is favoured by an alternating GC sequence

(c) Tends to be found at the 3 end of genes
(d) Is the most common conformation of DNA

1.10 Formation of a mature insulin does nor include
(a) Removal of a signal peptide
(b) Disulfide bond formation
(c) Removal of a peptide from an internal region
(d) Cyclization of a glutamate residue

1.11 How many energy bonds are expended in the formation of a peptide bond?
(a) 2 (b) 4 (c) 3 (d) 6

1.12 A technique for defining gene arrangement in very long stretches of DNA (50-100 kb) is,
(a) RFLP (b) Chromosome walking
(c) Nick translation (d) Southern blotting

1.13 Receptors for steroid hormones are found
(a) On cell membranes (b) In cytoplasm
(c) In mitochondria (d) On ribsomes

1.14 Platelet aggregation is
(a) Initiated at the site of injury by the conversion of fibrinogen to fibrin
(b) Causes release of serotonin
(c) Is inhibited by uninjured blood vessels by the secretion of prostacyclins
(d) Is inhibited by ADP and thromboxane

1.15 Oral administration of large amounts of tyrosine may interfere with the intestinal absorption of
(a) Leucine (b) Glycine (c) Aspartate (d) All of the above

1.16 In the 3-dimensional structure of immunoglobulin G,
(a) Free SH-group are conserved to function in binding of antigen
(b) C_H and V_H association form the complementary antigen binding sites.
(c) The predominant secondary structure is α-helix
(d) Hinge regions connect the globular domains

1.17 Change in protein conformation can be detected by
(a) Ultraviolet absorption spectroscopy (b) Fluorescence emission
(c) Circular dichrosim (d) All of these

1.18 Endonuclease is an enzyme that hydrolyzes,
(a) A nucleotide at the 3'end of an oligonucleotide
(b) A phosphodiester bond located in the interior of a polynucleotide
(c) A nucleotide from either terminii of an oligonucleotide
(d) A peptide bond located in the interior of a polypeptide

1.19 The K_m of an enzyme is

(a) One-half of the V_{max}
(b) A dissociation constant
(c) The substrate concentration that gives maximal velocity
(d) The substrate concentration that gives half maximal velocity

1.20 The class of immunoglobulins most abundant in body fluids is,

(a) IgM (b) IgG (c) IgA (d) IgE

1.21 Which of the following statements is correct in case of chaperone proteins?

(a) These do not prevent aggregation.
(b) They cleave incorrect S-S bonds.
(c) Act on fully synthesized polypeptides.
(d) Are involved in the transport of proteins across mitochondria and endoplasmic reticulum.

1.22 One of the following participates in phagocytic activites,

(a) Neutrophils (b) Master cells (c) T cells (d) Thrombocytes

1.23 The basal metabolic rate,

(a) Is not influenced by energy intake
(b) Increases in response to starvation
(c) Increases in direct proportion to energy expenditure
(d) Decreases during periods of starvation

1.24 Operons,

(a) Are of approximately uniform size
(b) Do not bind proteins
(c) Are found in all eukaryotic genes
(d) Are shorter and smaller in lower enkaryotes than higher eukaryotes

1.25 In humans, fatty acids

(a) Can be synthesized from excess dietary carbohydrate or protein
(b) Must be supplied entirely by diet
(c) Are not required at all in the diet
(d) Containing double bonds cannot be synthesized

2. (a) Blood groups are determined by the allelic genes A,B, and O. These genes code for different enzymes that result in the production of blood group antigens. What is the nature of these enzymes? (2)

(b) Where are Rh antigens found? (1)

(c) How many genes determine Rh antigens? (1)

(d) Name the immunodominant Rh antigen? (1)

3. Match the entries in Column I with one appropriate from Column II and write matching pairs in the answer books. (1× 5 = 5)

Column I	Column II
(a) Apoptosis	(i) CD3 complex
(b) T cell receptor	(ii) Caspases
(c) 5-Hydroxytrytamine	(iii) Adenylate cyclase
(d) β-adrenergic receptor	(iv) Mitotic spindle
(e) Colchicine	(v) Neurotransmitter
	(vi) immunosuppressive
	(vii) DNA binding protein

4. (a) What is the possible number of protein sequence that can be read from a genomic sequence? (1)

   (b) What are the different activities present in the Klenow fragment? Name the enzyme from which Klenow originates? (2)

   (c) In order to end-label a DNA molecule at its 5 end, what is the nature of the labeled nucleotide used? Name the enzyme used. (2)

5. (a) What effect would the removal of the bursa of Fabricius have on chickens? (1× 5 = 5)

   (b) Name the primary lymphoid organs in the adult human.

   (c) Define monoclonal antibodies.

   (d) Which are the two isotypes of immunoglobulins that can be simultaneously expressed on native B cells?

6. A purified enzyme preparation (3 mg/ml) was diluted 10 times. 10 μl of the diluted enzymes catalyzed the formation of 30 nmoles of product/min under standard assay conditions.

   (a) Express the velocity as μ moles product/min/mg. (2)

   (b) What does the velocity expressed in μ moles/min/mg signify? (1)

   (c) If the molecular weight of the enzyme is 30,000, calculate its turnover number. (2)

GATE-2003

Questions 1–10 carry one mark each

1. Which amino acid residue is most likely to be found in the interior of a water soluble globular protein?

(a) Ser (b) Arg (c) Val (d) Asp

2. Of the peptide sequences given below, which one is the digestive enzyme trypsin most likely to cleave?

(a) ——Val-Lys-Pro-Met—— (b) ——Arg-Val-Phe-Tyr——-
(c) ——Trp-Asp-Gln-Pro—— (d) ——Glu-Gly-Trp-Gly——

3. Which one of amino will have the highest absorbance at 280 nm? (Assume equimolar concentrations)

(a) Thr and His (b) Phe and Pro (c) Trp and Tyr (d) Phe and His

4. Which one of the following statements about protein secondary structure is correct?

(a) An α-helix is primarily stabilized by ionic interactions between the side chains of the amino acids.
(b) β–sheets exist only in antiparallel form
(c) β–turns often contain praline
(d) An α–helix is can be composed of more than one polypeptide chain

5. The enzymes where catalysis involves transfer of electrons are named as

(a) Isomerases (b) Transferases
(c) Oxidoreductases (d) Lyases

6. Vitamin D is derived from which of the following precursors by action of UV light?

(a) 7-Dehydrocholesterol (b) Lanosterol
(c) Glycocholate (d) Squalene epoxide

7. Molecular defect in familial hypercholesterolemia is due to the lack of functional

(a) VLDL receptor (b) IDL receptor (c) LDL receptor (d) HDL receptor

8. Alcaptonuria is an inborn error in metabolism, transmitted as a single recessive Mendelian trait where the enzyme that is absent is

(a) Phenylalanine hydroxylase (b) Ornithine decarboxylase
(c) Adenosine deaminase (d) Homogentisate oxidase

9. The prokaryotic RNA polymerase holoenzyme has the subunit structure
   (a) $\alpha_2\beta\ \beta'\sigma$ (b) $\alpha_2\beta_2'\sigma$ (c) $\alpha_2\beta_2$ (d) $\alpha_2\beta\ \beta'$
10. Given below are four enzymatic reactions involved in glycolysis. In which of the following steps is ATP generated?
    (a) 2-Phosphor glycerate to phosphoenol pyruvate
    (b) Glucose 6-phosphate to Fructose 6-phosphate
    (c) Phospoenol pyruvate to pyruvate
    (d) Glyceraldehyde 3-phosphate to 1,3-bisphosphoglycerate

Questions 11–30 carry one mark each

11. The correct decreasing order of permeability through a lipid bilayer of the molecules/ions Isoleucine, Tyrosine, O_2 and Na^+ is
    (a) O_2 > Na^+ > Isoleucine > Tyrosine
    (b) O_2 > iosleucine > Tyrosine > Na^+
    (c) Isoleucine > Tyrosine> O_2 > Na^+
    (d) Isoleucine > Tyrosine > Na^+ > O_2
12. For the reaction. Fructose 6-phosphate + PiFructose 1, 6- bisphosphate + H_2O the equilibrium constant at pH7 and 300 K is 10^{-3} The standard free energy change (in kcal per mole) for the reaction is approximately equal to (R = 2 cal deg^{-1} mol^{-1}
    (a) +4.1 (b) –4.1 (c) +2.2 (d) –2.2
13. Of the four statements given below only one is correct. Pick the correct one.
    (a) Progesterone is synthesized in the corpus luteum and it prepares the uterine lining for egg implantation and maintenance of pregnancy.
    (b) Progesterone is synthesized in the ovary and is responsible for female secondary sex characters.
    (c) Progesterone is synthesized in the adrenal cortex and promotes gluconeogenesis and glycogen formation.
    (d) Progesterone is synthesized in testis and is responsible for male secondary sex characters.
14. Which one of the following statements about lipoproteins is true?
    (a) Molecular mass of lipoproteins is directly proportional to their density.
    (b) The percent protein content in lipoproteins increases with molecular mass.
    (c) Density of a lipoprotein decreases with increase in protein content.
    (d) Molecular mass of lipoproteins is inversely proportional to their density.

15. Choose the correct commons sequence motif of Zn finger proteins from the choices given below. X stands for any amino acid.
    (a) X_3-Cys-$X_{2\text{-}4}$-Cys-$X_{1\text{-}2}$-His-$X_{2\text{-}4}$-His-X_4
    (b) X_3-Cys-Cys-$X_{2\text{-}4}$-His-His-X_4
    (c) X_3-Cys-Cys-His-His-X_4
    (d) X_3-Cys-$X_{2\text{-}4}$-His-$X_{1\text{-}2}$-His-$X_{3\text{-}4}$-Cys-X_4
16. Pick the correct statement,
    (a) In O-linked glycosylation, sugars are attached to the protein via O-glycosidic bonds to the carboxyl groups of Asp and Glu.
    (b) In O-linked glycosylation, performed oligosaccharides are attached to the relevant protein.
    (c) In O-linked glycosylation, N-acetylgalactosamine is added via O-glycosidic bonds to the OH groups of Ser and Thr after which other sugars are added sequentially.
    (d) O-linked glycosylation is inhibited by the passage of the newly synthesized protein through the Golgi complex.
17. Which of the following statements is NOT true with regard to photosynthesis?
    (a) The dark reactions use NADPH and ATP to drive the synthesis of carbohydrate from CO_2 and H_2O.
    (b) The principle phototeceptor, chlorophyll, is derived biosynthetically from protoporphyrin IX.
    (c) Photosystem II (PS II) generates a strong reductant capable of reducing $NADP^+$.
    (d) The components involved in the electron transport from H_2O to NADPH are largely organized into three thylakoid membrane-bound particles.
18. In the adjoining Ramachandran diagram, which type of secondary structure do the regions marked X,Y and Z represent?
    (a) X: right-handed α-helix
    Y: left-handed α-helix
    Z: β-sheet

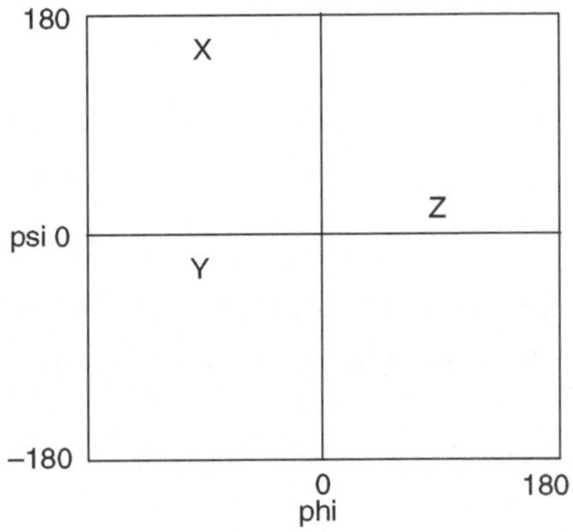

(b) X: left-handed α-helix
Y: right-handed α-helix
Z: β-sheet

(c) X: β-sheet
Y: right-handed α-helix
Z: left-handed α-helix

(d) X: β-sheet
Y: left-handed α-helix
Z: right-handed α-helix

19. The RNase A catalyzed hydrolsis of tRNA follows a two-step process with the intermediate formation of a 2'.3'-cyclin nucleotide. In these steps

(a) His 12 acts as a general base in the transphosphorylation step abstracting a proton from an RNA 2'-OH group

(b) His 119 acts as a general base in the transphosphorylation step abstracting a proton from an RNA'-OH group

(c) The 2', 3' – cyclic intermediate is hydrolyzed when His 12 acts as a general base

(d) His 12 acting as a general acid in the transphosphorylation step promotes bond scission by protonating the leaving group

20. Shown below is the autoradiogram of an electrophoresis gel obtained during the sequencing of a single-stranded DNA by Sanger's method. The base sequence of the DNA is

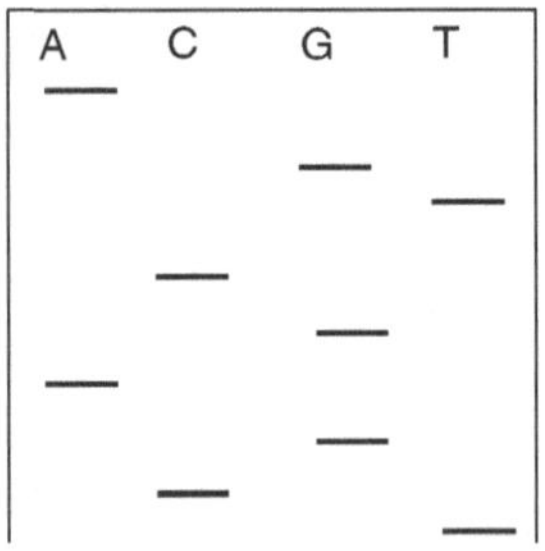

(a) 3'-AGTCGAGCT-5'
(b) 5'-TCAGCTCGA-3'
(c) 3'-TCAGCTCGA-5'
(d) 5'-AGTCGAGCT-3'

21. In an antibody interaction, in the zone of equivalence, the isolated antigen IgG complex was found to be in the molar ratio of antigen: IgG, 2:1. The number of epitope (s) present on the antigen is

(a) 1 (b) 2 (c) 4 (d) 10

22. Class switching occurs in a B cell to produce IgG from IgM. Which one of the following conclusions is correct?
    (a) Specificity on the IgG and IgM are different from each other.
    (b) Specificity or the IgG is the same as IgM.
    (c) The molecular weight of the new antibody is the same as the old one.
    (d) The valency of the new antibody is the same as the old one.

23. Two types of IgMs are produced by the same B cell, one that is secreted and the other that can bind to the membrane. Which one of the following statements is correct?
    (a) These two antibodies are coded by altogether two different genes.
    (b) The membrane anchor residues are added to the protein by post-translational modification.
    (c) The two different proteins associate, one of which provides the membrane anchor.
    (d) The proteins are produced by alternate splicing of its primary transcript.

24. During DNA replication, short RNA primers are synthesized which are then extended by DNA polymerase. These RNA primers in prokaryotes are removed by the enzyme
    (a) Primase (b) RNase H
    (c) DNA polymerase I (d) DNA polymerase III

25. A new antibiotic was discovered which strongly inhibited mRNA precursor transcripts and s_n RNA transcripts. This antibiotic was predicated to be an inhibitor of
    (a) RNA polymerase I (b) RNA polymerase II
    (c) RNA polymerase III (d) Helicase

26. Suppose [4-^{14} C] oxaloacetate is fed to mitochondria. After one turn of the citric acid cycle, which carbon(s) succinate would be labeled?
    (a) None
    (b) Equally distributed between C–1 and C–4
    (c) Equally distributed between C–2 and C–3
    (d) C–4

27. Two restriction enzymes A and B have eight and four base pairs as their recognition sites respectively. The ratio of the number of fragments that will generate on restriction digestion of a genomic DNA of *E. coli* is approximately
    (a) 4 : 8 (b) 8 : 4 (c) 1 : 64 (d) 1 : 256

28. A solution of tryptophan has an absorbance at e280 nm of 0.54 in a 0.5 cm path length cuvette. Given absorbance coefficient (ε) for tryptophan is 5.4×10^3 M^{-1} cm^{-1}, the concentration of the solution is
    (a) 0.mM (b) 20 μM (c) 1×10^{-3} M (d) 0.1 mM

29. From the data given below, identify the protein pair that would (a) give the least mobility band on a sodium dodecyl sulphate polyacrylamide gel electrophoresis (SDS-PAGE) experiment and (b) elute last on an anion exchange, e.g., DEAE column respectively.

Protein	pI	Subunit M.W.	Native M.W
A	9	10,000	20,000
B	8	35,000	35,000
C	6	15,000	90,000
D	5	20,000	80,000
E	3	30,000	30,000

(a) (a) Protein C (b) Protein A
(b) (a) Protein A (b) Protein E
(c) (a) Protein B (b) Protein E
(d) (a) Protein B (b) Protein A

30. The graph shows a Lineweaver–Burke plot for an enzyme-catalyzed reaction. Which of the following statements is correct?

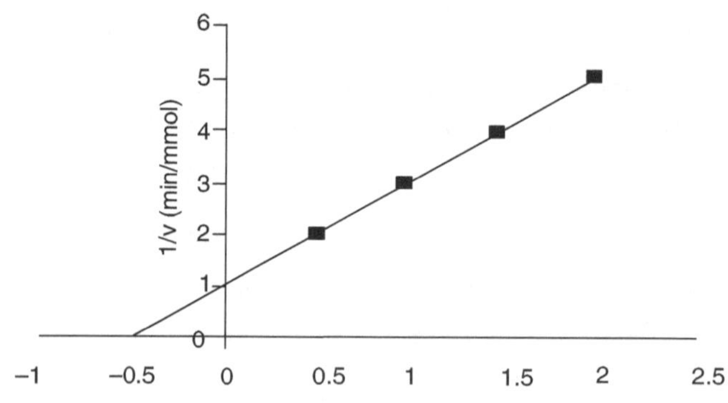

1/[S] (1/Mm)

(a) The V_{max} is 5 mmol/min and with competitive inhibition V_{max} remain unchanged
(b) K_m is 2 mmol/min and with competitive inhibition both K_m and V_{max} decrease
(c) K_m is 0.5 mM and with competitive inhibition V_{max} increases but K_m remains unchanged
(d) K_m is 2.0 mM and with competitive inhibition K_m increases but V_{max} remains unchanged

GATE-2004

Questions 1–10 carry one mark each

1. The most abundant atom in a living organism is
   (a) Hydrogen (b) Oxygen (c) Nitrogen (d) Carbon
2. The coenzyme involved in transfer of carboxyl group is
   (a) NADH (b) Coenzyme A (c) S-Adenosyl methionine (d) Biotin
3. The unit of length 'Angstrom'equal to
   (a) 10^{-6} m (b) 10^{-8}m (c) 10^{-10}m (d) 10^{-15}m
4. Which of the following triplet does NOT code for any amino acid
   (a) UUA (b) UCA (c) UAA (d) UGU
5. Erythrocytes resist shearing forces while traveling through narrow blood vessels because they contain the meshwork of
   (a) Glycophorin A (b) Spectrin (c) Hemoglobin (d) Myogobin
6. Tetrodotoxin, a highly poisonous toxin from puffer fish, acts on
   (a) Sodium channel (b) Potassium channel (c) Chloride channel (d) Calcium channel
7. Electron flow in cytochrome oxidase in the respiratory chain can be blocked by
   (a) Rotenone (b) Amytal (c) Cyanide (d) Cycloheximide
8. In case of humans, passive immunity observed in newborn child is due to the passage of the following immunoglobulin from the mother to the child through
   (a) IgD (b) IgE (c) IgM (d) IgG
9. In cukaryotic cells, around 5% cytosine residues are methylated, and the methylation most commonly occurs in the following sequence.
   (a) CpA (b) CptT (c) CpG (d) CpC
10. The following reagent will NOT release the integral proteins from the plasma amembrane
   (a) Sodium chloride (b) Triton X-100
   (c) Sodium dodecyl sulphate (d) NP-40 CpC

Questions 11–30 carry two mark each

11. The biochemical reactions leading from glucose to the generation of glyceraldehyde 3-phosphate consume ATP. The number of ATP molecules consumed for the conversion of one molecule of glucose to glyceraldehyde 3-phosphate is
   (a) 2 (b) 1 (c) 2 (d) 4

12. Monoclonal antibodies were raised against three different proteins containing unique epitopes. All the monoclonal antibodies were mixed together and placed in one of the wells in an Ouchterlony Double Diffusion plate. Similarly all three proteins were mixed together and placed in the adjacent well. The number of precipitin bands observed after incubation are

(a) 3 (b) 2 (c) 1 (d) 0

13. L-19 RNA from the protozoan *Tetrahymena* was demonstrated to have two enzymatic activities and these activities are

(a) RNA polymerase, DNA polymerase (b) Ribonuclease, RNA polymerase
(c) Ribonuclease, Deoxyribonuclease (d) Ribonulcease, Protease

14. A single-stranded DNA of unknown size CANNOT be distinguished from a double-stranded DNA by

(a) Analysis of base composition (b) Thermal denaturation
(c) Density gradient utracentrifugation (d) Gel Electrophoresis

15. Telomerase is an enzyme whose macromolecular composition is

(a) Lipoprotein only (b) Ribonucleoprotein only
(c) Ribonucleic acid only (d) Protein only

16. The molecular weight of IgG–Antigen complex at antigen excess is 300 Kd. The molecular weight of antigen is

(a) 75 (b) 50 (c) 100 (d) 150

17. The net reaction of the citric acid cycle is

(a) Acetyl CoA + 2NAD$^+$ + FAD + GDP + Pi + 2H$_2$O $\longrightarrow$ 2CO$_2$ + 2NADH + FADH$_2$ + GTP + 2H$^+$ + CoA

(b) Acetyl CoA + 3NAD$^+$ + FAD + GDP + Pi + 2H$_2$O $\longrightarrow$ 2CO$_2$ + 3NADH + FADH$_2$ + GTP + 2H$^+$ + CoA

(c) Acetyl CoA + 2NAD$^+$ + 2FAD + 2GDP + 2Pi + 2H$_2$O $\longrightarrow$ 2CO$_2$ + 2NADH + 2FADH$_2$ + 2GTP + 2H$^+$ + CoA

(d) Acetyl CoA + 2NAD$^+$ + FAD + GDP + Pi + H$_2$O $\longrightarrow$ 2CO$_2$ + NADH + FADH$_2$ + GTP + CoA

18. Genetic variation between individuals within a species involving different alleles at a locus is called

(a) Isotypic variation (b) Idiotypic variation
(c) Allotypic varitation (d) Haplotypic variation

19. A pentameric IgM molecule was electrophoresed on reducing SDA-polyacrylamide gel. How many bands can one expect?

(a) 1 (b) 2 (c) 3 (d) 4

20. Viral-encoded *Ras* oncogene transforms normal mammalian cells into cancer cells. Viral Ras protein differs from its normal counterpart by

(a) Diminished GTPase activity
(b) Increased GTPase activity
(c) Diminished ATPase activity
(d) Increase ATPase activity

21. Choose the correct match

Group I		Group II	
P	Immunodificiency disease	1.	Hexosaminidase-A
Q	Tay–Sach disease	2.	Hypoxanthine-guanine phosphoribosyl transferase
R	Lesch–Nyhan syndrome	3.	Chloride channel
S	Cystic fibrosis	4.	Adenosine deaminase

(a) P-1, Q-3, R-1, S-2
(b) P-3, Q-2, R-4, S-1
(c) P-4, Q-1, R-2, S-3
(d) P-2, Q-4, R-3, S-1

22. Choose the correct match

P	β-Ketoacyl-ACP synthase	1.	Reduces double bound forming Saturated acyl-ACP Hexosaminidase-A
Q	β-Ketoacyl-ACP reductase	2.	Reduces β-Keto group to β-hydroxy group
R	Enoyl-ACP-reductase	3.	Condenses acyl and malonyl Group
S	Cystic fibrosis	4.	Carries acyl group in thioester linkages

(a) P-2, Q-1, R-3, S-4
(b) P-3, Q-2, R-1, S-4
(c) P-4, Q-1, R-2, S-3
(d) P-3, Q-1, R-2, S-4

23. A DNA was labeled at its 5' end using γ^{32} P-ATP. Choose the correct combination from the following options to separate the free γ^{32} P labeled DNA.

P γ^{32} P-ATP can be removed by treating the sample with phosphatases.

Q The labeled sample can be electrophoresed and blotted onto nitrocellulose membrane and then labeled DNA can be recovered from nitrocellulose membrane.

R Sample can be passed through a gel filtration column to separate free γ^{32} P-ATP.

S DNA can precipitated using ethanol to separate it from contaminating γ^{32} P-ATP.

(a) P, Q (b) P, S (c) R, S (d) R, Q

24. A team of researchers had used the following eluting buffers to prepare an antialbumin IgG from an anti-IgG affinity column

(a) Buffer A : 0.1M Glycine-HCl, pH 3.0
(b) Buffer B : 0.1M Sodium Acetate, pH 6.0
(c) Buffer C : 0.1M Sodium phosphate, pH 7.0
(d) Buffer D : 0.1M Tris–HCl, pH 8.0

Which of the above buffers can be used for anti-albumin IgG elution?

(a) Buffer A (b) Buffer B (c) Buffer C (d) Buffer D

25. Oligosaccharide chain is added to the asparagine residue of a newly synthesized protein on the luminal side of the endoplasmic reticulum membrane. The asparagine of the newly synthesized protein that is linked to oligosaccharide chain is a part of the sequence (x, can be any amino acid other the proline)

(a) Asn-x-Asp (b) Asn-x-Thr (c) Asn-x-lle (d) Asn-x-Gly

26. The signal recognition particle binds to the signal peptide of the protein undergoing synthesis resulting into a pause in the translational process. The pause in the translational phenomenon is due to

(a) Inactivation of aminoacyl-tRNA synthetases
(b) Inhibition of the activity of elongation factor Tu
(c) Induced conformational change in the smaller ribosomal subunit
(d) Block of the entry of the next aminoacyl-tRNA for the reaction

27. Which of the following statement is NOT true with respect to the Photosystem II found in the chloroplasts? (Photosystem I = PS II)

(a) PSII is located in the grana
(b) Chlorophyll molecule in the reaction center for the PS II is bound to a 110 kD integral membrane protein
(c) A complex of Cytochromes b and f transports electrons from PS II to PS I
(d) PS II reaction center is bound to two Mn^{2+} ions

28. Choose the statement that is NOT correct with respect to the Protein Kinase C (PKC).

(a) Activated PKC can phosphorylate the specific tryptophan residue on the target protein.
(b) PKC is a Ca^{2+}-dependent kinase.
(c) PKC can be activated by diacylglycerol.
(d) Activation of PKC can lead to the activation of MAP kinase.

29. Enhancer elements are known to increase the rate of transcription when present at the upstream side of the promoter sequences. If the same enhancer element is placed in reverse orientation,

(a) The rate of transcription increases (b) The rate of transcription decreases
(c) The rate of transcription remains same (d) No transcription is observed

30. Choose the group containing only peptide harmones

(a) Vasopressin, Oxytocin, Epinephrine (b) Vasopressin, Testosterone, Glucagon
(c) Oxytocin, Vasopressin, Thyroxine (d) Oxytocin, Vasopressin, Somatostatin

GATE-2005

Questions 1–10 carry one mark each

1. Which pair of amino acid residues can interact in the interior of a protein only?

   (a) Arg, Thr (b) Ser, Thr (c) Glu, His (d) Val, Leu

2. Okazaki fragments ate joined by the enzyme

   (a) RNA polymerase (b) DNA polymerase
   (c) DNA ligase (d) Reverse transcriptase

3. The biosynthetic reaction in a cell mainly takes place in

   (a) Mitochondria (b) Lysosomes
   (c) Smooth endoplasmic reticulum (d) Golgi apparatus

4. Colchicine inhibits

   (a) DNA replication (b) Formation of spindle fibers
   (c) Condensation of chromosomes (d) Cytokinesis

5. The active form of Testosterone is

   (a) Dihydrotestosterone (b) Dehydrotestosterone
   (c) Dihydroxytestosterone (d) Dehydroepiandrosterone

6. The sensitivity of a radioimmunoassy depends primarily on

   (a) Titer of the antibody (b) Specificity of the antibody
   (c) Specific activity of the ligand (d) Purity of the antigen

7. In eukaryotes, introns can be found in transcripts which are precursors of

   (a) mRNA (b) rRNA (c) tRNA (d) All of the above

8. Homologous recombination can be employed to generate

   (a) Transgenic animals (b) Gene knock-out animal
   (c) Site specific mutagenesis (d) Specific promoter sequences

9. The mode of action of widely used anticancer drug methotrexate is to inhibit

   (a) Dihydrofolate reductase (b) Dihydroorotate dehydrogenase
   (c) Carbamoyl phosphate synthase-2 (d) Ribonucleotide reductase

10. For a double-stranded DNA, which one of the following base-ratios will always be equal to 1?

    (a) (A+T)/(G+C) (b) (A+G)/(C+T) (c) C/T (d) A/G

Questions 11–26 carry two marks each

11. Activated fatty acyl groups are transported into the mitochondria by
    (a) Coenzyme A (b) Oxaloacetate (c) Carnitine (d) Citrate
12. A mixture of Cytochrome-C (MW 11.7 kD) and Myoglobin (MW 17.2 kD) are to be separated by polyacrylamide gel electrophoresis. Their isoelectric pH (pl) values are 9.6 and 7.2 respectively. In which direction will each protein migrate at pH 8.5?
    (a) Myoglobin will migrate to anode and Cytochrome-C will migrate to cathode
    (b) Myoglobin will migrate to cathode and Cytochrome-C will migrate to anode
    (c) Both will migrate to anode
    (d) Both will migrate to cathode
13. Which one of the following fatty acids will have melting point higher than that of palmitic acid (16 : 0)?
    (a) Myristic acid (14 : 0) (b) Palmitoleic acid (16 : 1)
    (c) Oleic acid (8 : 1) (d) Stearic acid (18 : 0)
14. A diabetes mellitus patient excretes glucose in urine even when kept on carbohydrate-free diet. This is because
    (a) Fats are catabolised in the liver to form glucose
    (b) Amino acids are catabolised in liver to form glucose
    (c) of increased production of amino acids
    (d) of increased breakdown of glucogen
15. According to the second law of thermodynamics, molecules spontaneously move from a region of higher concentration to one of lower concentration. However, sodium ions are present at 143 mM outside the cell. Yet sodium cannot pass through the plasma membrane. Transport of sodium into the cell is achieved by
    (a) Facilitated diffusion (b) Release of acetylcholine
    (c) Release of norpinephrine (d) Sodium transporter
16. The choice of the enzyme used in ELISA depends on
    (a) Purity of the enzyme
    (b) Turnover number
    (c) Its absence in biological sample which is being analysed
    (d) Its availability in bulk
17. The T cell antigen receptor
    (a) Recognises conformational epitopes on the native molecule
    (b) Has Ig light chains
    (c) Is made up of heavy chain and β_2 microglobulin
    (d) Recognises epitopes on linear peptides associated with MHC determinants

18. The advantage of degeneracy in codons is that
    (a) It minimizes the deleterious effects of mutations
    (b) It provides more flexibility
    (c) It helps to code proteins resistant to proteases
    (d) It helps to code proteins of very high molecular weight
19. Which is the pK_a of lysine?
    (a) 7.07 (b) 9.50 (c) 6.10 (d) 5.60
20. The molecular weight of a bacterial DNA molecule is 2.64×10^6. The average molecular weight of a nucleotide pair is 660. Assume that the average protein is made up of a chain of 400 amino acid residues. What is the maximum number of proteins that can be coded by the bacterial DNA molecule?
    (a) 20000 (b) 3333 (c) 6667 (d) None of the above
21. The coenzymes involved in the formation of Acetyl CoA from Pyruvate are
    (a) Thiamine pyrophosphate, Lipoic acid and FAD
    (b) Pyridoxyl phosphate, Biotin and FAD
    (c) Vitamin B-12, Floic acid and Vitamin C
    (d) NADH, Lipoic acid and Vitamin E

Common Data Questions

Common Data for Questions 22–24

A restriction fragment, obtained with a type II endonuclease that recognizes a six base pair site, was subjected to Maxam-Gilbert sequence with results as shown in the autoradiogram below.

G	A+G	C	C+T
—	—		
	—		
			—
	—		
			—
—	—		
	—		
			—
	—		
—	—		
	—		
			—
		—	—

22. The DNA sequence is

(a) $^{3'}$CTAGATAGTATAG$^{5'}$ (b) $^{5'}$CTAGATAGTATAG$^{3'}$
(c) $^{3'}$GATCTATCATATC$^{5'}$ (d) $^{5'}$GATCTATCATATC$^{3'}$

23. The restriction site is

(a) TAGCTA
ATCGAT
(b) TAGGTA
ATCCAT
(c) ATCCAT
TAGGTA
(d) GATACC
CTATGG

24. Considering this DNA sequence as a template strand, the sequence of the corresponding mRNA is

(a) $^{5'}$CTAGATAGTATAG$^{3'}$ (b) $^{3'}$GATCTATCATATC$^{5'}$
(c) $^{3'}$GAUCUAUCAUAUC$^{5'}$ (d) $^{5'}$CUAGAUAGUAUAG$^{3'}$

Common Data for Questions 25 and 26

The kinetic data for an enzymatic reaction in the presence and absence of inhibitors are plotted in the following figure.

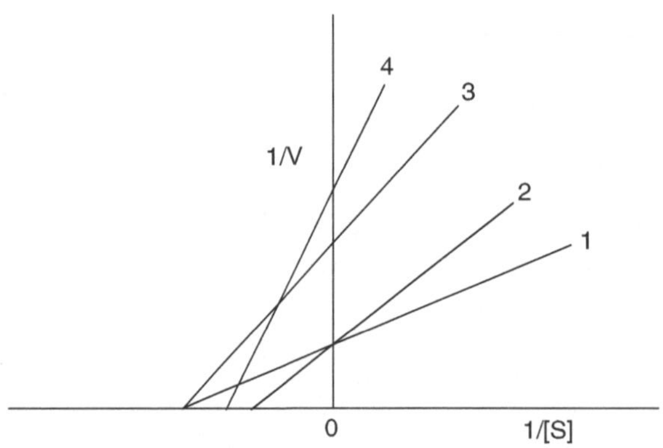

25. Which line represents the kinetics without inhibitor?

(a) Line 1 (b) Line 2 (c) Line 3 (d) Line 4

26. Which line represents kinetics of non-competitive inhibition?

(a) Line 1 (b) Line 2 (c) Line 3 (d) Line 4

Linked Answer Questions (Questions 27–30 carry two marks each))

Statement for Questions 27 and 28

Transmission at many synapses in the central nervous system is mediated by acetylcholine. Acetyl choline is cleaved to acetate and choline by the enzyme acetylcholine esterase which can be inhibited by diisopropyl phosphofluoridate (DIPF).

27. The mode of action of DIPF is by
    (a) Modifying histidine residue
    (b) Covalently modifying a crucial serine residue
    (c) Inducing a conformational change in the protein
    (d) Forming a complex with choline
28. Based on its mode of action, DIPF is used as
    a) A therapeutic agent to treat neurodegenerative diseases
    (b) A nerve gas
    (c) An agonist of acetylcholine
    (d) A reagent to determine the N-terminal amino acid

Statement for Questions 29 and 30

The following fragments are isolated on partial hydrolysis of a nonapeptide X:

Val-Arg-Pro-Gly, Lys-Phe-Val-Arg, Ala-Gly-Ser-Lys

29. The correct sequence of X is
    (a) Ala-Gly-Ser-Lys-Ala-Pro-Val-Arg-Gly
    (b) Val-Arg-Gly-Lys-Phe-Val-Arg-Ala-Pro
    (c) Lys-Phe-Val-Arg-Ala-Gly-Ser-Pro-Gly
    (d) Ala-Gly-Ser-Lys-Phe-Val-Arg-Pro-Gly
30. The number of fragments obtained when X is digested with trypsin is
    (a) 0 (b) 2 (c) 3 (d) 4

GATE-2006

Questions 1–10 carry one mark each

1. Which of the following amino acids does NOT contribute to fluorescence of protein?

   (a) Tyrosine (b) Phenylalanine (c) Cysteine (d) Tryptophan

2. Immunological memory is manifested during

   (a) Primary immune responses (b) Non-specific immune responses
   (c) Innate immune responses (d) Secondary immune responses

3. The main function of the pentose phosphate pathway is to

   (a) Supply ribose 5-phosphate and NADPH
   (b) Supply NADH and ATP
   (c) Provide a mechanism to use the carbon skeletons of excess amino acids
   (d) Provide carbon skeletons for oxidation of fatty acids

4. Which one of the following CANNOT be considered as weak interactions?

   (a) van der Waals forces (b) Peptide bonds
   (c) Hydrogen bonds (d) Ionic interaction

5. Polynucleotide kinase is used

   (a) To add a nitrogenous base at the 5’ end of DNA
   (b) To add a nitrogenous base at the 3’ end of DNA
   (c) To add a phosphate at the 5’ end of DNA
   (d) To add a phosphate at the 3’ end of DNA

6. The polymer shown below is a component of

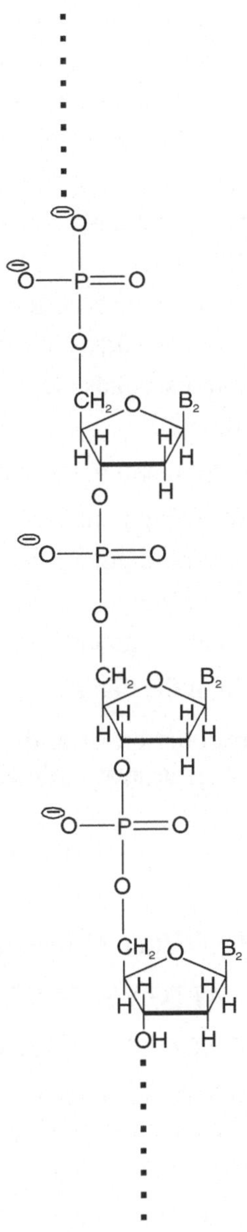

(a) Polysaccharide
(b) Polyribonucleotide
(c) Polyketide
(d) Polydeoxyribonucleotide

Questions 7–24 carry two marks each

7. FOS, Jun and MYC are
   (a) Proteins expressed on the surface of cancerous cells
   (b) Protein kinases that phosphorylate transcription factors in cancerous cells
   (c) Proteins involved in regulation of expression of genes involved in growth promotion
   (d) Proteins involved in ion transport in cancerous cells
8. Using microarray technique it was demonstrated that when a mammalian cell line was exposed to a drug, the expression of 10 genes is increased. Which of the following pairs of techniques could be used to validate microarray data?
   (a) Southern blotting and polymerase chain reaction (PCR)
   (b) Northern blotting and fluorescence in situ hybridization
   (c) Southern blotting and reverse transcriptase (RT)-PCR
   (d) Northern blotting and RT-PCR
9. Pepsin hydrolysis of IgG molecule will result in the production of
   (a) One Fc fragment and one F(ab') fragment
   (b) One Fhv fragment and one Fab fragment
   (c) One Fc fragment and one Fab fragment
   (d) One F(ab') fragment and one Fab fragment
10. Select the correct combination to fill in the blanks

    _______ are responsible for the production of antibody against free pathogens and soluble products from pathogens while _______ destroy pathogen and virally infected cells and abnormal cells.

    (a) Cytotoxic T cells and B cells
    (b) Macrophages and T cells
    (c) B cells and Helper T cells
    (d) B cells and Cytotoxic T cells
11. Select the correct primer pair for the PCR amplification of the following DNA fragment:

    5'TTTTTTTTTTAACTCGTAGGCATTGAACTATGCGGGGGGGGGG-3'

    3'AAAAAAAAAACATTGAGCCAGCATCCGTAACTTGATACCCCCCCCCC-5'

    (a) 5'-TTTTTTTTTT-3' and 5'GGGGGGGGGG-3'
    (b) 5'-AAAAAAAA-3' and 5'CCCCCCCCCC-3'
    (c) 5'-TTTTTTTTTT-3' and 5'CCCCCCCCCC-3'
    (d) 5'-AAAAAAAA-3' and 5'GGGGGGGGGG-3'
12. The double-reciprocal transformation of the Michaelis-Menten equation, also called the Lineweaver-Burk plot, is given by

    $1/V_0 = K_m/(V_{max}[S]) + 1/V_{max}$

    To determine K_m from a double-reciprocal plot, you would

(a) Take the reciprocal of the y-axis intercept
(b) Take the reciprocal of the x-axis intercept
(c) Multiply the reciprocal of the x-axis intercept by –1
(d) Mulitply the reciprocal of the y-axis intercept by –1

13. Which of the following statements is INCORRECT?
(a) Most of the eukaryotic mRNAs have a 7-methylguanosine cap at their 5' end
(b) TATA binding protein is involved in the synthesis of mRNA, tRNA and rRNA
(c) Histones have no other function except in chromatin organization
(d) Eukaryotic RNA polymerase II consists of more than 4 subunits

14. The organization of an eukaryotic gene expressed at high levels in liver is diagrammatically represented below?

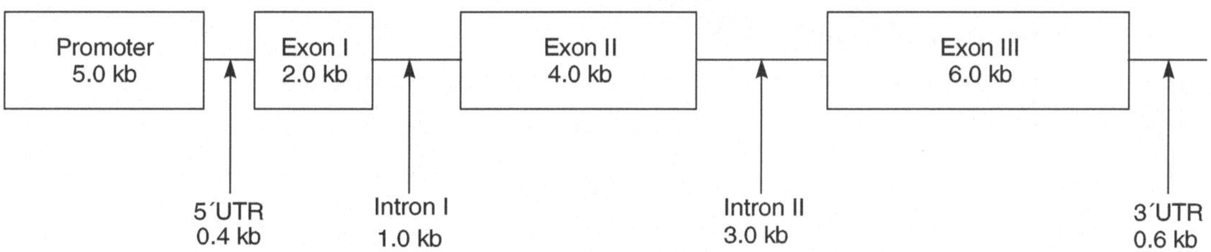

The size of the mature mRNA generated by the transcription followed by normal splicing of this gene will be (assume that this mRNA is not polyadenylated; 5' UTR and 3' UTR refer to the 5' and 3' untranslated regions, respectively)

(a) 12.4 kb (b) 13.0 kb (c) 12.0 kb (d) 12.6 kb

15. Incubation of a cell extract containing all enzymes of glycolysis with $\left[\gamma^{-32}P\right]$ ATP and unlabeled inorganic phosphate results in the formation of which of the following labeled compounds (assume that pyruvate kinase is inactivated)?
1. Glucose 6-^{32}Phosphate
2. (3-^{32}P)-Phosphoglycerate
3. (1-Phospho-3-^{32}Phospho)-Bisphosphoglycerate
4. (1-Phospho-6^{32}Phospho)-Fructose bisphosphate

(a) 1 and 3 (b) 1, 2 and 3 (c) 2 and 4 (d) Only 4

16. Electrphoresis of a purified protein named X in the presence of sodium dodecyl sulfate and 2-mercaptoethanol, shows a single band of 45 kDa. In gel filtration column chromatography, protein X elutes between alcohol dehydrogenase (160 kDa) and beta-amylase (190 kDa). How may identical subunits is protein X composed of?

(a) One (b) Two (c) Three (d) Four

17. Identify the correct pairs for the primary functions of the different enzyme classes

P. Kinases	1. Cleave bonds by elimination
Q. Lyases	2. Make large molecules from small molecules
R. Synthases	3. Transfer phosphate group to biomolecules
S. Phosphatases	4. Remove phosphate group from biomolecules

(a) P-4 Q-2 R-3 S-1
(b) P-3 Q-1 R-2 S-4
(c) P-4 Q-1 R-2 S-3
(d) P-3 Q-2 R-1 S-4

18. Which of the following statements about the mitochondrial proton gradient and membrane potential is correct?

(a) Either of them is sufficient to make ATP from ADP + Pi
(b) Both are required to make ATP
(c) Usually cancel one another since the system is at equilibrium
(d) Neither of them is required for ATP synthesis

19. The molecule shown is:

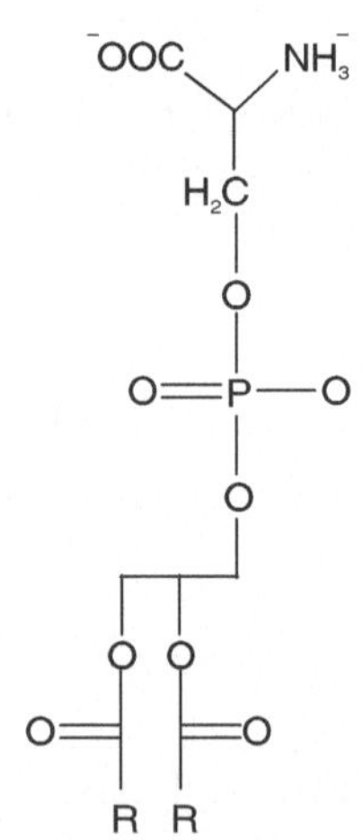

(a) Phosphatidylserine
(b) Phosphatidylcholine
(c) Phosphatidylethanolamine
(d) Phosphatidylinositol

20. Melting curve of two DNA specimens X and Y at the same pH and ionic strength have T_m values of 85°C and 80°C, respectively. This means that

(a) The AT content of Y is higher than X
(b) The GC content of Y is higher than X
(c) The AT content is same in X and Y
(d) The GC content of X is higher than Y

21. Part of the overall flow in amino acid catabolism is shown in the figure. Identify X and Y.

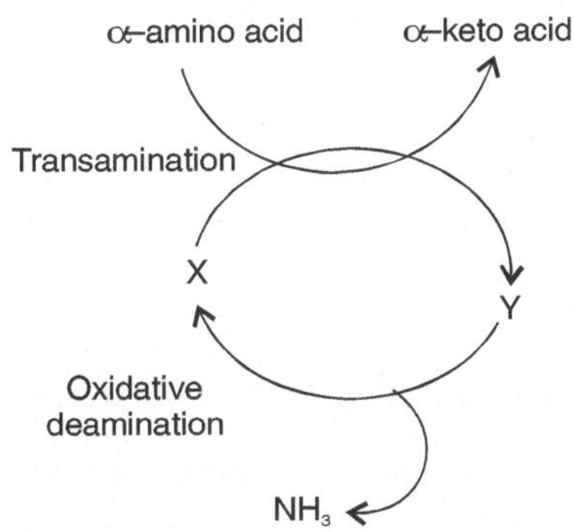

(a) X : L-Glutamate; Y: α -Ketoglutarate
(b) X : α -Ketoglutarate; Y: L-Glutamate
(c) X : α -Ketoglutamate; Y: L-Gltamine
(d) X : α -Ketoglutarate; Y: D-Glutamate

22. Which one of the following statements refers to glycogen, and which one refers to cellulose?

I. Branched molecule containing β-1, 4-glycosidic bond
II. Straight chain molecule containing β-1, 4-glycosidic bond
III. Branched molecule containing α-1, 6-glycosidic bond
IV. Straight chain molecule containing α-1, 6-glycosidic bond

(a) I = Glycogen; IV = Cellulose
(b) II = Glycogen; III = Cellulose
(c) III = Glycogen; II = Cellulose
(d) IV = Glycogen; I = Cellulose

Common Data Questions

Common Data for Questions 23 and 24

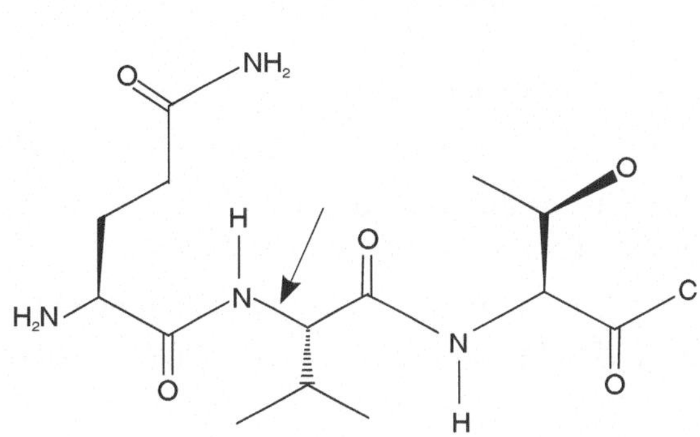

23. Dihedral angle indicated by an arrow in the tripeptide structure corresponds to the

    (a) Psi angle (b) Phi angle (c) Chi angle (d) Omega angle

24. The amino acid sequence of the above tripeptide is

    (a) Glutamine-valine-threonine (b) Asparagine-valine-serine
    (c) Glutamine-leucine-threonine (d) Asparagine-valine-threonine

Linked Answer Questions (Questions 27 –30 carry two marks each)

Statement for Question 25 and 26

Bind of glucagons to its receptor results in the generation of a specific second messenger.

25. Which of the following second messengers is generated in this case?

    (a) Calcium (b) cGMP (c) Phosphatidylinositol (d) cAMP

26. Which of the following enzymes is activated by this second messenger?

    (a) Protein kinase A (b) Protein kinase C
    (c) Phospholipase C (d) Protein phosphatase 2A

Statement for Linked Answer Question 27 and 28

The figure shows the oxygen binding curves for hemoglobin (Hb) and myoglobin (Mb).

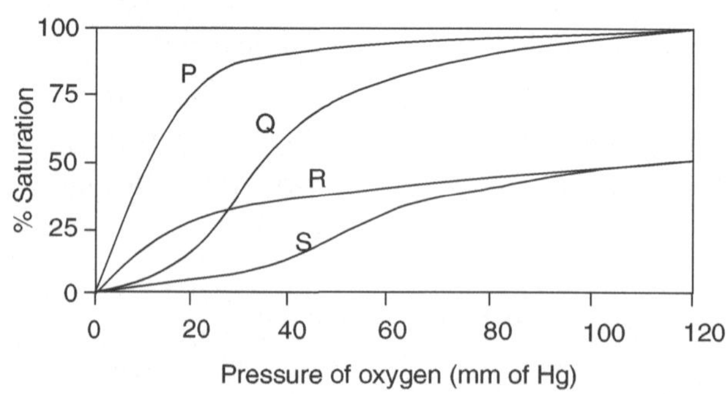

27. Identify the correct curves for Hb and Mb

    (a) P : Mb; Q : Hb (b) Q : Mb; P : Hb (c) R : Mb; Q : Hb (d) S : Mb; R : Hb

28. Sickle cell anemia arises due to the formation of a hydrophobic patch in one of the proteins shown in the above curves. This is due to the replacement of

    (a) Glu 6 by Val 6 in the β subunit of P (b) Glu 6 by Val 6 in the β subunit of Q
    (c) Glu 6 by Val 6 in the β subunit of R (d) Glu 6 by Val 6 in the β subunit of S

GATE-2007

Q.1 – Q.6 carry one mark each.

1. Deamination of cytosine produces

(a) Uracil (b) Pseudouracil (c) Hypoxanthine (d) 5-Methyluracil

2. Which of the following hormones binds to a cell surface receptor?

(a) Estrogen (b) Thyroid hormone (c) Insulin (d) Aldosterone

3. Systemic lupus erythematosus (SLE), an autoimmune disease, is characterized by the presence of

(a) Anti-DNA antibodies (b) Anti-thyroglobulin antibodies
(c) Anti-insulin antibodies (d) Anti-collagen antibodies

4. Optical density of 1 means

(a) 1% of the incident light is absorbed (b) 1% of the incident light is transmitted
(c) 90% of the incident light is absorbed (d) 90% of the incident light is transmitted

5. One of the carbon atoms of a glucose is [^{14}C]-labeled. If $^{14}CO_2$ is released during the conversion of pyruvate to acetyl coenzyme-A, which carbon atom of glucose was radiolabeled?

(a) C3 but not C4 (b) C3 or C4 (c) C1 or C6 (d) C1 but not C6

6. When yeast cells are shifted from a medium containing glycerol to glucose, an increase in the transcription of four genes involved in glucose metabolism was reported. Which of the following would be the most appropriate technique to demonstrate increased transcription of these genes?

(a) Southern hybridization (b) Northern hybridization
(c) Western hybridization (d) Fluorescence in situ hybridization

Q.7 – Q.24 carry two marks each

7. A mixture containing protein 1, 2, 3, 4 and 5 with molecular weights 5,000, 10,000, 25,000, 65,000 and 1, 00, 000 respectively, were separated on a Sephadex G-50 column. The order of elution of these proteins from the column will be

(a) Protein-1, Protein-2, Protein-3, Protein-4 and Protein-5
(b) Protein-5, Protein-4, Protein-3, Protein-2, and Protein-1
(c) Protein-1, 2, and 3 elute first, followed by protein-5 and 4
(d) Protein-4 and 5elute first, followed by protein-3, 2 and 1

8. The maximum number of hydrogen bonds that a molecule of water can form is

(a) 1 (b) 2 (c) 3 (d) 4

9. Match the techniques mentioned in Column A with their applications given in Column B.

Column A	**Column B**
P. PCR	1. Identification of transcription factor binding sites in chromatin
Q. DNA microarray	2. Identification of HIV infected patients using serum samples
R. ELISA	3. Isolation of mouse homologue of yeast Gene
	4. Analysis of differential gene expression in cancer and normal cells

(a) P-4 Q-1 R-3
(b) P-3 Q-4 R-2
(c) P-4 Q-1 R-2
(d) P-3 Q-2 R-1

10. A nonsense mutation in the gene encoding protein X leading to the synthesis of a truncated protein results in a slow-growing strain. Mutagenesis of this strain towards the isolation of extragenic suppressors led to the isolation of a strain which grew normally and synthesized the full-length protein X. The extragenic suppressor is likely to be a gene coding for

(a) rRNA (b) RNA polymerase (c) tRNA (d) Ribosomal protein

11. The total radioactivity in 1 ml solution of 0.25 mg of glycine is 1 mCi. The specific activity (mCi/millimole) of radiolabeled glycine will be

(a) 300 (b) 18.75 (c) 3000 (d) 1875

12. Ten grams of butter was saponified. The non-saponifiable fraction was extracted into 25 ml of chloroform. The absorbance of this solution in a 1 cm cuvette is 0.53 at 328 nm. If the extinction coefficient ($a_{1\%}$) of vitamin A at this wavelength is 1550, calculate the amount of vitamin A present

(a) 3.419×10^{-3} g/100 ml (b) 3.419×10^{-6} g/100 ml
(c) 3.419×10^{-5} g/100ml (d) 3.419×10^{-4} g/100 ml

13. Folate derivatives are required for the synthesis of which deoxynucleotides?

(a) Adenylate and guanylate (b) Cytidylate and thymidylate
(c) Adenylate, guanylate and thymidylate (d) Adenylate, guanylate and cytidylate

14. Cytochrome C reductase, also called as complex III or cytochrome bc_1 complex, localized on the inner mitochondrial membrane receives electrons from ubiquinol and donates to which of the cytochrome

(a) cyt *a* (b) cyt aa_3 (c) cyt *c* (d) cyt *b*

15. Match the biological functions mentioned in Column A with the enzymes given in B.

Column A	**Column B**
P. Diacylglycerol synthesis	(1) Protein kinase A
Q. CREB phosphorylation	(2) Ras
R. GTP hydrolysis	(3) Phospholipase C
	(4) Phospholipase D
	(5) Protein kinase G

(a) P-3 Q-1 R-5 (b) P-4 Q-1 R-2
(c) P-3 Q-1 R-2 (d) P-3 Q-5 R-2

16. How does haemoglobin carry carbon dioxide generated in tissues back to the lungs?
   (a) By coordination with heme
   (b) By forming N-terminal carbamate
   (c) By forming C-terminal carbamate
   (d) By linking to the epsilon-amino group of lysine

17. Which of the following enzyme activities can be detected in the supernatant obtained by centrifugation of liver homogenate at 100,000 g for 1 h at 4°C?
   (a) Succinate dehydrogenase
   (b) Glyceraldehyde 3-phosphate dehydrogenase
   (c) Glycogen synthetase
   (d) Aconitase

18. Which of the following statements about the enzyme complexes of the electron transport system is correct?
   (a) They interact with one another via mobile electron carriers
   (b) They are located in the mitochondrial matrix
   (c) They cannot be separated from one another in a functional form
   (d) They all have cytochromes

19. Match the DNA binding motifs mentioned in Column A with the proteins given in Column B

Column A	**Column B**
P. Zinc finger	(1) c-jun
Q. Leucine zipper	(2) Growth hormone receptor
R. Helix-turn-helix motif	(3) Glucocorticoid receptor
	(4) Histone H1
	(5) Lambda repressor

(a) P-4 Q-5 R-1 (b) P-2 Q-5 R-4
(c) P-2 Q-1 R-5 (d) P-3 Q-1 R-5

20. Which of the DNA polymerases listed below is primarily responsible for the de novo synthesis of new DNA strands?

    (a) DNA polymerase I (b) DNA polymerase II
    (c) DNA polymerase III (d) DNA polymerase IV

21. F_1F_0-ATPase in chloroplasts is located on the

    (a) Inner chloroplast membrane with F1 facing the stroma
    (b) Inner chloroplast membrane with F1 facing the inter-membrane space
    (c) Thylakoid membrane with F1 facing the stroma
    (d) Thylakoid membrane with F1 facing the thylakoid lumen

22. In addition to adjuvants, generation of anti-hapten antibodies will require injection of

    (a) The hapten to a mice
    (b) A mixture of hapten and protein to a mice
    (c) The hapten covalently linked to a protein to a mice
    (d) A mixture of hapten and lipid to a mice

Common Data Questions

Common Data for Questions 23 and 24

The number of protons translocated by the various vectorial proteins localized on the inner mitochondrial membrane of an organism was determined. They are as follows: NADH dehydrogenase: 4, cytochrome bc_1 complex: 2, cytochrome aa_3 complex : 4 and F_1F_0-ATPase: 3. One proton is also required for the transport of inorganic phosphate into the mitochondrial matrix.

23. The number of ATP molecules that can be synthesized by the oxidation of one NADH molecule is

    (a) 2 (b) 2.5 (c) 3 (d) 3.3

24. If the cytosolic NADH is transported to the matrix by the glyceraldehyde 3-phosphate shuttle, then the number of ATPs synthesized is

    (a) 1.5 (b) 2 (c) 2.5 (d) 3.3

Linked Answer Questions (Questions 25 –28 carry two marks each)

Statement for Question 25 and 26

Two mammalian cell lines were found to express either epidermal growth factor receptor (EGFR) alone (cell line A) or both EGFR and Ras (cell line B). These cell lines were treated with epidermal growth factor (EGF) and protein phosphorylation was examined in the membrane and cytosolic fractions using anti-phosphotyrosine and anti-phosphoserine antibodies.

25. EGF-dependent tyrosine phosphorylation will be detected in
    (a) Membrane and cytosolic fractions of both the cell lines
    (b) Only the membrane fraction of only cell line A
    (c) Only the membrane fraction of both cell lines
    (d) Only the cytosolic fractions of both cell lines
26. EGF-dependent serine phosphorylation will be detected in
    (a) Membrane and cytosolic fractions of both the cell lines
    (b) Only the membrane fraction of cell line A
    (c) Only the membrane fraction of cell line B
    (d) Only the cytosolic fraction of cell line A

Statement for Linked Answer Question 27 and 28

$\Delta G°$ is the symbol used to denote standard free-energy change of a chemical reaction system. The standard conditions are T = 298 K, concentration of water = 55.5 M, pHc : : 0 the reactants and products (other than water and proton) are initially present at 1 M come. ill or

27. Suppose ΔG denotes the free-energy change for the reaction A + B C + H^+ at pH 5, all other conditions being the same as the standard conditions specified above. Then
    (a) $\Delta G = \Delta G°'$
    (b) $\Delta G = \Delta G°' + 11.5$ RT
    (c) $\Delta G = \Delta G°' + 4.6$ RT
    (d) $\Delta G = \Delta G°' + 16.1$ RT
28. If $\Delta G°'$ for the reaction is –11.7 kJ/mol and R = 8.314 kJ/mol, the reaction is
    (a) Endergonic at both 37°C and 25°C
    (b) Endergonic at 37°C and exergonic at 25°C
    (c) Exergonic at both 37°C and 25°C
    (d) Exergonic at 37°C and endergonic at 25°C

GATE-2008

Questions 1–6 carry one mark each.

1. Which of the following inhibitor uncouples electron transport and oxidative phosphorylation?
   (a) Azide (b) Dinitrophenol (c) Oligomycin (d) Rotenone
2. The catalytic efficiency of an enzyme is represented by
   (a) V_{max} (b) K_m (c) k_{cat} (d) k_{cat}/K_m
3. Which of the following activate protein kinase C?
   (a) Inositol 1, 4, 5-triphosphate (b) Cyclic AMP
   (c) Inositol (d) Diacylglycerol
4. Transcription initiation sites can be determined by
   (a) Footprinting (b) Northern blotting
   (c) Primer extension (d) Nick translation
5. One common feature between B and T cells is that
   (a) Both cells produce antibodies
   (b) Both cells possess MHC class II
   (c) Both B cell receptor and T cell receptor undergo rearrangement
   (d) Both cells can produce cytokines
6. In hybridoma technology, the myeloma cells used
   (a) Lack HGPRTase
   (b) Lack the ability to produce lg
   (c) Lack both HGPRTase and the ability to produce lg
   (d) Lack thymidine kinase

Questions 7–24 carry one mark each.

7. Match the function in Column I with organelle in Column II

Column I	Column II
(P) Protein synthesis	(1) Endoplasmic reticulum
(Q) Protein degradation	(2) Golgi body
(R) Protein glycosylation	(3) Lysosome
	(4) Peroxisome

	P	Q	R
(a)	3	2	1
(b)	1	3	2
(c)	1	4	3
(d)	4	1	2

8. Match the polysaccharides in Column 1 with their constituent monosaccharide in Column II

Column I	**Column II**
(P) Chitin	(1) D- Glucose
(Q) Hemicellulose	(2) N-Acetyl glucosamine
(R) Glycogen	(3) D-Xylose
	(4) D-Galactose

	P	Q	R
(a)	1	3	4
(b)	2	4	1
(c)	4	2	3
(d)	2	3	1

9. The T_m of phosphatidyl choline A is higher than the T_m of phosphtidyl choline B because
   (a) A has shorter chain fatty acid and more unsaturated fatty acid than B
   (b) A has longer chain fatty acid and more saturated fatty acid than B
   (c) A has shorter chain fatty acid than B
   (d) A has more *cis*-unsaturated fatty acid than B

10. A mixture of proteins P, Q, R and S having molecular mass 50, 80, 120, and 150 kD is applied on the Sephadex-G 200 column. The order of their elution will be
   (a) P, Q, R, S (b) S, R, Q, P (c) Q, P, R, S (d) P, Q, S, R

11. Match the transition state or chemical entity of each enzyme that is responsible for their catalytic function.

(P) Ribonuclease	(1) Oxyanion
(Q) Lysozyme	(2) Pentacovalent phosphorus
(R) Chymotrypsin	(3) Carbonium ion
(S) Carboxypeptidase	(4) Mixed anhydride

	P	Q	R	S
(a)	3	2	4	1
(b)	2	3	1	4
(c)	2	1	3	4
(d)	4	3	2	1

12. Match the function of the following cofactors.

(P) Thiamine pyrophosphate	(1) Acyl group transfer
(Q) Coenzyme A	(2) Transfer of one carbon component
(R) Pyridoxal phosphate	(3) Group transfer to/or from amino acid
(S) Tetrahydrofolate	(4) Aldehyde transfer

	P	Q	R	S
(a)	4	3	1	2
(b)	4	3	2	1
(c)	4	1	3	2
(d)	3	1	4	2

13. Match the enzymes in Column I with their metabolic pathway in Column II

(P) Succinyl CoA synthetase	(1) β-Oxidation
(Q) Acyl CoA dehydrogenase	(2) Calvin cycle
(R) Transketolase	(3) Tricarboxylic acid cycle
(S) Ribulose 1,5-bisphosphate carboxylase	(4) Pentose phosphate pathway

	P	Q	R	S
(a)	1	2	3	4
(b)	3	1	2	4
(c)	2	4	1	3
(d)	3	1	4	2

14. Glycolysis and gluconeogenesis are reciprocally coordinated. Which of the following will activate pyruvate carboxylase in gluconeogenesis?

(a) Acetyl CoA (b) Fructose 2,6-bisphosphate
(c) ADP (d) ATP

15. The atoms of pyrimidine ring are derived from

(P) Carbamoyl phosphate (Q) Inosine monophosphate (R) Aspartate (S) Glutamate

(a) PQ (b) PR (c) PS (d) QR

16. Which of the following statements are true for steroid hormones?

(P) Increase the enzymatic activity of pre-existing target enzyme
(Q) Act at cell nucleus
(R) Interact with the plasma membrane receptor of target cell
(S) Form a complex with receptor and acts as transcriptional enhancers

(a) PR (b) QS (c) PQ (d) RS

17. Match the items on the left with the inhibitors on the right

(P) DNA polymerase α
(Q) RNA polymerase II
(R) Serine protease

(1) Phenyl methyl sulphonyl fluoride (PMSF)
(2) Aphidicolin
(3) α amanitin
(4) Actinomycin

	P	Q	R
(a)	2	3	1
(b)	3	1	2
(c)	2	1	2
(d)	1	2	4

18. A nucleic acid sample is resistant to digestion with λ exonuclease. When heated it does not show typical melting curve of a linear double stranded DNA. On CsCl-ethidium bromide equilibrium density centrifugation it settles at the bottom of the centrifuge tube. The nucleic acid is

(a) ccc pBR322
(b) Bacteriophage P22 DNA
(c) rRNA
(d) RFII M 13 DNA

19. The following 4 different solutions are prepared by mixing the components of the electron transport chain. Which among them is expected to cause a net transfer of electrons to cytochrome C?

(a) Reduced ubiquinone and reduced cytochrome C
(b) Reduced ubiquinone, cytochrome b-c_1 complex and reduced cytochrome c
(c) Oxidized ubiquinone and oxidized cytochrome c.
(d) Reduced ubiquinone, cytochrome b-c_1 complex and oxidized cytochrome c.

20. Nucleated cells tend to be more resistant to complement mediated lysis than RBC because

(a) Many nucleated cells can endocytose the membrane attack complex
(b) Membrane attack complex cannot get inserted in the nucleated cell membrane
(c) Membrane attack complex can gets inactivated by the nucleated cells
(d) Membrane attack complex gets inactivated hence cannot get inserted in the nucleated cell membrane

21. In a fluorescein labeled antibody to μ heavy chain and rhodamine labeled antibody to δ heavy chain, the fluorescent antibody staining pattern of the progenitor B cells (Pro-B cells) will be

(a) Anti-μ staining in cytoplasm and on membrane
(b) Anti-μ and anti-δ staining in cytoplasm and on membrane
(c) No cytoplasmic or membrane staining with either anti μ or δ antibody
(d) Anti-μ staining on the membrane

22. Serum IgM cannot activate the complement by itself because

(a) It does not have complement binding site
(b) It is planar in which complement binding site
(c) It gets degraded and hence unable to activate the complement
(d) It needs metal ions to activate complement

Common Data Questions

Common Data for Questions 23 and 24

A Caenorhabditis contig for one region of chromosome 2 contains contiguous locations marked 1, 2, 3, 4, 5, 6, 7, 8, and 9. Cosmid clones a, b, c, d and e overlap the locations 2-4, 3-5, 4-6, 5-8, 8-9 respectively. A cloned pBR322-x hybridize to cosmids b, c and d and pUC18-y hybridize to cosmids d and e.

23. The approximate locations of x and y are

(a) 4 and 7 (b) 5 and 8 (c) 4 and 8 (d) 5 and 7

24. Both pBR322-x and pUC18-y will bybridize to cosmids

(a) b (b) d (c) e (d) c

Linked Answer Questions (Questions 25 –28 carry two marks each)

Statement for Question 25 and 26

In animal cells, concentration of sodium ions is higher outside the cell and less inside the cell, yet sodium does not enter the cells.

25. The cellular environment is maintained by generating a gradient and transporting the Na^+ outside the cell through

(a) Diffusion process (b) Passive transport via Na^+-K^+pump
(c) Active transport via Na^+-K^+ Pump (d) Sodium ions not be transported

26. Digitoxigenin, a cardiotonic steroid that inhibits ATPase when applied on extracellular surface of membrane, helps in accumulation of Ca^{2+} inside the cardiac muscle cells by

(a) Activating Na^+-K^+ pump and blocking Na^+-Ca^{++} exchanger
(b) Inhibiting Na^+-K^+ pump and blocking Na^+-Ca^{++} exchanger
(c) Having no effort on Na^+-K^+ pump
(d) Increasing passive diffusion

Statement for Linked Answer Question 27 and 28

Nearly 46% of 45s ore-rRNA is unstable. The remaining portion of it forms mature 5.8s, 18s and 28s rRNA having lengths 160 bases, 1.9 kb and 5.1 kb respectively. The content of pre-rRNA per human genome is 7.8×10^{-15} g.

27. The mol.wt. of 45s pre-rRNA is

(a) 2×10^6 (b) 4.5×10^5 (c) 4.5×10^6 (d) 3.9×10^7

28. The number of pre-rRNA genes per genome is approximately

(a) 10 (b) 100 (c) 1000 (d) 10,000

21. In myasthenia gravis, a neuromuscular disorder, the acetylcholine receptor becomes dysfunctional because
    (a) The receptor is mutated
    (b) Antibodies to the receptor inhibit ligand binding
    (c) Of deficiency in acetylcholine transferase
    (d) Of excess of acetylcholine esterase

GATE-2009

Questions 1–7 carry one mark each.

1. Pernicious anemia is due to

(a) Blockage anemia is due to (b) Blockage of vitamin A absorption

(c) Deficiency of vitamin C (d) Deficiency of vitamin B_2

2. Which of these can be an antigen but cannot induce immune responses?

(a) Hemocyanin (b) Influenza virus

(c) Urea (d) Corynebacteria

3. In mammals, the second messenger Nitric Oxide (NO) is produced from

(a) Ammonium nitrate (b) Arginine

(c) Urea (d) Nitrous acid

4. Generally, the rate-limiting step of major metabolic pathway is a reaction

(a) In which the availability of the substrate is limited

(b) Catalyzed by an allosteric enzyme

(c) Catalyzed by an enzyme with very low K_m

(d) Whose products are not readily consumed by the subsequent step of the pathway.

5. Human DNA (3×10^6 kb) is replicated in 5 hrs at a rate of Kb/min. The number of origins of replication utilized are

(a) 1 (b) 3 (c) 300 (d) 10,000

6. Phospholipases A1 and A2

(a) Play no role in phospholipid synthesis

(b) Hydrolyze phosphatidic acid to diglycerate

(c) Remove fatty acid from sn-1and sn-2 of phospholipids

(d) Are involved in biosynthesis of phosphatidyl ethanolamine

7. Equal volumes of two buffers of pH 4 and pH 6 of identical ionic strengths are mixed. The resultant pH is

(a) Close to 4 (b) Close to 5 (c) Close to 6 (d) Exactly 5

Questions 8–21 carry two mark each

8. An insert (I) of 1 kb is ligated to a plasmid (p) of a Kb in a molar ratio of 4:1 respectively in a final DNA concentration 10 μg/ ml. The amounts of insert (I) and plasmid (p) required in μg are

(a) I = 8 and P = 2 (b) I = 2 and P = 8 (c) I = 5 and P = 5 (d) I = 1 and P = 4

9. The formation of ATP from ADP and phosphate is not a spontaneous reaction. A reason for this is
   (a) ATP readily ionizes
   (b) Electrostatic repulsion in ATP is lower than that in ADP
   (c) ATP is better hydrated than the total hydration levels of ADP and Pi
   (d) Resonance stabilization of P – O bonds is greater than in ATP.

10. A beam of light passes through 1 cm of a colored solution. Eighty percent of the incident light is transmitted. If the incident light passes through 2 cm of the same solution, the percentage of transmitted light is
    (a) 60 (b) 64 (c) 70 (d) 40

11. Lactose uptake in *E. coli* is an example of
    (a) Passive transport (b) Primary active transport
    (c) Secondary active transport (d) Simple diffusion

12. During receptor-mediated endocytosis of LDL bound to its receptor
    (a) Both receptor and ligand are degraded
    (b) The receptor is degraded and the ligand is recycled
    (c) Both are recycled
    (d) The ligand is degraded and the receptor is recycled

13. Choose the correct pairs from the following

A.	Isocitrate lyase	i	Conversion of amino into glucose
B.	PEP carboxykinase	ii	Biotin
C.	Pyruvate dehydrogenase complex	iii.	Synthesis of glucose from acetate
D.	Phosphofructokinase	iv.	Lipoic acid
E.	Pyruvate carboxylase	v.	An allosteric enzyme

    (a) A-i, B-ii, C-iv, D-v, E-iii (b) A-iii, B-i, C-iv, D-v, E-ii
    (c) A-iii, B-i, C-iv, D-ii, E-v (d) A-ii, B-v, C-i, D-iv, E-iii

14. The advantage of hemoglobin having a high histidine content is
    (a) Histidine binds to oxygen
    (b) Histidine carries oxygen to the tissues
    (c) Histidine imparts buffering capacity to hemoglobin
    (d) 'R' group of histidine has low pKa

15. IgG has four chains. Purified monoclonal IgG was subjected to electrophoresis. The number of bands visible is
    (a) Reducing SDS-PAGE (b) A-1; B-1; C-2
    (c) A-2; B-4; C-2 (d) A-4, B-2, C-4